Lecture Notes in Computer Science 8893

Commenced Publication in 1973
Founding and Former Series Editors:
Gerhard Goos, Juris Hartmanis, and Jan van Leeuwen

T0183204

Liqun Chen   Chris Mitchell (Eds.)

Security Standardisation Research

First International Conference, SSR 2014
London, UK, December 16-17, 2014
Proceedings

 Springer

Volume Editors

Liqun Chen
HP Labs
Bristol, UK
E-mail: liqun.chen@hp.com

Chris Mitchell
University of London
Information Security Group
Egham, UK
E-mail: me@chrismitchell.net

ISSN 0302-9743   e-ISSN 1611-3349
ISBN 978-3-319-14053-7   e-ISBN 978-3-319-14054-4
DOI 10.1007/978-3-319-14054-4
Springer Cham Heidelberg New York Dordrecht London

Library of Congress Control Number: 2014956396

LNCS Sublibrary: SL 4 – Security and Cryptology

Typesetting: Camera-ready by author, data conversion by Scientific Publishing Services, Chennai, India

Printed on acid-free paper

Springer is part of Springer Science+Business Media (www.springer.com)

Preface

The First International Conference on Research in Security Standardisation was held at Royal Holloway, University of London, UK during December 16–17, 2014. This event was the first in what is planned to become a series of conferences focusing on the theory, technology, and applications of security standards.

The conference program consisted of two invited talks, 14 contributed papers, and a panel session. We would like to express our special thanks to the distinguished keynote speakers, Charles Brookson (Zeata Security, UK) and Marijke De Soete (Security4Biz, Oostkamp, Belgium), who gave very enlightening talks. Special thanks are due also to the panel organizer, Joshua D. Guttman (MITRE Corporation), and the panel members, who included Karthikeyan Bhargavan (INRIA), Cas Cremers (University of Oxford), and Kenneth Paterson (Royal Holloway, University of London).

Out of 22 submissions from 12 countries, 14 papers were selected, presented at the conference, and included in the proceedings. The accepted papers cover a range of topics in the field of security standardisation research, including cryptographic evaluation, standards development, analysis with formal methods, potential future areas of standardisation, and improving existing standards.

The success of this event depended critically on the help and hard work of many people, whose help we gratefully acknowledge. First, we heartily thank the Programme Committee and the external reviewers, listed on the following pages, for their careful and thorough reviews. Each paper was reviewed by at least three people, and most by four. Significant time was spent discussing the papers. Thanks must also go to the hard-working shepherds for their guidance and helpful advice on improving a number of papers.

We must also sincerely thank the authors of all submitted papers. We further thank the authors of accepted papers for revising papers according to the various referee suggestions and for returning the source files in good time. The revised versions were not checked by the Programme Committee, and so authors bear final responsibility for their contents.

Thanks are due to the staff at Springer for their help with producing the proceedings. We must further thank the developers and maintainers of the Easy-Chair software, which greatly helped simplify the submission and review process.

December 2014

Liqun Chen
Chris Mitchell

Security Standardisation Research 2014

Royal Holloway, University of London, UK
December 16–17, 2014

General Chair

Chris Mitchell   Royal Holloway, University of London, UK

Programme Chair

Liqun Chen   Hewlett-Packard Laboratories, UK

Programme Committee

Ian Bryant	Trustworthy Software Initiative, UK
David W. Chadwick	University of Kent, UK
Takeshi Chikazawa	Information-technology Promotion Agency, Japan
Lizzie Coles-Kemp	Royal Holloway, University of London, UK
Cas Cremers	University of Oxford, UK
Riaal Domingues	South Africa
Andreas Fuchsberger	Microsoft
Aline Gouget	Gemalto, France
Phillip H. Griffin	Griffin Information Security, USA
Bridget Kenyon	University College London, UK
Eva Kuiper	Hewlett-Packard, Canada
Xuejia Lai	Shanghai Jiaotong University, China
Pil Joong Lee	Postech, Republic of Korea
Jiangtao Li	Facebook, USA
Peter Lipp	Graz University of Technology, Austria
Joseph Liu	Institute for Infocomm Research, Singapore
Javier Lopez	University of Malaga, Spain
Andrew Martin	University of Oxford, UK
Shin'ichiro Matsuo	National Institute of Information and Communications Technology, Japan
Jinghua Min	China Electronic Cyberspace Great Wall Corporation, China
Atsuko Miyaji	Japan Advanced Institute of Science and Technology, Japan

Table of Contents

Potential Future Areas of Standardisation

Improving Existing Standards

Unpicking PLAID

A Cryptographic Analysis of an ISO-Standards-Track Authentication Protocol

Jean Paul Degabriele[1], Victoria Fehr[2], Marc Fischlin[2],
Tommaso Gagliardoni[2], Felix Günther[2], Giorgia Azzurra Marson[2],
Arno Mittelbach[2], and Kenneth G. Paterson[1]

[1] Information Security Group, Royal Holloway, University of London, U.K.
[2] Cryptoplexity, Technische Universität Darmstadt, Germany
{j.p.degabriele,kenny.paterson}@rhul.ac.uk,
marc.fischlin@cryptoplexity.de,
{victoria.fehr,tommaso.gagliardoni,giorgia.marson,
arno.mittelbach}@cased.de, guenther@cs.tu-darmstadt.de

Abstract. The Protocol for Lightweight Authentication of Identity (PLAID) aims at secure and private authentication between a smart card and a terminal. Originally developed by a unit of the Australian Department of Human Services for physical and logical access control, PLAID has now been standardized as an Australian standard AS-5185-2010 and is currently in the fast track standardization process for ISO/IEC 25185-1.2. We present a cryptographic evaluation of PLAID. As well as reporting a number of undesirable cryptographic features of the protocol, we show that the privacy properties of PLAID are significantly weaker than claimed: using a variety of techniques we can fingerprint and then later identify cards. These techniques involve a novel application of standard statistical and data analysis techniques in cryptography. We also discuss countermeasures to our attacks.

Keywords: Protocol analysis, ISO standard, PLAID, authentication protocol, privacy.

1 Introduction

PLAID, the Protocol for Lightweight Authentication of Identity, is a contactless authentication protocol intended to be run between terminals and smartcards. The protocol was designed by Centrelink, an agency of the Australian government's Department of Human Services (DHS). According to the developers it is supposed to provide a cryptographically strong, fast, and private protocol for physical and logical access control, without exposing "card or cardholder identifying information or any other information which is useful to an attacker" [8,1,14].

PLAID was initially proposed for use in the internal ID management of Centrelink [19]. However, the intended scope of applications has since significantly

L. Chen and C. Mitchell (Eds.): SSR 2014, LNCS 8893, pp. 1–25, 2014.

broadened to include the whole of DHS and the Australian Department of Defence [26]. Indeed, the protocol's promoters aspire to broader commercial and governmental deployment, including on an international level [10]. Strategies that are mentioned to support these aspirations include freely available intellectual property and outreach to other governmental organizations. To the latter end, NIST organized a workshop to explore the potential of PLAID for U.S. Federal Agencies in July 2009 [22].

Another strategy that is being actively pursued is standardization. PLAID was previously registered as the Australian standard AS-5185-2010 [1] and was then entered into the ISO/IEC standardization process via the fast track procedure. At the time of writing, the current ISO/IEC version is draft international standard (DIS) 25185-1.2 [14] and is currently in the "Enquiry phase" 40.60 (close of voting). Minor changes in the original protocol to match the international standard have been applied. Reference implementations, based on the Australian standard, are available both from the Australian Department of Human Services (in Version 8.04) and of the Australian Department of Defence (in draft version 1.0.0).

The protocol. The main aim of the protocol is to perform mutual authentication and establish a shared key between the terminal (IFD) and the card (ICC). To this end the terminal and the card exchange nonces RND1 and RND2 in encrypted form and then derive the session key as part of the hash value of the two nonces. Encryption here uses both asymmetric RSA-based encryption (when the card transmits RND1 to the terminal) and symmetric AES-based encryption (when the terminal sends RND2 to the card). Authentication of the partner is presumably guaranteed by the fact that a party should know the secret key in order to be able to decrypt the other party's nonce. An overview of the protocol is depicted in Figure 1, where the encrypted nonces are exchanged with transmissions ^{e}STR1 and ^{e}STR2. The card confirms the receipt of RND2 by sending a string encrypted under the derived key in ^{e}STR3.

The role of the terminal's initial message KeySetIDs is as follows. Each PLAID deployment involves a set of key pairs consisting of an RSA key and an AES key. Each terminal and each card stores a certain subset of these pairs. More precisely, each terminal holds a set of RSA key pairs (both encryption and decryption key) and corresponding AES master keys, while each card holds a set of RSA public keys and card-specific AES keys, derived from the corresponding AES master keys using a card identity. The keys held by a card are intended to control what types of access the card should have, so each key represents a capability. The actually deployed pair of keys is negotiated during the protocol itself, by having the terminal send a sequence of supported RSA key identifiers KeySetID in the first message. Even though the encryption key in RSA is usually public, in PLAID it is kept secret to enhance privacy (since, for example, the set of RSA keys held by a card could be used to identify the card and track its use in a deployment).

One distinctive feature of the protocol, added for privacy reasons, is that the card switches to using a pair of so-called shill keys in case of an error. That is,

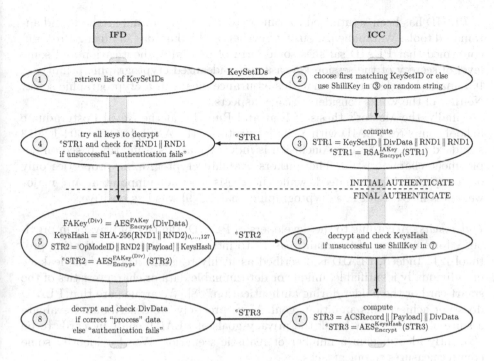

Fig. 1. PLAID protocol overview

if the card detects some potential error, then it uses its card-specific RSA shill key and AES shill key to encrypt random data. This mechanism is intended to hide information about failures from an adversary and thereby prevent leakage about which keys are possessed by a particular card.

Previous security analyses. Centrelink's accompanying description of PLAID [8] claims that PLAID is highly resistant against leakage of card or card holder identifying information, against various forms of active attacks, and provides mutual authentication. The document states as a goal that the protocol shall be "evaluated by the most respected cryptographic organisations, and the broader cryptographic community." For version 8 the document [8] refers to the input by various agencies like NIST and of "a number of independent cryptographic experts and consultants, a number of respected commercial cryptographic teams, as well as the internal Centrelink team."

However, we are not aware of any publicly available cryptographic evaluation of PLAID. None of the claimed security properties is backed up by arguments, nor matched against more precise formalizations in the description [8] or standards [1,14]. Some useful comments about the protocol's security have been given by the national representatives in the previous version of the ISO standard and documented there [13]. These comments refer partly to the points discussed in Section 5, where we asses them in a cryptographic context.

PLAID has been scrutinized to some extent by using formal methods and automated tools. Watanabe [28], using Scyther, and Sakurada [25], using ProVerif, confirmed that PLAID satisfies some form of mutual authentication and some level of secrecy of the session key, assuming idealized cryptographic primitives. It remained unclear to us what this assurance means in a cryptographic sense. Neither of the works considers privacy aspects.

Finally, the Master's thesis of Kiat and Run [18] at the Naval Postgraduate School compares PLAID with a similar protocol, the ANSI/INCITS 504-1-2013 standard OPACITY. The conclusion is indecisive and is primarily based on deployment characteristics. The authors evaluate cryptographic properties only on a superficial level. Indeed, while the thesis does not pinpoint at any major weakness in OPACITY, a cryptographic analysis [9] was less positive.

Our results. According to the developers of PLAID, the lack of privacy in previous efforts was one of the main reasons to introduce a new authentication protocol [24]. Indeed, PLAID is described as highly resistant against "the leakage of individually identifiable, unique or determinable data or characteristics of the smart card or the holder during authentication." [8]. We argue here that PLAID does not achieve this ambitious goal. More precisely, we describe and evaluate a suite of attacks that break the privacy goals of PLAID, enabling cards to be efficiently identified in a number of realistic scenarios. We also identify some countermeasures to our attacks.

In more detail, two of our attacks exploit PLAID's use of shill keys, which, being card-specific, can serve as a proxy for the card identity. While the shill keys themselves are not transmitted in the protocol, we show how they can be statistically estimated from RSA ciphertexts observed in protocol runs, enabling cards to be initially fingerprinted and then later re-identified. These two "shill key fingerprinting" attacks, presented in Section 3, deploy different techniques to perform the statistical estimation, and apply in different attack scenarios. Our first attack uses the standard solution to what is known as the "German Tank Problem", which concerns estimating the maximum of a discrete uniform distribution from a number of samples, while our second attack uses clustering techniques (and in particular the standard k-means clustering algorithm) to perform the estimation of the shill keys.

Our third attack, targeting the terminal's initial message KeySetIDs, is called "keyset fingerprinting" and is presented in Section 4. It exploits specific properties of the protocol flow to extract information about the set of keys held by a given card, potentially allowing us to draw conclusions about the card holder (e.g., via access authorizations). We show that this information can be efficiently extracted by interacting with a card a number of times and observing how the protocol proceeds (or fails to proceed). The information obtained in this attack may already be sufficient to identify individual cards from amongst a population, depending on the exact characteristics of a given deployment. The attack can also be combined with our first two attacks to increase their efficiency (by reducing the number of possible keys that need to be considered in the re-identification phase).

Finally, in Section 5, we make a number of other observations on cryptographic aspects of the PLAID protocol, focusing in particular on its lack of forward security, the use of weak RSA encryption, the lack of integrity protection for the symmetric encryption and a number of imprecisions in its specification. Some of the issues have already been briefly touched upon in the national body comments on the previous ISO standard version [13]; some aspects, like the lack of forward security, are new.

Reaction of the Responsible Authorities. We communicated our results to both the ISO 25185-1.2 project editor and to a contact person at the Department of Human Services. They are currently looking into our findings. The ISO project editor Graeme Freedman pointed out to us [11] that card-identifying information may be also available by other means, such as through the so-called Card Production Life Cycle (CPLC) data. The CPLC data contain information like serial numbers and manufacturers, uniquely identifying cards on a global scale. For privacy reasons access to the CPLC data must thus be restricted for PLAID. Indeed, the ISO draft standard itself already mentions this issue: "Consider switching off access to administrative applications from contactless interfaces, particularly ones which store unique card identification information such as the GlobalPlatform Card Production Life Cycle (CPLC) data." [14]. Our results show, however, that even if one restricts access to such administrative data, then PLAID still leaks card and cardholder identifying information.

2 PLAID Protocol Description

In this section we give a detailed description of PLAID according to the specification of the draft ISO/IEC DIS 25185-1.2 [14]. A more concise overview of the protocol flow is depicted in Figure 1. To make our description as close as possible to the original specification [14] we denote terminal and card by IFD (Interface Device) and ICC (Integrated Circuit Card) respectively. Table 1 provides a summary of the most important fields and objects occurring in the protocol.

2.1 PLAID Setup

In the setup phase PLAID initializes both terminals (IFDs) and cards (ICCs). PLAID supports up to 2^{16} key sets, each consisting of an RSA key pair $IAKey_i$ and an AES key $FAKey_i$. Each terminal and each card hold a subset of these overall possible key pairs, according to some access-control policy. However, the card only holds the public key part of the IAKey as well as a processed version of the original FAKey. More concretely, the card does not keep the FAKeys directly, but only a diversified version $FAKey^{(Div)} = AES_{Encrypt}^{FAKey}(DivData)$, where DivData is a 128-bit card identifier. The standard [14] highlights that these diversification data should be "random or unique". Using the diversified key instead of FAKey should retain security for other cards, in the case of a card being compromised and hence some of the (diversified) keys are disclosed. In addition to the RSA

Table 1. Most important fields and identifiers of PLAID

Variable	Description
ACSRecord	An access-control system record for each operation mode required for authentication.
DivData	A "random or unique" 16-byte ICC identifier.
FAKey	A 16-byte AES key which can be seen as master key to compute the diversified key used in the protocol (only known to the IFD).
FAKey$^{(\text{Div})}$	A 16-byte AES key derived from the FAKey and used in the FA phase.
IAKey	A 2048-bit pre-shared RSA key pair used in the IA phase. The ICC only knows the public key part.
KeySetID	A 2-byte index value identifying an IAKey and FAKey or FAKey$^{(\text{Div})}$, respectively.
OpModeID	A 2-byte index value identifying the operation mode. This value indicates which ACSRecord and payload the ICC needs to provide for authentication.
RNDi	A 16-byte random string for $i = 1, 2$.
KeysHash[1]	A 16-byte session key computed by IFD and ICC used in the FA phase.
ShillKey	A pair of 2048-bit RSA public key and 16-byte AES key of the ICC (randomly chosen per ICC during setup). These keys are to be used instead of error messages to simulate the next step of the protocol camouflaging that something went wrong.

keys, FAKey$^{(\text{Div})}$ and the value DivData, each card receives a pair of individual distress-keys (called ShillKey): a random encryption RSA key and a random AES key. These "shill keys" should be used to encrypt random data in case an error is detected, thus camouflaging errors or de-facto aborts on the card.

2.2   Initial Authenticate

The IA phase aims at exchanging the necessary information to compute the symmetric keys used in the FA phase as well as transferring DivData, the card-specific data later needed to guarante authenticity of the final message, securely to the terminal.

Step 1 (IFD) − IA Command: The interaction is initiated by the IFD, which transmits the complete sequence of supported KeySetIDs (in order of preference) to the ICC.

Step 2 (ICC) − IA Command Evalution: Upon receiving a set of KeySetIDs, the ICC traverses the entire list of indices to find the first KeySetID it supports, which determines the IAKey for RSA encryption. To prevent timing attacks it does not abort the search, even if a match has occurred. If no match is found, in Step 3 the ICC will encrypt a randomly generated string using its ShillKey.[2]

Step 3 (ICC) − IA Response: The ICC generates RND1, retrieves its DivData and derives string STR1, together with an encryption of it under IAKey, as follows:

---

[1] Note that in the original draft KeysHash refers to the entire 32 byte output of SHA-256(RND1 ‖ RND2) and the term *session key* is used to refer to the first 16 bytes which are used as secret key in the final message. For simplicity we refer to the session key as KeysHash in this paper.

[2] The standard neither specifies the exact format nor the length of this randomly generated string.

$$STR1 = KeySetID \parallel DivData \parallel RND1 \parallel RND1, \quad {}^eSTR1 = RSA^{IAKey}_{Encrypt}(STR1) \ .$$

The encrypted string ${}^eSTR1$ is sent to the IFD. Here PKCS#1 v1.5 padding is used.

Step 4 (IFD) – IA Response Evaluation: The IFD trial-decrypts ${}^eSTR1$ with all possible private IAKeys indexed by its KeySetID list and, for each valid decryption, it checks if the last two 16-byte blocks are equal. Again, to prevent timing attacks the IFD will continue the search even if a matching string has already been found. The (first) match is then used to extract KeySetID[3], DivData, and RND1. If no plaintext is of the anticipated format, authentication fails.[4]

2.3   Final Authenticate

The FA phase permits to specify the operation mode and to exchange data, like a PIN or biometrics, needed to complete the authentication. Here the diversified key $FAKey^{(Div)}$ (stored on the card and previously computed by the terminal during the IA phase) and a derived session key are used to secure the communication. The card authenticates by proving its ability to decrypt ${}^eSTR2$ as well as to include the correct DivData (transmitted in the previous IA phase) in the final message ${}^eSTR3$.

Step 5 (IFD) – FA Command: The IFD generates the 16-byte nonce RND2 and computes the unique session key KeysHash as the first 128 bits of

$$SHA\text{-}256(RND1 \parallel RND2) \ .$$

Next, using the master FAKey indexed by KeySetID, it computes the diversified AES key

$$FAKey^{(Div)} = AES^{FAKey}_{Encrypt}(DivData) \ ,$$

which corresponds to the AES key stored on the ICC under index KeySetID. The latter is used to encrypt

$$STR2 = OpModeID \parallel RND2 \parallel [Payload] \parallel KeysHash \ ,$$

using AES in CBC mode with the all-zero string as initialization vector, where Payload is an optional, variable-size field that depends on the operation mode. Concerning padding, the standard refers to the ISO/IEC 9797-1

---

[3] The standard is ambiguous in whether the trial KeySetID of the IFD or the value contained in ${}^eSTR1$ is stored.

[4] The standard does not specify what is meant by "authentication fails." We assume the protocol aborts in this case.

method 2, where one byte 0x80 is appended, followed by blocks of 0x00 bytes until the length is a multiple of the block length.[5] The resulting string

$$^e\text{STR2} = \text{AES}_{\text{Encrypt}}^{\text{FAKey}^{(\text{Div})}}(\text{STR2}) \ ,$$

is then transmitted to the ICC.

Step 6 (ICC) – FA Command Evaluation: The ICC decrypts ^{e}STR2 with FAKey$^{(\text{Div})}$ and retrieves RND2. It computes the session key as described above as first half of SHA-256(RND1 $\|$ RND2) and compares the result to the value KeysHash extracted from the decrypted ^{e}STR2. If they do not match the ICC encrypts a random byte string[6] using its AES ShillKey in the FA Response. Else Payload, if given, should be processed as specified by the implementation.

Step 7 (ICC) – FA Response: The ICC retrieves the Payload data specified by the operation mode (if necessary) and encrypts

$$\text{STR3} = \text{ACSRecord} \| [\text{Payload}] \| \text{DivData} \ ,$$

using AES in CBC mode with the all-zero string as initialization vector. Again, Payload is an optional, variable-size field which may (and usually will) differ from the Payload in Step 5. The resulting ciphertext

$$^e\text{STR3} = \text{AES}_{\text{Encrypt}}^{\text{KeysHash}}(\text{STR3}) \ ,$$

is transmitted as final message to the IFD.

Step 8 (IFD) – FA Response Evaluation: The IFD decrypts ^{e}STR3 and checks whether the recovered DivData matches the one received in the IA phase: if so, then the other data is considered authenticated and processed according to the implementation, otherwise authentication fails.

3   ShillKey Fingerprinting – Tracing Cards in PLAID

According to the developers of PLAID, privacy was one of the main reasons to introduce a new authentication protocol. In this and the next section we present three attacks on the privacy of PLAID contradicting the claims that no static information is available to be exploited. In this section we focus on the *traceability* of cards, that is, we consider an adversary who learns some information about one or more cards and then tries to identify these cards at a later time.

We consider two distinct attack scenarios, each consisting of a fingerprinting phase and then an identification phase. The difference is roughly that in the first scenario the fingerprinting is a supervised learning phase in the sense that we can attribute execution traces to cards, whereas the second setting corresponds to unsupervised learning where we get a set of random traces. More precisely:

---

[5] Though referring to ISO/IEC 9797-1 method 2, the standard explicitly describes a different padding method and thus makes unambiguous decoding impossible (cf. Section 5.4).

[6] Again, the standard does neither specify the exact format nor the length (note that STR3 in Step 7 contains a variable sized field Payload) of this random byte string.

- In the first scenario we allow the adversary to first interact in turn with each and every card in the system in a number of protocol runs (the fingerprinting phase). We then draw a card at random and let the adversary interact with this specific card a number of times, with the adversary's goal being to identify which of the cards was selected. The adversary's ability to interact with each card in the system in turn in the identification is not wholly realistic. However, given the high success rates of this attack that we will report below, we believe that good success rates would still be achieved in the more realistic scenario where the adversary does not have the guarantee of being able to interact with each distinct card in turn in a first phase, but instead must build up its picture of the system as it goes along.
- In the second scenario, which is much more challenging for the adversary, we do not allow the adversary to interact in turn with every card in a number of protocol runs, but simply present it with a sequence of transcripts of individual protocol executions, each execution involving a randomly chosen card. The identification phase and the adversary's goal are the same as before. This much more demanding attack scenario models a situation where the adversary cannot interact many times with each distinct card during fingerprinting, but only in one protocol run at a time with a random card.

In Section 4 we consider a different attack scenario and show how an adversary can learn the capabilities of a card (that is, it learns *which* keys are stored on a card). Besides being a serious breach of privacy on its own, this attack can also be combined with the attacks described in this section to gain better performance.

Our attacks in this section specifically target the ShillKey values used by PLAID. A ShillKey pair, generated for every card, contains an RSA public key and an AES key that are to be used in the IA and in the FA phases respectively in place of the actual keys should an error in the terminal message be detected. Intended as a security measure—to prevent attackers from exploiting potential information leaked by error messages—the use of the ShillKey turns out to drastically weaken the anonymity properties of PLAID.

Before explaining the details of the attacks, we note that in order to run the attacks in this section, we need to be able to force each card into replying with RSA ciphertexts generated using its ShillKey in the first phase of the protocol. This can, however, easily be arranged by sending the card a first message containing an empty sequence of KeySetIDs, or a set of KeySetIDs containing a single and particularly high index that is not in use in any card on the system. Thus we may assume that the adversary is able to gather samples of ShillKey ciphertexts from cards at will.

3.1   Tracing Cards via ShillKey Ciphertexts

We consider the following situation: we assume the system has t cards with corresponding ShillKey moduli $N_1, \ldots, N_t$, where each N_i is an n-bit RSA modulus (the current draft version gives $n = 2048$ [14]). Our basic attack considers the following scenario. In a first phase the adversary learns, for every card in the system, k_1 encryptions of a random message under the card's ShillKey (N_j, e_j); then, in a challenge phase the adversary is given k_2 fresh ciphertexts (again for random messages) computed under ShillKey (N_{j^*}, e_{j^*}), for j^* chosen uniformly at random from $\{1, \ldots, t\}$. The adversary's goal is to identify from which card the challenge ciphertexts come, that is, to output the correct index j^*. We define the adversary's advantage as its success probability bounded away from the guessing probability $\frac{1}{t}$.

The idea behind our first attack is that, although each ShillKey (N_j, e_j) is meant to be kept private, each of the k_1 ciphertexts $X_{i,j}$ computed using (N_j, e_j) leaks some information about the modulus N_j. Specifically, we learn that $X_{i,j} < N_j$ for each i. Similarly, in the challenge phase, where we have k_2 ciphertexts computed using (N_{j^*}, e_{j^*}), each ciphertext leaks some information about the challenge ShillKey modulus. Starting from this observation, we now seek a procedure to obtain a good estimate of the ShillKey moduli given only a certain number of corresponding ciphertexts for each modulus.

The problem can be reposed as follows. Notice that each ciphertext $X_{i,j}$ can be regarded as a uniformly random integer in the range $[1, N_j - 1]$. We are then faced with the task of estimating N_j, which is one more than the size of the interval from which the sample comes. This is essentially an instance of a classical statistical problem that is known as the *German Tank Problem*[7]. A naive approach would be to use twice the mean values of the samples $X_{i,j}$ as an estimate for N_j. A statistically strictly better approach is to use as an estimator for N_j the value

$$\tilde{N}_j = m_j + \frac{m_j}{k_1} \, ,$$

where m_j is the maximum value of the observed samples $X_{i,j}$ and k_1 is the number of samples. It basically corresponds to the maximum plus the average distance of observed samples. This estimator arises from a frequentist interpretation of the problem, and has the benefit of providing what is known as a Minimum-variance Uniform Estimator (MVUE). It can be replaced by a more appropriate Bayesian estimator, but the estimator above is sufficient for our purposes.

Our first attack proceeds using this estimator as follows. In the first phase, we use it to produce estimates $\tilde{N}_j$ for each of the ShillKey moduli N_j. In the challenge phase, we again use it to produce an estimate $\tilde{N}^*$ for the challenge ShillKey modulus (now with parameter k_2, representing the number of samples available in that phase). We finally output as our guess for the challenge index j^* the index j for which $\tilde{N}^*$ is closest in absolute value to $\tilde{N}_j$, that is,

$$\arg\min_j \left| \tilde{N}^* - \tilde{N}_j \right| \, .$$

This concludes the description of our first attack.

---

[7] See [17] for a good introduction. The name stems from the problem initially being posed as that of estimating the total number of tanks in the German army from observing a subset of their serial numbers.

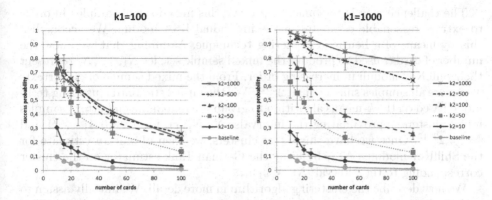

Fig. 2. Simulations of the ShillKey attack. On the left with k_1 (the number of samples during the fingerprinting phase) set to 100 and on the right with k_1 set to 1000. The success probability is averaged over ten runs with one hundred repetitions of the identification phase and the simulation was done with $t = 10, 15, 20, 25, 50, 100$ cards. The baseline indicates the success probability of an adversary that tries to win the game by purely guessing.

Simulation results. We have conducted extensive simulations of the first attack above for various values of t (the number of cards), k_1 (the number of ciphertext samples per card available in the first phase) and k_2 (the number of ciphertext samples in the challenge phase). A selection of our results are depicted in Figure 2.

It can be seen that even with many cards in the system ($t = 100$, say), our attack significantly outperforms simply guessing the card's identity. This is particularly so when the number of samples k_2 available in the challenge phase is large. However, even with k_2 as low as 50, the attack performance is still significantly better than guessing; given the target execution time of the PLAID protocol of 300 ms, this many samples could be gathered from a card within 15 s.

It is also the case that increasing the number of samples available in the first phase of the attack, k_1, improves the attack performance (compare the left and right sub-figures in Figure 2, which correspond to values of k_1 equal to 100 and 1000, respectively). This is particularly accentuated at higher levels of k_2, where the attack's success probability is better than 75%, even for a system with $t = 100$ cards.

3.2 Tracing Cards from a Mixed Set of ShillKey Ciphertexts

For the basic attack in Section 3.1, we assumed that during the initial phase the attacker was able to identify ciphertexts computed from the same key. In our second attack, we relax this assumption: we now give the attacker a large mixed set of $k_1 \times t$ ciphertext samples, each sample coming from a randomly selected card. The challenge phase of the attack proceeds as before where the attacker obtains a small sample of k_2 ciphertexts computed by the same card, and the attacker's goal is to identify this card.

The challenge now is to somehow process this mixed set of samples in order to extract reasonable estimates of the individual RSA moduli. We accomplish this by means of a heuristic clustering technique. Assuming that we know the number of cards t used to produce the mixed sample set, let $N_1, \ldots, N_t$ represent their ShillKey moduli in increasing order. From the mixed sample of ciphertexts we ignore all samples smaller than 2^{2047}. We then use a standard clustering technique based on the k-means algorithm to group the remaining ciphertext samples into t clusters approximating the intervals $[N_j, N_{j+1})$, for $j \in \{0, 1, \ldots, t\}$ and $N_0 = 2^{2047}$. Once we have this set of clusters, we then obtain an estimate for the ShillKey modulus N_{i+1} by using the German Tank estimator on the cluster corresponding to the interval $[N_i, N_{i+1})$.

We now describe the clustering algorithm in more detail. We initially assign to each of the t clusters a uniformly random value in the range $(2^{2047}, 2^{2048})$. This value is called the *centroid* of the cluster. For each ciphertext sample greater than 2^{2047} we calculate its distance from each of the cluster centroids, and assign that ciphertext to the cluster to whose centroid it is closest. The distance metric is merely the absolute value of the arithmetic difference. Once that every ciphertext sample has been assigned to a cluster we ensure that no cluster is empty. If an empty cluster is found, we pick another cluster at random whose size is greater than one and move its largest element to the empty cluster. We then set the centroid of each cluster to be the mean of the ciphertext samples contained in that cluster, as per the standard k-means algorithm. We iterate this process of assigning ciphertext samples to clusters and recalculating their centroids until the centroids converge to stable values, or the maximum number of iterations is exceeded.

In the challenge phase of the attack, the attacker is given k_2 ciphertexts computed by the same card. Here, our attack proceeds identically to the previous one: the attacker uses the estimator to produce an estimate $\tilde{N}^*$ for the challenge ShillKey modulus and outputs as its guess the index of the modulus from the first phase that is closest to $\tilde{N}^*$.

Simulation results. We ran simulations of the above clustering attack for a mixed sample set of size $t \times k_1$ for various values of t (the number of cards) and values of k_1 equal to 100 and 1000. The results of our simulations are depicted in Figure 3.

Being a more ambitious type of attack, the success probabilities for this attack are considerably lower than for the previous attack. Nonetheless we see that our clustering approach is able to correctly identify a card with a probability of roughly 4 times the guessing probability. Notably, as we increase the values of k_1 and k_2 beyond 100 we do not get a corresponding increase in performance as in the previous attack. On the other hand, for low parameter values its performance is comparable to the previous attack. In fact if we compare Figures 2 and 3 we see that for $k_2 = 10$ the two attacks perform almost identically. Therefore if the attacker is limited to a small number of samples (≈ 10) during the identification phase, he can trace cards as effectively without requiring a sorted set of ciphertext samples during fingerprinting.

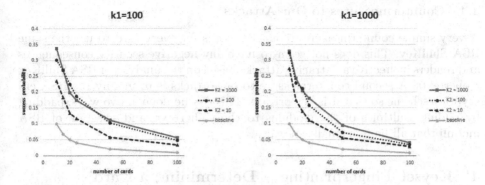

Fig. 3. Simulations of the clustering attack for k_1 equal to 100 and 1000, and varying values of k_2 for each. The success probability is averaged over fifty runs, and the simulation was done with $t = 10, 15, 20, 25, 50, 100$ cards. The baseline indicates the success probability of an adversary that tries to win the game by pure guessing.

3.3   Connection to Key Privacy of RSA Encryption

We remark that our two ShillKey fingerprinting attacks only consider properties of RSA moduli and are, thus, of independent interest in the study of *key privacy* (or *key anonymity*) of RSA encryption, a security notion introduced by Bellare *et al.* in [2]. In the key privacy security model of [2], an adversary plays against two key pairs and is given both the public keys. Security is modelled in terms of key indistinguishability, requiring that it is infeasible for any efficient adversary, which can request encryptions of messages of its choice under one of the two public keys, to tell which key was chosen with probability higher than guessing. As already pointed out in [2], the RSA cryptosystem does *not* provide key privacy. Security is trivially broken when the two key lengths are different. However, RSA keys of the same bit length are easy to tell apart, too: let $N_0 < N_1$ be two RSA moduli: independently of the underlying plaintext, a ciphertext c computed under one of the two corresponding keys satisfies $c < N_b$. A single-query attack which succeeds with non-negligible advantage simply requests to encrypt an arbitrary message and then compares the resulting ciphertext c with the smallest modulus: if $c < N_0$ then it returns 0 and else guesses 1.

This attack is not directly applicable to the PLAID setting because there the RSA encryption keys are kept secret. Moreover, a realistic adversary against card untraceability should play against a number of cards $t \gg 2$. Still, our ShillKey fingerprinting attacks on PLAID can be seen as similar in spirit to, but obviously harder to perform than, the above single-query attack. Moreover, if the encryption scheme used in PLAID were to enjoy key privacy, then the attacks presented would be completely thwarted (and the RSA moduli would no longer need to be kept secret).

3.4 Countermeasures to Our Attacks

A very simple countermeasure to our attacks is for every card to use the same RSA ShillKey. This does not seem to have any negative security consequences and renders ineffective any tracing attacks based on the analysis of RSA ShillKey ciphertexts. A second countermeasure to our attacks is to modify the RSA encryption scheme so that it is key private. This can be done in two ways: padding by adding multiples of the modulus to the ciphertext, and selection of RSA moduli that all lie in a small interval.

4 Keyset Fingerprinting – Determining a Card's Capabilities

In this section we present another type of attack on PLAID's privacy, which we call *keyset fingerprinting*. This attack reveals the exact set of keys a card knows[8], thereby determining its *capabilities* in terms of which keyset it can use, i.e., which specific terminal it is able to talk to. In order to mount the attack, we exploit the following observations: (i) the KeySetIDs list sent by the terminal (in the clear) in the IA Command contains all keys known by the terminal [14, 6.1], and is *not* authenticated[9], (ii) in its IA Response, the card is required to use the first key of the received KeySetIDs list it knows [14, 6.2], and (iii) if the card uses its ShillKey, in the IA Response, then the terminal aborts [14, 6.4].

4.1 The Attack in a Nutshell

We first explain the core idea of our attack by describing a concrete attack scenario. Assume an adversary observes a successful protocol run between a card and a terminal where the latter had sent (in the clear) KeySetIDs $= (2, 5, 8)$. From this, the attacker not only learns that the ICC holds at least one of the keys with IDs $\{2, 5, 8\}$, but it can also determine *all* of the keys the ICC supports, independently of the identifiers announced in KeySetIDs. To this end, the adversary can trigger a protocol run and mount a man-in-the-middle attack as described below.

In a first phase, the attacker sequentially replaces the IFD's original initial message by one containing only a single identifier from the original list of KeySetIDs, that is, 2, 5 or 8 in our example; by observing the subsequent protocol run, the attacker deduces that the ICC supports the selected key if and only if the protocol execution reaches the third step, i.e., if the terminal

---

[8] Recall that terminals announce their supported keysets by sending corresponding KeySetIDs in the clear. As a consequence, any observer can see which keys are related to which resource/terminal.

[9] We note that the unauthenticated nature of the PLAID protocol messages has already been criticized in the national body comments on an earlier ISO draft [13]. In our attack we exploit this weakness, refuting the claim of the current ISO draft [14, Annex H.1.1] that sending KeySetIDs in clear is "of no use to an attacker."

responds with a third message. In a second phase, the attacker sequentially prepends to the IFD's original initial message all key identifiers that were not contained in KeySetIDs, e.g., $(1, 2, 5, 8)$, $(3, 2, 5, 8)$, ..., $(65536, 2, 5, 8)$ in our example. Then, from each of the subsequent protocol runs, the attacker learns that the ICC knows the inserted key if and only if the IFD does *not* respond with a third message. This is because of observation (ii) above about the first matching key in the list to be used. At the end of the two phases, the attacker knows the identifiers of all keys supported by the ICC.

We stress the attack above can be performed in a remote fashion where two attackers, placed in physical proximity to the terminal (respectively, the card) relay the exchanged messages between each other, playing the role of a card (respectively, a terminal). Moreover, this attack can be mounted independently of the values announced in KeySetIDs, as long as the attacker observes a single, successful protocol execution.

Note that knowledge of all the keys supported by a card also reveals its capabilities (e.g., access authorizations), thereby potentially disclosing highly sensitive information. While this is not, in general, sufficient to identify a card uniquely, it effectively allows to derive capability classes, containing cards with the same capabilities. Moreover, in certain scenarios, capabilities like access authorizations might even leak the identity of a card's owner, hence breaking its anonymity, as some keys might be used exclusively to access security-critical infrastructure [14, Annex C] such as, e.g., server rooms or the CEO's office. The impact of keyset fingerprinting is furthermore increased by the remote nature and the low cost of the attack (in terms of the number of interactions between terminal and card). Even in large-scale, realistic scenarios, the attack requires only few seconds (and no physical proximity of card and terminal) to determine a card's capabilities. See Section 4.2 for a more detailed discussion.

We remark that keyset fingerprinting can, in addition, be used as a prefilter for the tracing attacks discussed in Section 3 based on ShillKey Fingerprinting. Recall that the performance of these attacks heavily depends on the number of cards in the system that have to be distinguished. By first performing keyset fingerprinting on the card(s) in question, this number can potentially be reduced substantially (thereby improving the overall efficiency), as the ShillKey Fingerprinting in a second step only has to discriminate amongst the smaller number of cards belonging to the same capability class. Finally, we note that there are cases where the cheaper keyset fingerprinting attack on its own is actually already sufficient for a tracing attack: whenever a traced card has a *unique* set of supported keys (i.e., is the only member in its capability class), this attack is able to uniquely (re)identify that card. Furthermore, keyset fingerprinting suffices to distinguish two cards as long as there is a key supported by only one of the cards.

4.2 The Attack Details

Suppose that we observe a successful authentication between an honest terminal IFD and an honest card ICC. In the course of the protocol execution, the

IFD starts by sending the list KeySetIDs $= (\text{KeySetID}_{i_1}, \ldots, \text{KeySetID}_{i_\ell})$ of (all) KeySetIDs it supports. The keyset fingerprinting attack proceeds in two phases, focusing first on the keys supported by the IFD and then on the remaining keys.

Phase 1. In the first phase, we replace the initial KeySetIDs list with a list containing only one of the keys supported by the IFD at a time, i.e., we replace the first message by (KeySetID_{i_j}) for $j = 1, \ldots, \ell$ in ℓ sequential interactions. We relay the response of the ICC unmodified to the IFD. If the IFD replies with a third message in the jth interaction, we can infer that the ICC knows the key with KeySetID$_{i_j}$. Otherwise, the ICC did not support this key and hence used its ShillKey, leading the IFD to abort.

Phase 2. In the second phase, we prepend the initial KeySetIDs list with one (or multiple, see below) values $\text{KeySetID}_j \notin \{\text{KeySetID}_{i_1}, \ldots, \text{KeySetID}_{i_\ell}\}$ at a time. We relay the response of the ICC unmodified to the IFD. If the IFD replies with a third message, we can infer that the ICC knows none of the prepended keys. Otherwise, the ICC did know at least one of these keys (which the IFD does not support), leading the IFD to abort. This relies on observations (ii) and (iii) above.

We measure the attack costs in terms of the number of interactions between IFD and ICC needed to extract the keys supported by the ICC. In the first phase, which requires ℓ interactions between the IFD and the ICC, we are able to determine exactly which of the keys $\{\text{KeySetID}_{i_1}, \ldots, \text{KeySetID}_{i_\ell}\}$ supported by the IFD the ICC knows. The second phase aims at determining which of the remaining $2^{16} - \ell$ KeySetIDs are known by the ICC. There are different strategies to proceed in Phase 2:

1. The *basic approach* is to simply prepend each one of the $2^{16} - \ell$ KeySetIDs not supported by the IFD one at a time, resulting in $2^{16} - \ell$ interactions in order to determine exactly which of the keys the ICC knows. Together with the first phase, this approach leads to 2^{16} interactions to fingerprint a card.

2. In the *binary search approach*, the set of KeySetIDs is partitioned along a binary tree with the full set of all $2^{16} - \ell$ KeySetIDs at the root, the first half of them as the left child, the second half of them as the right child, etc. In Phase 2, first the root (i.e., all $2^{16} - \ell$ KeySetIDs) is prepended. If the IFD replies, the ICC knows none of these keys and we have thus completed the keyset fingerprinting for the card. Otherwise, both the left half and the right half are prepended (sequentially) and, again, if the IFD replies then the ICC knows none of the prepended keys. This process can be repeated recursively until the IFD replies for each branch.

    Using this approach, we can quickly rule out those parts of the KeySetID space where the ICC does not know any key. More precisely, denote by n the number of keys the ICC knows in total and by ℓ' the number of keys the ICC knows amongst the ℓ keys supported by the IFD. Then we can upper bound the number of interactions needed to fingerprint a card by $(n - \ell') \cdot \log(2^{16}) + \ell = (n - \ell') \cdot 16 + \ell$, since a traversal of the binary search tree in order to pinpoint a single key requires at most $\log(2^{16})$ (i.e., height of the tree) additional interactions.

3. The *binary search with known maximum approach* is a further optimization which is applicable in scenarios where the highest KeySetID in the system, MaxID, is known in advance. In this case, the binary tree can be reduced to the tree having only the $\text{MaxID} - \ell$ remaining unknown KeySetIDs (instead of all $2^{16} - \ell$) as leaves. The number of interactions to fingerprint a card therefore is reduced to $(n - \ell') \cdot \lceil \log(\text{MaxID}) \rceil + \ell$.

When comparing the strategies for the second phase, in the (unlikely) worst case where the card knows all 2^{16} possible keys, the basic approach requires 2^{16} interactions whereas the binary search approach takes approximately 2^{20} interactions. We observe however that the binary search is more efficient as long as the card holds less than 2^{12} keys (which we assume to be the case in any practical scenario).

A practical example. For the sake of providing the reader with some estimates on a more realistic, but still large-scale example, consider a scenario where MaxID = 5000 keys are deployed (enough for, e.g., a large building or a small campus) and the considered terminal and card both hold $\ell = n = 10$ keys, from which $\ell' = 1$ key is known by both.

We chose these parameters in the light of the targeted execution time for PLAID and the resource restrictions imposed by the terminal and card hardware. First, in every execution of the PLAID protocol, the terminal has to perform ℓ RSA decryptions, which is an expensive cryptographic operation for a computationally constrained embedded device.[10] But since the previous ISO draft aims at an overall protocol execution time of less than 300 ms [13, p. 27], this means that ℓ cannot be too large. Second, the card has to store all of its n keys in a protected, tamper-proof memory [14, Annex B]. As this kind of memory is very expensive, it is reasonable to assume that a card can store only a small number of keys. With these parameters, the binary search approach would require $(10 - 1) \cdot 16 + 10 = 154$ interactions to fingerprint the card which, knowing the highest KeySetID MaxID, can be optimized to $(10 - 1) \cdot \log(5000) + 10 \approx 121$ interactions using the binary search with known maximum.

4.3   Potential Countermeasures against Our Attack

As the keyset fingerprinting attack relies heavily on the malleability of the initial KeySetIDs message sent by the terminal, tamper-protecting this message is the obvious way to prevent this attack. One potential and immediate remedy to detect and to prevent tampering with the initial message would be to let the ICC include a hash value of the KeySetIDs value in the plaintext of STR1. The terminal could then check whether the ICC obtained the unmodified initial message by comparing the hash value that it receives with the hash of the original KeySetIDs value. However, a rigorous analysis would be required to put this idea on a profound foundation.

---

[10] For 2048-bit RSA decryptions or signatures, [23] reports times of over 100 ms for mobile devices, while our simulations on an Intel Core i7 2.4 GHz are around 10 ms.

5 Further Security Considerations

Here we discuss further security considerations, and mainly the secrecy of the established keys which, according to the standard, can be optionally used "as a secure messaging, session or encryption key in subsequent sessions." We also point out that the design of PLAID also deviates in several ways from good cryptographic practice. We observe that some of these issues have already been pointed out in the comments on the previous ISO draft version [13].

5.1 Forward (In)security

Forward security [3] demands that one cannot recover session keys generated in the past, even if the long-term secrets of a party become known. In the case of PLAID, the long-term secrets correspond to the secret RSA keys and the FAKey on the terminal side, and to the public RSA keys, DivData, and the diversified keys FAKey$^{(Div)}$ on the card side. The loss of keys of either party immediately reveals all past session keys, and also of future sessions, even if they are executed honestly between the parties and the adversary merely observes these execution traces. Furthermore, revealing a card's secrets also allows the identification, *a-posteriori*, of traces belonging to that card and so breaches privacy in this sense.

Assume first that a terminal's long-term secrets become known to the adversary, and consider the trace of an execution between this terminal with an arbitrary card: the adversary can, analogously to the genuine server, try to decrypt the ciphertext encrypting string ^{e}STR1 under all possible RSA private keys of the terminal, until it succeeds with one key. It then obtains DivData, hence can compute FAKey$^{(Div)}$ by executing AES$^{\text{FAKey}}_{\text{Encrypt}}$(DivData) and then decrypt ^{e}STR2 sent by the honest terminal to recover the session key KeysHash.

Next, suppose that the adversary gets hold of the diversification data DivData and the diversified key FAKey$^{(Div)}$ of a card. It can then try to decrypt ^{e}STR2 with this key to obtain some candidate KeysHash for the session key. The adversary can verify the validity of this candidate by checking that ^{e}STR3 decrypts under the candidate key to the given DivData. This way, the adversary is able to identify traces belonging to the specific card and to determine correct session keys of the card.

Most importantly, any such breach would lead to the disclosure of the payload data which may be highly sensitive (for example, a user's biometric data).

5.2 Key (In)security in the Bellare–Rogaway Model

The PLAID protocol specifies the option of reusing the negotiated session key KeysHash for subsequent secure communication. We comment on possible consequences of doing so. Our starting point is the widely-used Bellare–Rogaway (BR) security model [4] for key exchange protocols. This model demands that all session keys should look random to the adversary. Neglecting technical details, this is formalized by presenting the adversary either the genuine session key or an independent random key and challenging it to decide which is the case.

This immediately requires of a protocol that its session keys are not themselves used in a non-trivial way in the key exchange steps, otherwise the adversary can try to test the given key against a protocol execution trace. In the specific case of PLAID, the adversary can try to decrypt ^{e}STR3 with the given key, and will recover a meaningful plaintext with overwhelming probability if and only if this key equals the genuine key KeysHash. Thus PLAID *cannot* achieve security in the BR model.

Note that the lack of security in the BR sense does not necessarily imply that a protocol is insecure. It merely means that other models must be used to assess its security. PLAID is not unique in this respect: a prominent example of a protocol not achieving BR security is TLS, leading researchers to investigate various alternative security evaluations [15,20,7,12,5]. The usage of the session key in the exchange step is often alleviated by the fact that messages in this part and in the channel protocol differ in format, e.g., if a counter value is used and incremented with each application. This form of "domain-separation", however, is not necessarily given in case of PLAID, because the subsequent channel message format has not been specified.

Interestingly, PLAID could easily avoid the problems with the session key being used in the key exchange phase. Recall that the session key KeysHash for AES (with 128 bits) is derived as the first 128 bits of the hash value SHA-256(RND1 ‖ RND2). Since the hash value has 256 bits one could easily use the remaining 128 bits as the AES-128 key for the final message in the key exchange step, and then switch to KeysHash as before in the channel protocol. In the original protocol the card in some sense demonstrates knowledge of FAKey$^{(\text{Div})}$ by being able to decrypt the terminal's message and answer under the derived key. This would still be true with the proposed modification. Note however that this modification still requires a formal security treatment.

5.3   On the Applicability of Bleichenbacher's Attack

Recall that PLAID uses PKCS#1 v1.5 padding for RSA encryption. The accompanying protocol description [8] argues that there is no need to use OAEP padding, because "PLAID doesn't expose the modulus or any other RSA primitive" and that "there is a significant performance advantage in using PKCS#1 v1.5 padding." While we do not feel inclined to comment on the performance related issue, the first part of the argument is debatable in light of the fact that exposure of a card's secrets does reveal the public keys. Further, our attacks in the previous sections show that some information about the moduli is revealed, and the exponent e may be fixed. We note that the comments section in the previous ISO version of PLAID [13] also asks for investigations of the possibility of mounting Bleichenbacher's attack.

Once the RSA public key is known one can in principle mount Bleichenbacher's attack [6] on PKCS# v1.5 padding. In this attack the adversary takes a ciphertext $c \in \mathbb{Z}_N^*$ of some unknown padded message m and "shifts" the message by multiplying c with a random $s^e \bmod N$. With sufficiently high probability the derived "message" $sm \bmod N$ is PKCS#1 v1.5 padding compliant.

The adversary could thus potentially deduce information about m in case of an error message[11] indicating correct or incorrect padding, and given sufficiently many error messages, recover m. The attack has been significantly improved in a series of papers, e.g., [16,21].

For PLAID, the message format carries some redundancy in terms of repeating RND1. Hence, most likely the shifted message $sm \mod N$ will not be accepted by the terminal in any case, independently of the padding. However, the detailed behaviour is implementation-specific. For example, the current implementation is based on the JavaCard framework and the decryption procedure of PKCS#1 v1.5 merely throws an exception in case of incorrect padding and leaves it up to the higher level program to treat this exception.

5.4   CBC-Mode Encryption

PLAID proposes to use CBC-mode encryption based on AES. The standard explicitly demands that the initialization vector IV is set to the all-zero string for both ^{e}STR2 (from the terminal to the card) and ^{e}STR3 (from the card to the terminal). This usage does not conform with standard practice, which demands the use of random IVs to achieve security against chosen plaintext attacks. As remarked before, the PLAID specification states that padding is only applied "if necessary" and is thus not compliant with ISO/IEC 9797-1 padding method 2, where padding is always applied. Indeed, this imprecision makes the standard unimplementable as currently specified, since there will be cases arising during decryption where it is not possible to discern whether padding should be removed or not. It is well-known that CBC-mode encryption is especially vulnerable to padding oracle attacks [27], and that careful implementation is needed to avoid them. The lack of precision in this aspect of the PLAID specification does not bode well.

It is also now well-understood in the cryptographic community that CBC-mode encryption does not offer sufficient integrity guarantees on its own to provide adequate security against active attacks. The usual solution is to add explicit integrity protection through the application of a MAC algorithm to the CBC-mode ciphertext. PLAID does not do so, and a justification for why this lack of integrity does not endanger security was requested in the comments of the previous ISO standard draft [13], but was not addressed in the latest version [14].

PLAID does offer mild forms of plaintext integrity. For example, STR2 contains the session key KeysHash computed as the hash of RND1 ∥ RND2 while STR3 contains DivData. These elements can be checked for after decryption by the relevant party, and this would detect some forms of adversarial plaintext manipulation through simple bit-flipping in the corresponding ciphertexts ^{e}STR2 and ^{e}STR3. However, it is easy to see that an attacker can still manipulate other fields in STR2 and STR3 by bit-flipping in ciphertexts (even with a fixed IV).

---

[11] The protocol explicitly notes that no error messages should be issued, but wrong implementations or side-channel attacks may reveal such information.

While this lack of integrity has not led us to the discovery of specific attacks on PLAID, it is a worrying feature that could be easily avoided through the application of mainstream cryptographic design principles.

5.5 Entity Authentication

Note that both parties, IFD (reader) and ICC (card), basically authenticate one another by proving knowledge of a secret key. For the terminal this is done via the secret RSA key, whereas the card uses its unique DivData and therefore unique key $\text{FAKey}^{(\text{Div})} = \text{AES}^{\text{FAKey}}_{\text{Encrypt}}(\text{DivData})$. The standard mechanism to do so would be either to compute a signature or a message authentication code for a random challenge, or to return, in clear, a nonce encrypted under the party's key.

PLAID follows the encryption-based approach. Yet the usual security argument for this type of authentication requires chosen-ciphertext security for the deployed encryption scheme. PLAID, on the other hand, uses two encryption schemes which are known not to provide this level of security, i.e., RSA-PKCS#1 v1.5 and plain AES-CBC-encryption. This does not mean that the protocol is insecure and does not provide any form of entity authentication. However one cannot infer security from known results but would instead need carefully constructed *de novo* arguments.

5.6 Payload Insecurity

During the ISO standardization process the PLAID protocol was changed to introduce an optional payload field in the third protocol message, the second message from the IFD to the card (see Step 5 in Figure 1) [13]. The standard motivates the purpose of this payload field—this field should not be confused with the payload field in the last message by the card in Step 7 (Figure 1)—for scenarios where, for example, a user enters a PIN and the verification should be done on the card. In this case, the PIN is to be sent to the card within the payload field [14, Annex G]. The problem is that sensitive information send by the IFD can always be intercepted via a simple man-in-the-middle attack, assuming an adversary has corrupted any card. Breaking a card allows the attacker to learn the diversified $\text{FAKey}^{(\text{Div})}$ of the card, the card's DivData field as well as the public part of one (or more) IAKeys. Thus, an adversary can simply replace the second message during an honest execution with one corresponding to the broken card. This will lead to the IFD encrypting the third message under the $\text{FAKey}^{(\text{Div})}$ of the broken card and hence, if a PIN (or any other payload) is included, the adversary can trivially learn it by a simple decryption operation. Note that this problem is not due to the payload being sent in the third message, but that a user, when entering a PIN, cannot tell whether or not the terminal is actually communicating with his/her card. Therefore, PIN comparison on the card as proposed [14, Annex G] is generically insecure due to the given attack scenario.

5.7   On the Impossibility of Key Revocation

Although PLAID uses a public-key encryption system (i.e. RSA) during the initial authentication phase, the overall setup resembles more a symmetric setting where all static keys used by parties are exchanged during system setup (abstracting away the diversification procedure of PLAID). As a consequence, it is not possible to revoke any compromised keys within PLAID. In order to exemplify the resulting consequences, assume that an attacker is able to break into an IFD (terminal). The IFD contains a list of IAKeys and a list of FAKeys which thereby are revealed to the attacker. With this information, the attacker can generate arbitrary new cards with the capabilities of any of the KeySetIDs known by the broken IFD. Furthermore, there is no way to revoke the compromised keys in the system without issuing new cards, as the keys known by IFDs are hardcoded into the cards. Thus, even the break of a single IFD can lead to an entire PLAID setup becoming insecure.

5.8   Key Legacy Attack

A *key legacy attack* is related to the same issues allowing for the KeySet fingerprinting attack, namely, the lack of authentication in the list of keys used by a card and a terminal to establish a connection. Recall that the protocol specifies that the first commonly shared key in the list has to be used, even if there are other shared keys. This means that an adversary could force the card to use one particular key (among those supported by both card and terminal) by *reordering* the list of keys sent by the terminal in a man-in-the-middle fashion. This could be dangerous in case one or more of the keys in the system are compromised, or turn out to provide inferior security for any reason, even if the use of these keys is de-prioritized (e.g., by having the terminals set them always last in order of preference). We note that this type of attack was already mentioned in the national body comments to the first ISO draft [13], but remained unconsidered in the current version [14].

6   Conclusion

Our results show that PLAID has significant privacy weaknesses. The shill key attack and the keyset fingerprinting attack reveal card identifying information and, via access authorizations, information about the card holder. As for entity authentication and the secrecy of established keys for subsequent communication, in several places the design of PLAID follows some uncommon strategies and reveals potential attack vectors, such as the lack of forward security. The case of PLAID also shows that standards should specify details thoroughly, in order to avoid vulnerable implementations. An example here is the ISO/IEC 9797-1 non-compliant CBC padding in PLAID, which potentially enables padding attacks (see our remark in Section 5).

We do not recommend the indiscriminate usage of PLAID in its current form, especially not for privacy-critical scenarios. While our proposed countermeasures seem to thwart our attacks on privacy, a more comprehensive analysis of the protocol in light of clearly stated security goals would be necessary. The PLAID description promises that the protocol should be scrutinized by "the most respected cryptographic organisations, as well as the broader cryptographic community" [8]. Unfortunately, we are not aware of any available documents in this regard. Indeed, standardization processes in general would benefit if supporting material, arguing the security of a proposal, was available at the time of evaluation.

Acknowledgments. We thank Pooya Farshim for his contributions during the early stages of this paper and the anonymous reviewers for valuable comments. Marc Fischlin is supported by the Heisenberg grants Fi 940/3-1 and Fi 940/3-2 of the German Research Foundation (DFG). Tommaso Gagliardoni and Felix Günther are supported by the German Federal Ministry of Education and Research (BMBF) within EC SPRIDE. Felix Günther and Giorgia Azzurra Marson are supported by the DFG as part of the CRC 1119 CROSSING. Giorgia Azzurra Marson and Arno Mittelbach are supported by the Hessian LOEWE excellence initiative within CASED. Kenneth G. Paterson and Jean Paul Degabriele are supported by EPSRC Leadership Fellowship EP/H005455/1.

References

1. Standards Australia: AS 5185-2010 Protocol for Lightweight Authentication of IDentity (PLAID). Standards Australia (2010)
2. Bellare, M., Boldyreva, A., Desai, A., Pointcheval, D.: Key-privacy in public-key encryption. In: Boyd, C. (ed.) ASIACRYPT 2001. LNCS, vol. 2248, pp. 566–582. Springer, Heidelberg (2001)
3. Bellare, M., Pointcheval, D., Rogaway, P.: Authenticated key exchange secure against dictionary attacks. In: Preneel, B. (ed.) EUROCRYPT 2000. LNCS, vol. 1807, pp. 139–155. Springer, Heidelberg (2000)
4. Bellare, M., Rogaway, P.: Entity authentication and key distribution. In: Stinson, D.R. (ed.) CRYPTO 1993. LNCS, vol. 773, pp. 232–249. Springer, Heidelberg (1994)
5. Bhargavan, K., Fournet, C., Kohlweiss, M., Pironti, A., Strub, P.-Y., Zanella-Béguelin, S.: Proving the TLS Handshake Secure (as it is). In: Garay, J.A., Gennaro, R. (eds.) CRYPTO 2014, Part II. LNCS, vol. 8617, pp. 235–255. Springer, Heidelberg (2014)
6. Bleichenbacher, D.: Chosen ciphertext attacks against protocols based on the RSA encryption standard PKCS #1. In: Krawczyk, H. (ed.) CRYPTO 1998. LNCS, vol. 1462, pp. 1–12. Springer, Heidelberg (1998)
7. Brzuska, C., Fischlin, M., Smart, N.P., Warinschi, B., Williams, S.C.: Less is more: relaxed yet composable security notions for key exchange. Int. J. Inf. Sec. 12(4), 267–297 (2013)

8. Centrelink: Protocol for Lightweight Authentication of Identity (PLAID) — Logical Smartcard Implementation Specification PLAID Version 8.0 - Final (December 2009), http://www.humanservices.gov.au/corporate/publications-and-resources/plaid/technical-specification

9. Dagdelen, Ö., Fischlin, M., Gagliardoni, T., Marson, G.A., Mittelbach, A., Onete, C.: A cryptographic analysis of OPACITY (extended abstract). In: Crampton, J., Jajodia, S., Mayes, K. (eds.) ESORICS 2013. LNCS, vol. 8134, pp. 345–362. Springer, Heidelberg (2013)

10. Department of Human Services: Protocol for Lightweight Authentication of Identity, PLAID (2014), http://www.humanservices.gov.au/corporate/publications-and-resources/plaid/

11. Freedman, G.: Personal communication by e-mail (July 2014)

12. Giesen, F., Kohlar, F., Stebila, D.: On the security of TLS renegotiation. In: Sadeghi, A.R., Gligor, V.D., Yung, M. (eds.) ACM CCS 2013, pp. 387–398. ACM Press (November 2013)

13. ISO: Draft International Standard ISO/IEC DIS 25185-1 Identification cards — Integrated circuit card authentication protocols — Part 1: Protocol for Lightweight Authentication of Identity. International Organization for Standardization, Geneva, Switzerland (2013)

14. ISO: Draft International Standard ISO/IEC DIS 25185-1.2 Identification cards — Integrated circuit card authentication protocols — Part 1: Protocol for Lightweight Authentication of Identity. International Organization for Standardization, Geneva, Switzerland (2014)

15. Jager, T., Kohlar, F., Schäge, S., Schwenk, J.: On the security of TLS-DHE in the standard model. In: Safavi-Naini, R., Canetti, R. (eds.) CRYPTO 2012. LNCS, vol. 7417, pp. 273–293. Springer, Heidelberg (2012)

16. Jager, T., Schinzel, S., Somorovsky, J.: Bleichenbacher's attack strikes again: Breaking PKCS#1 v1.5 in XML encryption. In: Foresti, S., Yung, M., Martinelli, F. (eds.) ESORICS 2012. LNCS, vol. 7459, pp. 752–769. Springer, Heidelberg (2012)

17. Johnson, R.: Estimating the size of a population. Teaching Statistics 16(2), 50–52 (1994), http://www.mcs.sdsmt.edu/rwjohnso/html/tank.pdf

18. Kiat, K.H., Run, L.Y.: An Analysis of OPACITY and PLAID Protocols for Contactless Smart Cards. Master's thesis, Naval Postgraduate School, Monterey, CA, USA (September 2012)

19. Kline, R.: Improving contactless security is goal of emerging PLAID project, secureIDNews (January 2010), http://secureidnews.com/news-item/improving-contactless-security-is-goal-of-emerging-plaid-project/

20. Krawczyk, H., Paterson, K.G., Wee, H.: On the security of the TLS protocol: A systematic analysis. In: Canetti, R., Garay, J.A. (eds.) CRYPTO 2013, Part I. LNCS, vol. 8042, pp. 429–448. Springer, Heidelberg (2013)

21. Meyer, C., Somorovsky, J., Weiss, E., Schwenk, J.: Revisiting SSL/TLS Implementations: New Bleichenbacher Side Channels and Attacks. In: 23rd USENIX Security Symposium (USENIX Security 2014). USENIX Association, San Diego (2014), https://www.usenix.org/conference/usenixsecurity14/technical-sessions/presentation/meyer

22. National Institute of Standards and Technology: Protocol for Lightweight Authentication of Identity (PLAID) Workshop (July 2009), http://csrc.nist.gov/news_events/plaid-workshop/

23. Rifà-Pous, H., Herrera-Joancomartí, J.: Computational and energy costs of cryptographic algorithms on handheld devices. Future Internet 3(1), 31–48 (2011)

24. Risky.biz: Risky Business 106 — Centrelink's new PLAID auth protocol (May 2009), http://risky.biz/netcasts/risky-business/risky-business-106-centrelinks-new-plaid-auth-protocol

25. Sakurada, H.: Security evaluation of the PLAID protocol using the ProVerif tool (September 2013), http://crypto-protocol.nict.go.jp/data/eng/ISOIEC_Protocols/25185-1/25185-1_ProVerif.pdf

26. Taylor, J.: Centrelink ID protocol still in trial phase, zDNet (May 2012), http://www.zdnet.com/centrelink-id-protocol-still-in-trial-phase-1339336953/

27. Vaudenay, S.: Security flaws induced by CBC padding - applications to SSL, IPSEC, WTLS ... In: Knudsen, L.R. (ed.) EUROCRYPT 2002. LNCS, vol. 2332, pp. 534–546. Springer, Heidelberg (2002)

28. Watanabe, D.: Security analysis of PLAID (September 2013), http://crypto-protocol.nict.go.jp/data/eng/ISOIEC_Protocols/25185-1/25185-1_Scyther.pdf

The SPEKE Protocol Revisited

Feng Hao and Siamak F. Shahandashti

Newcastle University, UK
{feng.hao,siamak.shahandashti}@ncl.ac.uk

Abstract. The SPEKE protocol is commonly considered one of the classic Password Authenticated Key Exchange (PAKE) schemes. It has been included in international standards (particularly, ISO/IEC 11770-4 and IEEE 1363.2) and deployed in commercial products (e.g., Blackberry). We observe that the original SPEKE specification is subtly different from those defined in the ISO/IEC 11770-4 and IEEE 1363.2 standards. We show that those differences have critical security implications by presenting two new attacks on SPEKE: an impersonation attack and a key-malleability attack. The first attack allows an attacker to impersonate a user without knowing the password by engaging in two parallel sessions with the victim. The second attack allows an attacker to manipulate the session key established between two honest users without being detected. Both attacks are applicable to the original SPEKE scheme, and are only partially addressed in the ISO/IEC 11770-4 and IEEE 1363.2 standards. We highlight deficiencies in both standards and suggest concrete changes.

1 Introduction

Password Authenticated Key Exchange (PAKE) is a protocol that aims to establish a secure communication channel between two remote parties based on a shared low-entropy password without relying on any external trusted parties. Since the seminal work by Belloven and Merrit in 1992 [2], many PAKE protocols have been proposed, and some have been standardised [10, 11].

The Simple Password Exponential Key Exchange (SPEKE) protocol is one of the most well-known PAKE solutions. It was originally designed by Jablon in 1996 [7]. Although some concerns have been raised [8, 9], no major flaws seems to have been uncovered. Over the past decade, SPEKE has been included in the IEEE P1362.2 [10] standard draft[1] and ISO/IEC 11770-4 [11]. Furthermore, SPEKE has been deployed in commercial applications – for example in Black-Berry devices produced by Research in Motion [6] and in Entrust's TruePass end-to-end web products [5].

In this paper, we revisit the original SPEKE protocol and review its specifications in the two standardisation documents: IEEE P1363.2 and ISO/IEC 11770-4. We observe that the original protocol is subtly different from those

---

[1] At the time of writing, the latest draft available on the IEEE P1363.2 website is D26. See http://grouper.ieee.org/groups/1363/passwdPK/draft.html

L. Chen and C. Mitchell (Eds.): SSR 2014, LNCS 8893, pp. 26–38, 2014.

defined in the standards. The reason for the difference, or deviation from the original specification, is not justified clearly in the standards.

During this investigation, we have identified several issues with SPEKE that have not been reported before. Our findings are summarised below:

1. We show that the original SPEKE protocol is subject to an impersonation attack when the victim is engaged in two parallel sessions with an active attacker. The attacker is able to achieve mutual authentication in both sessions without knowing the password.
2. We show that the original SPEKE protocol is subject to a key-malleability attack. The attacker, sitting in between two honest users, is able to manipulate the session key without being detected.
3. While both attacks clearly succeed against the original SPEKE protocol, we show they are partially addressed in IEEE P1363.2 and ISO/IEC 11770-4, but not in any rigorous manner. We propose explicit and concrete changes to both standards.

Details of our findings are explained in the following sections.

2   The Original SPEKE Scheme

First, we define the original SPEKE scheme based on Jablon's 1996 paper [7]. Let p be a safe prime, $p = 2q + 1$ where q is also a prime. Assume two remote parties, Alice and Bob, share a common password s. SPEKE defines a function $f(\cdot)$ to map a password s to a group element: $f(s) = s^2 \bmod p$. We use g to denote the result returned from $f(s)$, i.e., $g = f(s)$. The SPEKE protocol provides implicit authentication in one round, which is defined below. (Unless stated otherwise, all modular operations are performed modulo p, hence the explicit $\bmod p$ is omitted for simplicity.)

SPEKE (one round). *Alice selects* $x \in_R [1, q - 1]$ *and sends* g^x *to Bob. Similarly, Bob selects* $y \in_R [1, q - 1]$ *and sends* g^y *to Alice.*

Upon receiving the sent data, Alice verifies that g^y is within $[2, p - 2]$. This is to ensure the received element does not fall into the small subgroup of order two, which contains $\{1, p - 1\}$. Alice then computes a session key $\kappa = H((g^y)^x) = H(g^{xy})$, where H is a secure one-way hash function. Similarly, Bob verifies that g^x is within $[2, p - 2]$. He then computes the same session key $\kappa = H((g^x)^y) = H(g^{xy})$.

To provide explicit key confirmation, the SPEKE paper defines the following procedure. One party sends $H(H(\kappa))$ and the other party replies with $H(\kappa)$. The paper does not specify who must initiate the key confirmation and hence leaves it as a free choice for specific applications to decide.

3 Previously Reported Attacks

In 2004, eight years after SPEKE was initially designed, Zhang presented an exponential-equivalence attack [8]. The attack is based on the observation that some passwords are exponentially equivalent. Hence, an active attacker can exploit that equivalence to test multiple passwords in one protocol execution. This is especially problematic when the password is digits-only, e.g., a Personal Identification Numbers (PIN). As a countermeasure, Zhang proposed to hash the password before taking the square operation. In other words, he redefined the password mapping function as: $f(s) = (H(s))^2 \mod p$. The hashing of passwords makes it much harder for the attacker to find exponential equivalence among the hashed outputs. Zhang's attack is acknowledged in IEEE P1363.2 [10], which adds a hash function in SPEKE when deriving the base generator from the password.

In 2005, Tang and Mitchell presented three attacks on SPEKE [9]. The first attack is similar to Zhang's [8] – an on-line attacker tests multiple passwords in one execution of the protocol by exploiting the exponential equivalence of some passwords. The second attack assumes that the user shares the same password with two servers, say $S1$ and $S2$. By relaying the messages between the client and $S2$, the attacker may trick the client into believing that she shares a key with $S1$, but actually the key is shared with $S2$. The authors call this an "unknown key-share" attack. They suggest to address this attack by including the server's identifier into the computation of g. (However, we note that this suggested countermeasure has the side-effect of breaking the symmetry of the original protocol.) The third attack indicates a generic vulnerability. In this scenario, two honest parties launch two concurrent sessions. The attacker can swap the messages between the two sessions to exchange the two session keys. The two communicating parties will be able to decrypt messages successfully but they may get confused about which message belongs to which session.

4 New Attacks

In this section, we describe two new attacks: an impersonation attack and a key-malleability attack. The first attack indicates a practical weakness in the original design of SPEKE, while the second attack has an unfavourable implication on the theoretical analysis of the protocol.

4.1 Impersonation Attack

The impersonation attack works when the user is engaged in several sessions in parallel with another user. This is a realistic scenario in practice as two users may want to run several concurrent SPEKE key exchange sessions and use each established channel for a specific application, as explained by Tang and Mitchell [9].

We assume Alice and Bob share a common password. Their respective identities are denoted by $\hat{A}$ and $\hat{B}$. Without loss of generality, we assume Alice

Alice		Mallory (impersonating Bob)
Select $x \in_R$ [1,q-1]	$1.\ g^x, \hat{A}$ $\longrightarrow$	
Compute $\kappa = H(g^{xyz})$	$4.\ g^{y \cdot z}, \hat{B}$ $\longleftarrow$	Choose arbitrary z (Session 1)
Start key confirmation	$5.\ H(H(\kappa)), \hat{A}$ $\longrightarrow$	
Verify key confirmation	$8.\ H(\kappa), \hat{B}$ $\longleftarrow$	
		$\{g^{xz}, H(H(\kappa))\} \downarrow\uparrow \{g^y, H(\kappa)\}$
Select $y \in_R$ [1,q-1]	$2.\ g^{x \cdot z}, \hat{B}$ $\longleftarrow$	
Compute $\kappa = H(g^{xyz})$	$3.\ g^y, \hat{A}$ $\longrightarrow$	(Session 2)
Verify key confirmation	$6.\ H(H(\kappa)), \hat{B}$ $\longleftarrow$	
Reply key confirmation	$7.\ H(\kappa), \hat{A}$ $\longrightarrow$	

Fig. 1. Impersonation attack on SPEKE

initiates a SPEKE session – which we call Session 1 – with Bob by sending g^x (see Figure 1; we append the sender's purposed identity in the key exchange flow to make the illustration of the attack clearer). But the message is intercepted by Mallory. Mallory chooses an arbitrary z from $[2, p-2]$ and raises the intercepted g^x by the power of z to obtain g^{xz}. Pretending to be "Bob", Mallory initiates another SPEKE session – which we call Session 2 – with Alice by sending g^{xz}. The use of z serves to make the messages different between the two sessions. In the second session, Alice replies with g^y. Mallory raises this item to the power of z to obtain g^{yz}, and sends the result to Alice as the reply in Session 1. Following the key confirmation procedure as in the original SPEKE paper, Alice provides the first key confirmation challenge in Session 1 $H(H(\kappa))$, which is subsequently relayed to Session 2 as Bob's key confirmation challenge. In Session 2, Alice answers the key confirmation challenge by replying with $H(\kappa)$, which is then relayed in Session 1 to complete the mutual authentication in both sessions. Recall that in a Password Authenticated Key Exchange protocol, the notion of "authentication" is defined based on the knowledge of a secret password. However, without knowing the password, Mallory has been successfully authenticated by Alice as "Bob", someone who supposedly shares the exclusive knowledge with Alice about a secret password.

In essence, this impersonation attack follows the "wormhole attack" [3], in which the attacker relays the sender's message back to the sender in order to pass authentication. However, the "wormhole attack" presented in [3] works in a PKI-based key exchange setting while the attack reported here occurs in a password-based key exchange setting. The two settings are distinct. Nonetheless, both attacks highlight the importance of including explicit user identities in the authenticated key exchange process.

To some extent, this impersonation attack is similar to the "unknown key-share" attack described in Tang-Mitchell's paper [9]. However, our attack seems to be more feasible and more harmful than theirs. The main difference is that in our attack, the attacker changes the user's message and sends the modified message back to the user herself (instead of to a third party as in [9]). At the end of the key establishment process, Alice thinks she is sharing a session key with the real "Bob", but she is actually sharing the key with another instance of herself. This confusion of identity in the key establishment can cause problems in some scenarios. For example, using the derived session key κ in an authenticated mode (e.g., AES-GCM), Alice may send an encrypted message to "Bob": *"Please pay Charlie 5 bitcoins"*. But Mallory can relay the message back to Alice in the second session. Since the message is verified to be authentic from "Bob", Alice may follow the instruction and pay Charlie instead (in a practical application, Alice is likely an automated program that follows the protocol). Thus, although Alice's initial intention is to make "Bob" pay Charlie 5 bitcoins, she ends up paying Charlie instead. In this attack, the supposed "Bob" seems to be liable but the real Bob is actually never involved.

4.2 Key-Malleability Attack

A second attack is called the key-malleability attack. In this attack, the attacker sits in the middle between two honest users (see Figure 2). The attacker chooses an arbitrary z within the range of $[2, q - 1]$, raises the intercepted item to the power of z and passes it on. The users at two ends are still able to derive the same session key $\kappa = H(g^{xyz})$, but without being aware that the messages have been modified.

We do not claim there is a direct practical harm caused by this attack. However, the fact that an attacker is able to manipulate the session key without being detected may have significant implications on the theoretical analysis of the protocol. In the original SPEKE paper, the protocol comes with no security proofs[2]. However, it is heuristically argued that the security of the session key in SPEKE depends on either the Computational Diffie-Hellman assumption (i.e., an attacker is unable to compute the session key) or the Decisional Diffie-Hellman assumption (i.e., an attacker is unable to distinguish the session key from random). The existence of such a key-malleability attack suggests that a tight reduction to CDH or DDH is impossible. The attacker's ability to inject randomness into the session key without being noticed can significantly complicate the theoretical analysis. As an example, let us assume the attacker chooses

---

[2] In 2001, MacKenzie published a manuscript "On the Security of the SPEKE Password-Authenticated Key Exchange Protocol" on IACR ePrint 2001/057. Although an ePrint manuscript is usually not regarded as a formal peer-reviewed publication, it is cited in IEEE P1363.2 [10] to support the argument that the SPEKE protocol has been formally proved to be secure based on some number theoretical assumptions. We observe that the formally analysed SPEKE is substantially different from the original SPEKE: e.g., 1) the derivation of the generator is different, and as a result, the protocol is two-round instead of one-round as in the original SPEKE paper or the standards; 2) the definition of the key confirmation function is different from that in the original paper or standards; 3) it regards the key confirmation as mandatory rather than optional.

Alice ($\hat{A}$)	MITM	Bob ($\hat{B}$)
Select $x \in_R [1, q-1]$ $\quad \xrightarrow{g^x, \hat{A}}$	Select $z \in [2, q-2]$	$\xleftarrow{g^y, \hat{B}}$ Select $y \in_R [1, q-1]$
Check $(g^y)^z \in [2, p-2]$ $\xleftarrow{(g^y)^z, \hat{B}}$ Raise to power z $\xrightarrow{(g^x)^z, \hat{A}}$ Check $(g^x)^z \in [2, p-2]$
Compute $\kappa = H(g^{xyz})$		Compute $\kappa = H(g^{xyz})$

Fig. 2. Key-malleability attack on SPEKE

z as a result of an arbitrary function with the incepted inputs, i.e., $z = f(g^x, g^y)$. Because of the correlation of items on the exponent, standard CDH and DDH are no longer applicable here (recall that in the CDH/DDH assumption, the secret values on the exponent are assumed to be independent).

5   Discussion

While the two attacks clearly work on the original SPEKE protocol [7], it may be arguable whether they are applicable to the variants of SPEKE defined in IEEE P1363.2 and ISO/IEC 11770-4. In this section, we explain the difference between the original protocol and its variants in the standards in relation to the two attacks.

5.1   Explicit Key Confirmation

First of all, we observe that the key confirmation procedure of SPEKE defined in the standards is different from that in the original SPEKE paper [7]. For example, in ISO/IEC 11770-4, the key confirmation works as follows [11] (the procedure in IEEE P1363.2 [10] is basically the same).

$$\text{Alice} \rightarrow \text{Bob} : H(\text{"0x03"} \| g^x \| g^y \| g^{xy} \| g)$$
$$\text{Bob} \rightarrow \text{Alice} : H(\text{"0x04"} \| g^x \| g^y \| g^{xy} \| g)$$

As explicitly stated in the ISO/IEC 11770-4 standard, there is no order in the above two steps[3]. Either party is free to send out the key confirmation message without waiting for the other party.

Effect on impersonation attack We observe that the above key confirmation procedure does not prevent the impersonation attack. The attacker is still able to relay the key confirmation string in one session to another parallel session to accomplish mutual authentication in both sessions without being detected. The attack works largely because the session keys are identical in the two sessions.

---

[3] In the same standard, it is also stated that there is no order during the SPEKE exchange phase. We find the two statements contradictory: the fact that g^x comes before g^y in the definition of key confirmation implies there is an order during the key exchange phase.

Effect on Key-malleability attack The key-malleability attack no longer works with the key confirmation procedure defined in ISO/IEC 11770-4 (and IEEE P1363.2). However, it is worth noting that the key confirmation procedures in both standards are marked as "optional". Hence, the key-malleability attack is not completely addressed.

5.2   Definition of Password

In the original SPEKE paper, the mapping of a password s to a group element over the prime field is simply achieved by $f(s) = s^2$. To prevent Zhang's exponential-equivalence attack, it is necessary to add a hash function before performing the squaring operation, i.e., $f(s) = (H(s))^2$. This is essentially the mapping function defined in ISO/IEC 11770-4 and IEEE P1363.2 (for the case that p is a safe prime). However, the definition of the shared secret is subtly changed in both standards. For example, in IEEE P1363.2 [10], the shared low-entropy secret (denoted π in the standard document) is defined as follows:

> *"A password-based octet string, used for authentication. π is generally derived from a password or a hashed password, and may incorporate a salt value, identifiers for one or more parties, and/or other shared data."*

It is worth nothing that in the above definition, the incorporation of "a salt value, identifiers for one or more parties, and/or other shared data" is not mandatory (as indicated by the use of the word "may").

In ISO/IEC 11770-4 [11], the shared low-entropy secret is defined as follows with an additional note:

> *"A password-based octet string which is generally derived from a password or a hashed password, identifiers for one or more entities, an identifier of a communication session if more than one session might execute concurrently, and optionally includes a salt value and/or other data.*
>
> *NOTE - it is required to include one or more the entity identifiers and a unique session identifier into the value of π, in order to avoid that a key establishment mechanism might be vulnerable to an unknown key share attack addressed in [TC05]."*

The above definition seems to include the "identifiers for one or more entities" as part of the shared secret. However, the standard does not provide any formula. It is not even clear if one or both entities' identifiers should be included, and if only one identifier needs to be included, which one and how. Furthermore, the word "generally" weakens the rigour in the definition and makes it subject to potentially different interpretations. The note below the definition states that: the inclusion of the entity/session identifiers is required to address the UKS attack in [TC05] [9]. However, the UKS attack reported in [9] works under the assumption that the user shares the same password with two different servers. In many applications, it is often considered reasonable to exclude that assumption from the threat model (otherwise, the solution may become overly complex). In those cases, the justification becomes no longer valid. Therefore, on whether the entity/session identifier "should", "must" or "may" be included as part of the shared secret, we find the current ISO/IEC 11770-4 standard not sufficiently clear.

Effect on impersonation attack. Strictly speaking, if the entity identifiers (or the session identifiers) are included in the definition of the shared secret, the impersonation attack presented in Section 4 will not work. In the IEEE definition, the inclusion of the entity identifiers is clearly not mandatory, hence the impersonation attack should be applicable to the IEEE variant of SPEKE. On the other hand, we cannot state the same for the ISO/IEC variant of SPEKE, because its definition of the shared secret is not sufficiently clear. Neither standard provides any clear formula about the definition of the shared secret. This is unsatisfactory, especially because the detail here has critical security implications.

For a more concrete discussion, let us denote Alice's identifier as $\hat{A}$, Bob's identifier as $\hat{B}$ and the session identifier as SID. One straightforward way to include all these identifiers is: $s = H(\text{Password}\|\hat{A}\|\hat{B}\|\text{SID})$. But this implies a preferred order of the parties' identifiers, which need to be agreed beforehand. A slightly better definition is as follow: $s = H(\text{Password}\|\min(\hat{A}, \hat{B})\|\max(\hat{A}, \hat{B})\|\text{SID})$. Yet it remains questionable how the SID should be defined and by whom. In the general case, the unique session ID is decided by both parties as part of the key exchange process, but this usually requires extra rounds of communication. The requirement for extra rounds is undesirable as it would remove the most notable advantage of SPEKE in terms of its optimal one-round efficiency. The way that the two standards address this extra-round issue is by defining the session ID (together with the entity identifiers) as part of the "prior shared parameters" before the key exchange. Hence, the SPEKE protocol remains one-round.

However, we believe the above solution in the standards is inappropriate, as it does not address the real problem. It is not a safe assumption that the user must know the other party's identifier or a session identifier before any communication is started. It is often difficult enough for a user to remember her own user name (identifier) and password; requiring the user to remember the other entity's (exact) identifier will only add to the burden on the user's memory and consequently make a PAKE protocol less useful. When the PAKE session fails, it will be no longer clear if that is due to the mismatch of the password, or simply because the user misremembered the identifiers. The user identifiers, as well as the session ID, should be determined as part of the key exchange process. In Section 5.3, we will present a solution that addresses the identified attacks without requiring any extra memory burden, in the meanwhile still keeping the SPEKE protocol one-round only.

Effect on Key-malleability attack. The inclusion of identifiers for one or more entities and the specific session into the definition of the password-based string has no effect in preventing the key-malleability attack.

Alice ($\hat{A}$)		Bob ($\hat{B}$)
Select $x \in_R [1, q-1]$. Compute $M = g^x$	$\xrightarrow{\hat{A}, M = g^x}$	Check $M \in [2, p-2]$
Check $N \in [2, p-2]$	$\xleftarrow{\hat{B}, N = g^y}$	Select $y \in [1, q-1]$. Compute $N = g^y$

Alice Computes: $\kappa_a = H\left(\min(\hat{A}, \hat{B}), \max(\hat{A}, \hat{B}), \min(M, N), \max(M, N), N^x\right)$

Bob Computes: $\kappa_b = H\left(\min(\hat{A}, \hat{B}), \max(\hat{A}, \hat{B}), \min(M, N), \max(M, N), M^y\right)$

Fig. 3. Patched SPEKE

5.3   Countermeasures and Suggested Changes to Standards

There are several reasons to explain the cause of the two attacks. First, there is no reliable method in SPEKE to prevent a sent message being relayed back to the sender. Second, there is no mechanism in the protocol to verify the integrity of the message, i.e., whether they have been altered during the transit. Third, no user identifiers are included in the key exchange process. It may be argued that all these issues can be addressed by using a Zero Knowledge Proof (ZKP) (as done in [4]). However, in SPEKE, the generator is a secret, which makes it incompatible with any existing ZKP construction. Since the use of ZKP is impossible in SPEKE, we need to address the attacks in a different way.

Our proposed solution is to redefine the session key computation. Assume Alice sends $M = g^x$ and Bob sends $N = g^y$. The session key computation is defined as follows:

$$\kappa = H\left(\min(\hat{A}, \hat{B}), \max(\hat{A}, \hat{B}), \min(M, N), \max(M, N), g^{xy}\right) \qquad (1)$$

The patched SPEKE protocol is summarized in Fig. 3. When the two users are engaged in multiple concurrent sessions, they need to ensure the identifiers are unique between these sessions. As an example, assume Alice and Bob launch several concurrent sessions. They may use "Alice" and "Bob" in the first session. When launching a second concurrent session, they should add an extension to make the identifier unique – for example, they may agree at the protocol level to start the extension from "1" and increment by one if a new concurrent session is created. Thus, the actual user identifiers become "Alice-1" and "Bob-1" in the second session. In the third session, the user identifiers become "Alice-2" and "Bob-2", and so on. As long the user identifiers are unique between concurrent sessions, the use of the extra session identifier does not seem needed.

The new definition of the session-key computation function in Eq. 1 should address the impersonation and key-malleability attacks in Section 4 (and also the "unknown-key share" attack and the generic attack reported by Tang and Mitchell [9]). This is achieved without having to involve explicit key confirmation, so the key confirmation can remain as "optional" as it is in the current standards. Furthermore, this countermeasure preserves the optimal one-round efficiency of the original SPEKE protocol.

There is an alternative solution, which is to make the definition of a shared low-entropy secret more explicit in the standards. One way is to define the shared secret as below:

$$s = H\left(\text{Password}\|\min(\hat{A}, \hat{B})\|\max(\hat{A}, \hat{B})\right) \qquad (2)$$

In the above definition, the session identifier SID is not included, as the concept seems to have been absorbed in the user identifiers as long as they are ensured to be unique between concurrent sessions.

Comparing the two solutions, we recommend the first solution in Eq. 1 (also see Fig. 3) for the following reasons.

- The first solution is more flexible to accommodate pre-computation of g^x and g^y. In the second solution, the user must know the identifier of the other party before the key exchange, which effectively prevents pre-computation.
- The first solution is more round-efficient. Alice and Bob do not have to know the exact identifier of the other party before starting the key exchange. But in the second solution, Alice and Bob may need an extra round before they are able to compute the generator g.
- The first solution is computationally more efficient. Because the generator g is unchanged for the same password, it only needs to be computed once. In comparison, the generator needs to be re-computed with any change in the user identifiers. (This may not make much difference in terms of computation if a safe prime is used, but it can significantly decrease performance in some other group settings.)

A further suggestion we would like to make for both standards is to reconsider the definition of the key confirmation method. The existing method, as defined in ISO/IEC 11770-4 and IEEE 1363.2, breaks the symmetry of the protocol (the key confirmation cannot be completed within one round). The key confirmation method in the original SPEKE paper [7] has the same limitation.

Our rationale for suggesting this change is not based on security considerations, but on the grounds of round efficiency. The key confirmation method defined in the original SPEKE paper [7] and the two standards [10, 11] cannot be completed in one round. We use the method defined in [7] as an example. If both parties attempt to initiate the explicit key confirmation at the same time, i.e., Alice sends $H(H(\kappa))$ and without receiving Alice's message, Bob also sends $H(H(\kappa))$. In that case, they may enter a deadlock and may have to abort the session and restart a new one. The chance of such an occurrence would be non-negligible in a high-latency network.

The solution we propose is based on the key confirmation defined in NIST SP 800-56A Revision 1 [1]. It works as follows:

$$\text{Alice} \rightarrow \text{Bob} : \text{HMAC}(\kappa, \text{``KC\_1\_U''}\|\hat{A}\|\hat{B}\|g^x\|g^y)$$
$$\text{Bob} \rightarrow \text{Alice} : \text{HMAC}(\kappa, \text{``KC\_1\_U''}, \|\hat{B}\|\hat{A}\|g^y\|g^x)$$

Table 1. Summary of results

SPEKE protocol variants	Round efficiency	Impersonation attack	Key-malleability attack
Original SPEKE with KC	3	Yes	Yes
Original SPEKE without KC	1	Yes	Yes
SPEKE in IEEE P1363.2 with KC	3	Yes	No
SPEKE in IEEE P1363.2 without KC	1	Yes	Yes
SPEKE in ISO/IEC 11770-4 with KC	≥ 3	Maybe	No
SPEKE in ISO/IEC 11770-4 without KC	≥ 1	Maybe	Yes
SPEKE in IETF I-D with KC	3	Yes	No
SPEKE in IETF I-D without KC	1	Yes	Yes
Patched SPEKE with KC	*2*	*No*	*No*
Patched SPEKE without KC	*1*	*No*	*No*

In the above key confirmation method, HMAC is a hash-based MAC algorithm and the string "KC_1_U" refers to unilateral key confirmation [1]. There is no dependence between the two flows, so Alice and Bob can send messages in one round.

5.4 Summary of Results

We summarize the applicability of the reported attacks on various variants of SPEKE in Table 1. For the completeness of discussion, we also include the version of the SPEKE protocol defined in the IETF Internet Draft[4], authored by the original SPEKE designer Davlid Jablon. In this Internet Draft, the entity identifiers are not included into the default definition of the shared secret. From the Draft, "... *in a peer-to-peer application using SPEKE, both parties may compute* $<g>$ *directly from the shared password.*". The key confirmation function defined in this Internet Draft is basically the same as that in IEEE P1363.2.

With the exception of ISO/IEC 11770-4, all previous versions of SPEKE are vulnerable to the impersonation attack regardless of whether the key confirmation is in place. In ISO/IEC 11770-4 [11], we cannot determine the applicability of the same attack because the wording in the standard is not sufficiently clear. Hence, we mark "Maybe" instead of "Yes" in the table.

The key-malleability attack is applicable to the original SPEKE (patched against the exponential-equivalence attack [8]) regardless of whether the key confirmation is used. In both the IEEE and ISO/IEC standards (and also in the submitted IETF Internet Draft), the key confirmation is modified to include the key exchange messages. Hence, the key-malleability attack no longer works when the modified key confirmation is used. But the key confirmation is marked as "optional" in these standards. Therefore, the versions that do not use explicit key confirmation are still vulnerable to the key-malleability attack.

In this paper, we propose two changes to the standardized SPEKE protocol: one is to redefine the session key computation based on Equation 1 and the

---

[4] The latest draft version is "02", dated 22 October, 2003. See
http://www.ietf.org/archive/id/draft-jablon-speke-02.txt

other one is to redefine the key confirmation function based on NIST SP 800-56A Revision 1 [1]. The first change addresses both the impersonation and the key-malleability attacks. The second change allows the key confirmation to be completed in one round.

Our patched SPEKE preserves the overall round-efficiency in the optimal manner. By comparison, in the original SPEKE paper, the IEEE 1363.2 standard and the IEFT Internet Draft, the specified protocol is one-round without explicit key confirmation, and is three-round with explicit key confirmation. In ISO/IEC 11770-4, it is not clear if the entity/session identifiers must be included into the definition of the shared secret. If such an inclusion is mandatory, it would generally need an extra round to send the entity/session identifiers before the key exchange.

Finally, we examine how the SPEKE protocol is actually implemented in practice, particularly with regard to whether or not the entity/session identifiers are included. In practice, SPEKE has been used by Blackberry for secure messaging. In this implementation, only the password (no entity/session identities) is used to derive the generator of the designated group. From the on-line documentation about the derivation of the generator[5]: *"The function applies the ECREDP-1 primitive to the password to derive a generator point."* Hence, the impersonation attack is in principle applicable to the protocol that underpins the Blackberry application. This does not necessarily mean the Blackberry application must be insecure, since it also depends on the context of an application and other implementation details (e.g., if the application supports parallel sessions). We leave this as a subject for further investigation.

6 Conclusion

In this paper, we present two new attacks on SPEKE, a protocol that has been included in the IEEE P1363.2 and ISO/IEC 11770-4 standards, and deployed in commercial products. The first attack indicates a practical flaw that needs to be addressed, while the second attack has an unfavourable theoretical implication. We explain the differences between the original SPEKE protocol and its variants defined in both standards and show how these differences are critically relevant to the presented attacks. We suggest concrete changes to both standards to address the issues identified in this paper.

Acknowledgement. We thank the anonymous reviewers from SSR'14, Brian Randell, Liqun Chen and Chris Mitchell for many useful comments. This work is supported by the European Research Council (ERC) Starting Grant (No. 106591).

---

[5] http://developer.blackberry.com/native/reference/core/
com.qnx.doc.crypto.lib_ref/topic/hu_ECSPEKEKeyGen.html

References

1. Barker, E., Johnson, D., Smid, M.: Recommendation for pair-wise key establishment schemes using discrete logarithm cryptography (revised), NIST Special Publication 800-56A (March 2007),
   http://csrc.nist.gov/publications/nistpubs/800-56A/
   SP800-56A_Revision1_Mar08-2007.pdf
2. Bellovin, S., Merritt, M.: Encrypted Key Exchange: password-based protocols secure against dictionary attacks. In: Proceedings of the IEEE Symposium on Research in Security and Privacy (May 1992)
3. Hao, F.: On robust key agreement based on public key authentication. In: Sion, R. (ed.) FC 2010. LNCS, vol. 6052, pp. 383–390. Springer, Heidelberg (2010)
4. Hao, F., Ryan, P.Y.A.: Password authenticated key exchange by juggling. In: Christianson, B., Malcolm, J.A., Matyas, V., Roe, M. (eds.) Security Protocols 2008. LNCS, vol. 6615, pp. 159–171. Springer, Heidelberg (2011)
5. Entrust TruePass Product Portfolio: Strong Authentication, Digital Signatures and end-to-end encryption for the Web Portal, Technical Overview, Entrust Inc. (July 2003), http://www.entrust.com/wp-content/uploads/2013/05/
   entrust_truepass_tech_overview.pdf
6. BlackBerry Bridge App and BlackBerry PlayBook Tablet, Security Technical Overview - Version 2.0, Research in Motion Ltd. (February 2012), Available online through Blackberry Knowledge Base at
   http://btsc.webapps.blackberry.com/btsc/microsites/searchEntry.do
7. Jablon, D.: Strong password-only authenticated key exchange. ACM Computer Communications Review 26(5), 5–26 (1996)
8. Zhang, M.: Analysis of the SPEKE password-authenticated key exchange protocol. IEEE Communications Letters 8(1), 63–65 (2004)
9. Tang, Q., Mitchell, C.J.: On the security of some password-based key agreement schemes. In: Hao, Y., Liu, J., Wang, Y.-P., Cheung, Y.-M., Yin, H., Jiao, L., Ma, J., Jiao, Y.-C. (eds.) CIS 2005. LNCS (LNAI), vol. 3802, pp. 149–154. Springer, Heidelberg (2005)
10. IEEE P1363 Working Group, P1363.2: Standard Specifications for Password-Based Public-Key Cryptographic Techniques. Draft available at
    http://grouper.ieee.org/groups/1363/
11. International Standard on Information Technology, Security Techniques, Key Management, Part 4: "Mechanisms based on week secrets", ISO/IEC 11770-4:2006

Analyzing Proposals for Improving Authentication on the TLS/SSL-Protected Web

Christopher W. Brown[1] and Michael Jenkins[2]

[1] National Security Agency / U. S. Naval Academy, Annapolis, MD USA
wcbrown@usna.edu
[2] National Security Agency, Maryland, USA
mjjenki@tycho.ncsc.mil

Abstract. "Secure" web browsing with HTTPS uses TLS/SSL and X.509 certificates to provide authenticated, confidential communication between web clients and webservers. The authentication component of the system has a variety of weaknesses, which have led to a variety of proposals for improving the current environment. In this paper we survey, analyze, compare and contrast three prominent proposals. To do this, we attempt to systematically capture the properties one might require of such a system: authentication properties, forensics/privacy properties, usability properties, and pragmatic properties. Enumerating these properties is an important part of understanding these proposals and the nature of the authentication problem for the secure web. Finally, we offer a few conclusions and suggestions pertaining to these proposals, and possible future directions of research.

Keywords: web security, authentication, TLS, HTTPS, certificates.

1 Introduction

This article provides a summary and analysis of three proposals for improving the authentication component of the current environment for the TLS/SSL protected web (HTTPS) — specifically the client's authentication of the server. As a part of this analysis, we identify the properties the overall system would, ideally, satisfy, and describe the various proposals in terms of the degree to which they do or do not provide/have these properties. The paper's real contributions, hopefully, are a clearer picture of what the authentication problem for the TLS/SSL protected web really is, and a framework for evaluating new proposals both individually, and in combination with one another.

The current environment for secure web-browsing is based on TLS (see RFC5246 [1]). TLS provides a mechanism by which a web client (browser) and a webserver establishing a connection between one another make use of public key cryptography to agree on a shared secret key, which is then used to encrypt communication using symmetric encryption for the rest of that session. Typically the process begins with the client being furnished a domain name by the user, which is then translated to an IP address via DNS name resolution, after

L. Chen and C. Mitchell (Eds.): SSR 2014, LNCS 8893, pp. 39–56, 2014.

which the client sends the ClientHello TLS message to the server presumed to be listening on port 443 at that IP Address. The client then receives messages from the server, one of which contains a public key. The client chooses a secret key (more accurately, a value that determines the secret key), encrypts it with the public key the client received, and sends the resulting ciphertext back to the server. At this point, both parties' have the same secret key, and encrypted communication can commence.

This process guarantees that the client and the owner of the public key — i.e. the entity in control of the associated private key — are the only ones who know the secret key, assuming that the private key is kept private. It does not, however, guarantee anything about the identity of the owner of the private key. There is no assurance that the party the user of the client wanted to contact is the one with whom the client now has a secure connection. Routing to the IP address could have gone wrong. DNS resolution could have produced the wrong IP address. The domain name itself might be wrong — e.g., amazen.com instead of amazon.com. Thus, there is a critical authentication problem to be dealt with!

The current environment for secure web-browsing generally handles this authentication problem in the following (highly simplified!) way: 1) It is the user's responsibility to ensure that the domain name provided to the browser is in fact the correct domain name for the entity they are trying to contact. 2) The webserver sends its public key to the client in an X.509 certificate (see RFC5280 [2]) as part of a certificate chain in the TLS Server Certificate message, and it is the browser's responsibility to validate the certificate chain, and verify that it chains to one the certificates in the browser's list of trust anchors. 3) It is the responsibility of the CA that issued the trust anchor at the end of the certificate chain to ensure that the public key in the certificate really belongs to the entity that owns the given domain name.[1]

It is assumed that the reader of this paper is familiar with this process, including details like certificate revocation lists, self-signed certificates, X.509 validation, etc. not covered in the very brief description above. It is also assumed that the reader is familiar with the manifold problems inherent in this system — for instance, that every CA trusted by the browser represents a single point of failure for the whole system, or realities of how users usually bypass the whole system [3,4]. The key point is that the authentication component of the current secure web-browsing environment has problems both in principle and in practice.

The many problems that exist in the current environment have prompted a wide variety of proposals for improvements. The proposals considered in this paper are limited to ones with a reasonable level of pragmatism (full scale replacement of the current environment with something new is extremely unlikely!), and a reasonable degree of visibility or momentum behind them. User interface improvements, while important, are also outside of the scope of this analysis. The proposals examined here are: DANE, Certificate Transparency and HTTP Pinning. We looked at the Sovereign Keys proposal closely, but it is not yet

---

[1] The responsibility for managing trust anchors falls to some combination of browser vendors, OS vendors, users and, perhaps, IT support departments.

mature enough to really analyze to the same extent as the above. Space restrictions forced us to remove our coverage of TACK [5], which is, however, similar in many respects to HTTP Pinning. Perspectives [6] and similar projects like Convergence, Google's now defunct Certificate Catalog project, and the Berkeley ICSI project, are interesting as well, but space precludes covering them here.

This paper consists of three parts. The first describes the set of properties that might be desirable for the authentication component of the "secure web". The second provides summaries of three proposals, along with some commentary and analysis — all of which is done in terms of the properties from the previous section. The last focuses on comparisons of different proposals and their potential to work in combination with one another.

2   Desirable Properties for the Authentication Component of the "Secure Web"

It is easier to understand and compare these various proposals if we first describe what it is we really want. In other words, What properties do we really desire of the authentication component of the "secure web"? What follows is a list of such properties, grouped into four categories. The fourth category is more accurately described as a list of properties we'd like to see in a proposal for improving on the current state of affairs, for example that the proposal is realistic. These properties were in large part deduced from reading a number of proposals and commentary on those proposals.

Authentication Properties[2]
continuity: that when a client connects to a host with name X, one can be sure
    that, in some meaningful sense, it is communicating with the same entity
    as it was communicating with the last time it connected to a host with
    name X. Note: it could make sense to think of continuity on two levels, the
    individual client level and the community level. Pinning proposals are about
    individual clients observing continuity. Notary proposals like Perspectives or
    the Berkeley ICSI, and gossiping proposals are about a community of clients
    observing continuity.
domain-name authentication: that the client is connected to the server au-
    thorized, intended or allowed to run under that name-and-port by the legit-
    imate owner of that domain name.
higher-level authentication: that the client is connected to the server au-
    thorized, intended or allowed to run under that name-and-port by an entity
    described by some notion of identity beyond merely ownership of a given
    domain name. (e.g., Southwest Airlines, the U. S. Postal Service) There are
    actually different classes of certificates — EV (extended validation) [7], OV
    (Organization Validation), DV (Domain Validation) — that seek to provide
    some higher-level authentication. NOTE: Many companies have made their

---

[2] This article is only concerned with *server* authentication, so client-authentication
    properties are not addressed.

domain name their identity, e.g., amazon.com, so that domain-name identity and higher-level identity are one and the same.

attribute authentication: that the client is connected to the server authorized, intended or allowed to run under that name-and-port by an entity with certain attributes (e.g., FDIC insured bank, NASDAQ listed company).

Forensics and Privacy Properties

client privacy: that third parties cannot, without the willing participation of the client or the server, deduce what websites a given client has been connecting to (without special access to the client machine).

impostor discoverability: that the legitimate owner of a given domain name should be able to discover what servers are presenting themselves as belonging to that name.

server privacy: that third parties cannot, without the willing participation of the client or the server, deduce the existence of a given server without actually attempting a connection themselves.

local privacy: that someone with access to the client machine after a website has been visited cannot deduce facts about what sites have been visited by the client. Of course, if a proposal includes client-side data, a user should be able to "clear history" as they can with current mechanisms like cookies. However, this presumably degrades the improvements to authentication. So this property refers to privacy concerns assuming that this clearing of authentication history hasn't happened.

Usability Properties

minimal false positives: that the client seldom refuses a connection or presents the user with an error message when the server it is connected to is indeed the right one. Ideally the client never decides to disbelieve that the entity with which it is communicating has the proper identity when, in fact, it does. This is the ideal. In reality, we can only try to minimize the number of false positives. As noted elsewhere, encountering many false positives conditions users to simply override-and-accept. Conversely, users that err on the side of caution are actually discouraged from connecting to legitimate servers. Note that this is described as a "false positive" because of the analogy to medicine — the test says there is a problem when, in fact, everything is OK.

maximal actionable information: that the client should have good, actionable information to present to the user, information that allows for good decisions. This includes information that allows for a reasonable plan of action in case the user decides not to override-and-accept.

minimal user trust: that the trust users must place in other entities is minimized. Note: this is a complex issue, since it involves not only how many entities need to be trusted, but also to what extent they are trusted, and how bad things could be if some number of them colluded.

user control: that the user is allowed to override decisions, determine who or what to trust, etc. This means the system must allow for it. Particular clients might not.

minimal server trust: that the trust server operators must place in other entities is minimized.

minimal server roadblocks: that setting up a TLS server is not overly burdensome. Already, lots of people and organizations have difficulty doing it right. Ideally improved authentication mechanisms shouldn't set up too many barriers, technical, logistical or financial, to organizations that want to set up websites that use them.

Implementation/Infrastructure/Pragmatics Properties

incremental: that there are benefits (progress towards other properties) for those that opt-in without requiring that everyone participates.

minimal-impact: that big changes to the internet architecture are not required. Client-only solutions, for instance are very low impact, as are server-only. When both have to change, when third-parties are involved, and so forth, the impact grows.

no-break: that things that currently work relying on relatively common practices (e.g., local network TLS connections, changing domain ownership, webhosting services, user-level TLS servers on hosts, etc.) do not break under the new proposal.

scalable/maintainable/robust: that the system works at internet scale, and can function over a long period of time. For example, if keys are lost or no longer secure: can they be changed? The system should function reasonably in the face of outages and attacks.

resource-friendly: that adopting the proposal does not slow communication too much, or require too much CPU time or memory. Resource-constrained mobile devices must be able to participate.

realistic: that the proposal does not make unreasonable assumptions or demands on individuals or society. For example, expecting organizations with no suitable financial or other incentive to run big servers might not be realistic.

In each of the following three sections we briefly describe a well-known proposal for improving authentication on the secure web, and analyze that proposal in terms of the properties described above. Each property is listed, along with commentary on the proposal in terms of that property. A small "gauge" icon accompanies each property to give a quick indication of whether the proposal positively or negatively affects a certain property, or has no substantial impact. For the first three categories — Authentication, Forensics and Privacy, and Usability — the gauge value is to be understood as relative to the current secure web environment. For the last category — Implementation/Infrastructure/Pragmatics — gauge values are to be understood as relative to all the other proposals for improvements. The gauge values are simply manifestations of qualitative judgments, not true quantitative data. The hope is that when combined in tables as in the following sections, they are a concise means to provide insight into the strengths, weaknesses, and purpose of any given proposal. However, it is the accompanying commentary that provides the actual analysis.

3   DNS-Based Authentication of Named Entities (DANE)

DNS-Based Authentication of Named Entities (DANE) uses DNSSEC to make assertions that constrain valid certificate chains. These assertions can specify the end entity certificate or public key, the trust anchor certificate or public key, or an intermediate certificate or public key. By using DNSSEC to distribute these assertions, clients can guarantee that the assertions really belong to the domain name in question. Thus, DANE is a mechanism that provides very strong *domain name authentication*. DANE depends on DNSSEC, and DNSSEC adoption seems to be proceeding very slowly. DANE (see RFC6698 [8]) adds a new record type to DNS, the TLSA resource record, which allows the nameserver that is authoritative for a given domain name to make assertions tied to a pairing of the name with a port-and-protocol. These assertions are of the following form:

(usage, selector, matching type, certificate association data)

where usage $\in \{0, 1, 2, 3\}$, selector $\in \{0, 1\}$, matching type $\in \{0, 1\}$, and the "certificate association data" can be a certificate, the Public KeyInfoField of a certificate, or a hash value.

The semantics of DANE assertions are essentially this: The certificate chain is constrained in usage 0 to contain a given certificate/public key, and in usage 1 to start with a given certificate/public key (which specifies the end-entity). For both, the client is still required to validate the certificate chain up to a client-trusted anchor. Usages 2 and 3 are similar, except that the client's role in determining trust is eliminated. The certificate chain is constrained in usage 2 to contain a given certificate/public key as trust anchor, and that trust anchor must be trusted by the client for purposes of that validation, regardless of the client's current trust store. The certificate chain is constrained in usage 3 to have a given end-entity certificate/public key, and if it does the client must accept the connection without doing certification path validation of the chain.

Because it is tied to DNSSEC, DANE's pragmatic outlook is tied to DNSSEC adoption. It's worth clarifying the sense in which DANE offers something more than DNSSEC alone would offer. A client relying on DNSSEC to resolve a given hostname to an IP address has a strong guarantee that the IP address it uses for that domain name is correct. However, it has no guarantees that the entity it connects to is a really a host properly associated with that IP address. After all, the attacker could be corrupting routing rather than DNS. With DANE, however, the client has stronger guarantees that the host it is connected to is properly associated with the given domain name.

Usage 2 & 3 assertions are potentially problematic. With usage 3, certification path validation does not occur, i.e. if the end-entity certificate presented to the client matches the certificate in the DANE assertion the connection is accepted. A usage 2 DANE response mandates a certain trust anchor for validation, and mandates that it be trusted — regardless of whether it is currently in the client's trust store. Both deny the user (or system administrator) the opportunity to make trust decisions. The security issue here is that without usages 2 & 3 both DNSSEC/DANE and the CA system have to be defeated in a successful attack, while with them an attacker that can successfully subvert DNSSEC can successfully pull off a man-in-the-middle attack. Defense in depth has been lost.

- -⌒+ **continuity:** —

- -⌒+ **domain-name auth.:** this is DANE's strength.

- -⌒+ **higher-level auth.:** a usage 2&3 response bypasses certification path validation, so information in certificates is less trustworthy than in the current system.

- -⌒+ **attribute auth.:** —

- -⌒+ **client privacy:** generally, having to contact a third party server is a client privacy concern. However, clients would almost always be contacting a DNS server for name resolution anyway, so it's not really a concern here.

- -⌒+ **impostor discoverability:** —

- -⌒+ **server privacy:** normally a DNS server would store the IP address associated with a given name, and nothing more. DANE records include a port as well as IP, so the fact that a particular host is running a TLS server at a particular port number is then known to the DNS server.

- -⌒+ **local privacy:** —

- -⌒+ **minimal false positives:** DANE provides mechanisms (usages 3 & 4) by which a certificate that does not chain to a trust anchor would be accepted without any error or warning, which reduces false positives ... although it can also defeat authentication, as pointed out above.

- -⌒+ **maximum actionable info:** if a site uses DANE and the client issues an error, the DANE assertion itself provides extra information about what public key / certificate should have been expected.

- -⌒+ **minimal user trust:** the user doesn't need to trust the CAs as much, but they put more trust in DNS.

- -⌒+ **user control:** with usages 2 & 3, there are situations in which a connection will be accepted regardless of the user's trust anchor settings.

- -⌒+ **minimal server trust:** with DANE, the webserver operator puts even more trust in the DNS, however CAs don't need to be trusted as much.

- -⌒+ **minimal server roadblocks:** to use DANE the webserver operator requires the cooperation of the DNS administrator. To also ensure that clients that do not support DANE aren't locked out, a certificate from a CA would also be required.

- -⌒+ **incremental:** both web clients and sites (though not really webservers) need to change for DANE to work, any pair that supports DANE gain the security of the system, regardless of whether it's adopted elsewhere. However, it's not enough for just one of the two parties to elect to participate.

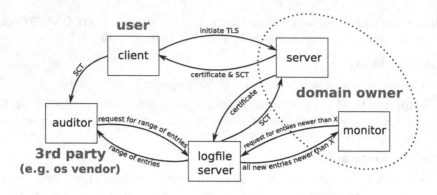

Fig. 1. Diagram illustrating CT. A full picture would show multiple logfile servers.

-⟨1⟩+ **minimal-impact:** it's not enough for client and server to change, the infrastructure has to change if DNSSEC is not already available.

-⟨—⟩+ **no-break:** —

-⟨—⟩+ **scal./maint./robust:** —

-⟨2⟩+ **resource-friendly:** DNSSEC has some overhead (see [9]) and more signed info will be sent when DANE is used than would be sent using DNSSEC solely for name resolution.

-⟨1⟩+ **realistic:** given that DNSSEC is in use, DANE is quite realistic. The question is whether it's realistic to expect significant DNSSEC adoption anytime soon.

4   Certificate Transparency

Certificate Transparency is described in RFC6962 [10]. Its primary purpose is to provide "impostor discoverability". The basic idea is this: If there was a public logfile of all certificates issued, then domain name holders could view the public logfile to root out bogus certificates for their domain names and, as the proposal puts it, "invoke existing business mechanisms for dealing with misissued certificates". If TLS clients all agree to reject any certificate not recorded in the public logfile, attackers would be forced to record their forged certificates in the logfile where, hopefully, server/domain owners would observe the bogus certs and do something about it. While this basic idea is straightforward, realizing it in a secure way is non-trivial. The Certificate Transparency proposal is somewhat complicated in terms of the number of entities involved: in addition to servers and clients, there are logfile servers, trusted auditors, and logfile monitors (see Figure 1). So the description that follows adds these various pieces in small steps.

Step 1: We first consider how clients determine whether a certificate record is in the logfile. Of course the client could contact the logfile server and ask.

Even if that could be done in an utterly secure and authenticated manner, there are still two issues: 1) contacting the third party has a performance and availability concern, and 2) letting the logfile server know every domain name you want to contact has privacy concerns. Therefore, the proposal calls for a different approach. Clients have preloaded/out-of-band-received public keys for the logfile server. The TLS server is supposed to send a "Signed Certificate Timestamp" (SCT) along with the certificate, which is essentially a hash of the certificate concatenated with a timestamp, signed by the logfile server. This gives the client something that it can verify quickly, without any third-party communication, so it addresses both concerns 1 and 2. IANA has issued a value for the TLS SCT extension. For technical reasons beyond the scope of this overview, the SCT is issued before the certificate is logged. However, the SCT contains an additional field with a value called the Maximum Merge Delay (MMD). Implicit in the SCT is a promise by the logfile server that the time between when the SCT was issued and when it is logged will not exceed that MMD value.

Step 2: If the logfile is misbehaving, or if it has been compromised, or its private key stolen or broken, clients could get forged SCTs. In other words, they could be accepting certificates that weren't actually logged. To address this, the proposal calls for "trusted auditors" that clients are supposed to submit SCTs to, in order to keep tabs on the logfile server and make sure it really is reporting the submitted SCT as part of the log. The RFC mentions having the client do this (asynchronously, so as not to take the performance hit), but 1) that has all the same privacy concerns, and 2) the logfile server could systematically lie to that one host. So it makes more sense to introduce trusted auditors into the system. It's unclear who or what auditors are notifying in case they detect a misbehaving logfile server, nor is it clear what the plan would be were a logfile server discovered to be misbehaving. Recovering from that situation could be quite challenging.

Step 3: Another way that a logfile server could misbehave is by modifying past entries in the log. For instance, maybe a bogus certificate gets a real SCT from the logfile server and is in the logfile (so the auditor doesn't see any trouble) but then after the attack the logfile entry gets erased, and all this happens before the domain name owner gets a chance to check for any new entries in the logfile for his domain name. This would defeat the whole purpose of certificate transparency. Therefore, the logfile is *append-only* — once an entry is there, it's there forever. This is done with Merkle trees, which provide a mechanism whereby anyone observing the logfile server could detect modifications or erasures of past entries. Of course, some entity has to bother to make these checks, so the proposal calls for "logfile monitors", which would periodically query the logfile server and and check that it was really operating in append-only mode. These might do double duty by also checking for new logfile entries for domains the host is interested in.

Step 4: Finally, the proposal envisages not one, but multiple logfile servers. To protect against Denial of Service attacks, in which the attacker floods a logfile server with bogus certificates to be logged, the proposal suggests that each logfile server would publish a list of root CAs, and it would only log entries that validate via a chain up to one of the CAs in the list. The proposal also calls for "gossiping" to root out misbehaving logfile servers. However, no details on the gossiping protocol are given.

Certificate Transparency provides impostor discoverability. This is a benefit to server/domain-name owners, but only secondarily a benefit to users. It provides no benefit for the initial targets of attacks, but it does offer a potential benefit to the larger user community, in as much as a vigilant server/domain-name owner may notice the attacker and take steps to shut him down. The proposal takes great pains to ensure that entities that care to do so can monitor the activities of logfile servers in order to ensure that they are being honest, so that logfile operators don't need to be trusted. There are, however, some issues to consider.

The purpose of the logfile monitors is to ensure that the logfile servers behave properly. However, once again, it is unclear how to deal with a logfile server that has misbehaved. It could be blacklisted somehow, but it's not clear what to make of the SCTs it had previously issued. Webservers would be sending them out, potentially unaware that it was no longer trusted. Perhaps a bigger question is what to do with logfile servers that are not purposefully acting badly, but fail to meet an obligation — for example a logfile server that does not not get the SCT into the log within the window specified by the MMD because of an attack, or a simple programming or administrative error. Simply blacklisting such a server seems highly undesirable. An alternative would be to rollback the log to the point of the error, but that's a problem because all the legitimate SCTs that had been issued in between issuing this SCT and noticing that the merge deadline had been missed would then be invalidated. Some mechanism is required to deal with this gracefully.

The proposal doesn't address how logfile public keys are distributed and updated. It seems that we end up in a similar situation as with CAs, namely that some arbitrary list of trusted logfile servers is preloaded into the browser. There is then the potential for even more certificate-related error messages, since a client could receive a certificate that is actually OK, but receive an SCT along with it that refers to a logfile server that is not trusted by the client (or for which the public key stored in the client is too old or too new).

It is not clear why users/clients would opt to submit SCTs to auditors. Collectively, there is the benefit that attackers could be discovered and eventually dealt with. But for the individual user there is little short-term benefit, and there is definitely a risk to privacy. If the intended model is that browser vendors would run their own trusted auditors, the privacy issue is mitigated, since their users are essentially trusting them completely anyway ... at least for Google, Microsoft and Apple. Less so for open source projects like Firefox, where users may put their faith in the "many eyes" that are supposedly on the source code. How a "trusted auditor" run by the Mozilla foundation is set up would not be subject to all those eyes, making it easier for a single individual or small group to misuse it than to introduce errors into the Firefox codebase. Below is a summary analysis of CT.[3]

- ⌢+ **continuity:** no real first-order effect.

- ⌢+ **domain-name authentication:** no real first-order effect.

- ⌢+ **higher-level authentication:** —

- ⌢+ **attribute authentication:** —

- ⌣+ **client privacy :** the auditor sees every secure site the client connects to.

- ⌣+ **impostor discoverability :** this is the whole point of CT!

- ⌢+ **server privacy :** a legitimate server has to announce its presence by submitting to a logfile server.

- ⌢+ **local privacy:** no additional data is stored locally.

- ⌣+ **minimal false positives :** with CT users could be faced with errors when valid certificates aren't logged, or when SCTs are sent to clients that don't have that logfile server's public key.

- ⌢+ **maximal actionable information :** when a cert's trust anchor is untrusted by the client, but the cert is logged, the user at least knows that the cert has been available for scrutiny and for how long. Otherwise, the user will know that it is unlogged (which is more suspicious).

---

[3] Draft revisions of the RFC address some of the issues raised here. This evaluation notes server privacy as an issue — a legitimate server needs to announce its presence by submitting its certificate to a public logfile server. To address this, draft revision 3 of the RFC includes a mechanism for redacting portions of the domain name in the certificate information submitted to a logfile server. For example, if the domain name in the certificate was `super.secret.example.com`, the information submitted to the logfile server might be `(PRIVATE).example.com`. Another mechanism added in the draft that addresses this problem is logging a name-constrained intermediate authority, along with a field that explicitly allows the SCT for the intermediate authority to stand in lieu of an SCT for the end-entity certificate. Thus, the situation above might be handled be having `super.secret.example.com` send an SCT for an intermediate CA constrained to `.example.com`. Concerns raised here regarding the Minimal Impact and Minimal Server Roadblocks Properties are addressed in draft revision 3 by providing a mechanism for including a server's SCT in its certificate. This way, the server/server owner does not necessarily need to change or do anything different in order for CT to be used. Instead, CA's could make sure SCTs are bundled in the certificates they issue, and servers simply send those certificates as they always do. Of course, this creates a chicken-and-egg problem: the CA needs the SCT to create the certificate, but the logfile server needs the certificate to create the SCT. To deal with this, draft revision 3 allows CA's to submit "pre-certificates" to the logfile server, which contain enough information for the logfile server to create an SCT. The SCT gets sent back to the CA, which then can complete the certificate. Because these mechanisms are only described in draft revisions under very active development, we are not including them in our analysis.

-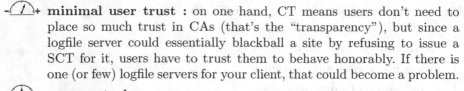+ **minimal user trust** : on one hand, CT means users don't need to place so much trust in CAs (that's the "transparency"), but since a logfile server could essentially blackball a site by refusing to issue a SCT for it, users have to trust them to behave honorably. If there is one (or few) logfile servers for your client, that could become a problem.

-+ **user control:** —

-+ **minimal server trust** : server/domain owners don't need to place as much trust in CAs.

-+ **minimal server roadblocks** : server/domain owners have to submit their certs to a logfile server, and have to find one that supports their trust anchor.

-+ **incremental** : adoption is a major issue. If clients do participate, all sorts of legitimate sites will suddenly stop working, and users will get swamped in false positives.

-+ **minimal-impact** : both clients and webservers need to change in order for CT to work, and a lot of additional infrastructure and new kinds of servers needs to be created.

-+ **no-break** : how will this work in local network only situations? Will organizations be forced to run logfile servers inside their local networks? How about enterprise trust anchors? None of the usual logfile servers will support them, of course, so would such an enterprise need to run its own logfile server?

-+ **scal./maint./robust** : lots of questions: what happens when logfiles make errors or are found to be acting improperly? How can logfile server keys change? How are logfile server keys distributed to clients? There are some significant maintenance problems!

-+ **resource-friendly:** not a lot of extra burden on client or server; although clients have to report to auditors, they do it asynchronously.

-+ **realistic** it is unclear what would motivate operators to stand up logfile servers, monitors or auditors.

5   An HTTP Extension for Public Key Pinning (HPKP)

At its most basic, "pinning" just means hard-coding or caching cert/public key (or the hash of the cert/public key) in a client, and requiring the cert/public key received at connection time to match what is currently "pinned". More flexibly, the client might pin a set of certs/public keys, or pin the cert/public key of an intermediate element of the certificate chain, both of which allow the end entity cert/public key to change in a controlled manner. Essentially, pinning is a commitment that the user won't allow certs/public keys to change. What's interesting is looking at the question of who controls whether, when and what pinning takes place. Pinning could be directed by 1) the user, 2) the client (e.g., hard-coded pins, or pinning that would be updated by the client "calling home" or calling an external service, or a policy of caching certs/public keys after an initial unpinned connection), or 3) the server (directing pinning for itself or for subdomains). An example of (1) is when a user preloads or chooses to accept an ssh public key. An example of (2) is when applications (like Chrome) have preloaded pins or call back home to get new pinning directives. If a website were to send pinning directives to the client, that would be an example of (3). Pinning, obviously, is a mechanism for providing authentication *continuity*.

The IETF draft document draft-ietf-websec-key-pinning (at the time of this writing in revision 12) [11] proposes an HTTP extension (HPKP) that allows the server to direct the pinning performed by the client — i.e. it is an example of the category (3) type of pinning described above. In this proposal, the server sends clients HTTP directives (the proposed extension) providing (hash-algorithm,hash-of-public-key) pairs that are to be pinned. The client saves this pin information indexed by the domain name it used in creating the connection. On subsequent connections to the same name, the client then checks whether any hash value in the set of pins is matched by a hash of any of the public keys in the certifying chain. If so, the client continues as normal. If not, there is an error. A hash-algorithm + hash-of-public-key pair must be accompanied by a "max-age" value and may be accompanied by a "report-uri" value. The max-age value instructs the client to keep the pin for a certain time. The report-uri gives a URI that is to be used by clients to report pinning errors for that domain name. An additional assertion may be sent that directs the client to apply the pin not only to the server's domain name, but to all of its subdomains as well.

The obvious benefit of HPKP is the *continuity authentication* it provides. When a user connects to a server often enough (meaning that the time between visits is less than the max age) with the same client, man-in-the-middle attacks should be detected. Because the server directs the pinning, and because sets of pins are allowed and intermediate public keys can be pinned, servers can pull off planned transitions to new public keys gracefully. As will be described in more detail below, the proposal is very good in terms of the Usability Properties. Among the Implementation/Infrastructure/Pragmatics properties, the only real concern is the extra resources required by a participating client — namely that a potentially large number of of pins will have to be stored, which could be problematic for resource-constrained clients. There is an especial concern that a malicious site could flood the pin store and use up all the available space. The specification could perhaps be modified to bound the storage given a (non-top-level) domain, or reclaim space from the non-top-level domain with the largest storage footprint. Maintainability questions surrounding unplanned key transitions are answered by requiring servers to pin a "backup key", which is a key to be used in case the current public key is compromised and needs to be revoked and its use discontinued.

Next we consider Forensics & Privacy Properties. What should be another obvious benefit of HPKP is *impostor discoverability*. This is, after all, the point of the reporting mechanism provided by the report-uri directive. But to what degree will HPKP really provide this property? In the case of a man-in-the-middle attack (which is what an "impostor" really has to do), the client will be provided with a domain name X, and the attacker will somehow arrange things so that the client will think it is communicating with the host properly referred to by that name when, in fact, it is communicating with some other host — for example by disrupting routing. If the client receives a certificate chain and it doesn't match what is pinned for the name X, the client is supposed to send information about the pinning error to the URI in the report-uri directive. However, it seems quite likely that this message will never arrive at its destination given that the attacker is already subverting network traffic to carry out the man-in-the-middle attack. So the only case in which this would actually have its desired effect is when the attacker was unwary enough to allow the reporting message through. We note that this could be remedied by having clients send these reports at exponentially decaying intervals — perhaps until a signed acknowledgment is received. As long as the attack is not permanent, the report should eventually get through. To avoid flooding-style attacks, a carefully analyzed approach that looks at domain relationships and drops multiple error reports from the same (non top level) parent domain should be investigated.

There are a variety of ways clients may end up with pins that don't match the public key presented by a legitimate server. A domain name may change hands without the willing cooperation of the party losing the domain. Both primary and backup keys could be lost. An attacker manages to pull off a successful man-in-the-middle attack for a period of time on a site that doesn't use pinning, and puts a "poison pin" in the browser of all clients that connect during that time, with a very big max-age. In all these cases, there's actually a hole in the DNS namespace — a domain name that, for a large number of clients, is unusable for HTTPS connections. This is a potentially serious problem.

Finally, HPKP breaks with the general design principle of separating concerns, and the specific cryptographic principle that different security properties should be safeguarded by different keys (see, for example, Section 5.2 of [12]). TACK, which space precludes us from covering here, is a similar proposal, but it uses a separate key (the "Tack Signing Key") to provide continuity authentication, and the certificate chain keys provide (as they are supposed to) domain name and higher-level authentication.

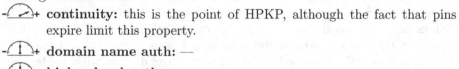

- + **continuity:** this is the point of HPKP, although the fact that pins expire limit this property.
- + **domain name auth:** —
- + **higher-level auth:** —
- + **attribute auth:** —
- + **client privacy:** attacks referenced above.
- + **impostor discoverability :** the report-uri directive provides this but, for reasons described above, it's unclear how effectively.
- + **server privacy:** —

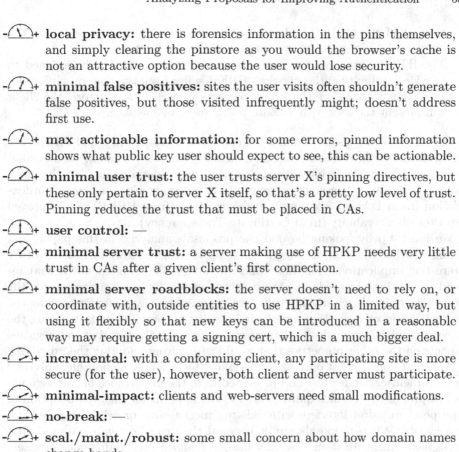

- **local privacy:** there is forensics information in the pins themselves, and simply clearing the pinstore as you would the browser's cache is not an attractive option because the user would lose security.

- **minimal false positives:** sites the user visits often shouldn't generate false positives, but those visited infrequently might; doesn't address first use.

- **max actionable information:** for some errors, pinned information shows what public key user should expect to see, this can be actionable.

- **minimal user trust:** the user trusts server X's pinning directives, but these only pertain to server X itself, so that's a pretty low level of trust. Pinning reduces the trust that must be placed in CAs.

- **user control:** —

- **minimal server trust:** a server making use of HPKP needs very little trust in CAs after a given client's first connection.

- **minimal server roadblocks:** the server doesn't need to rely on, or coordinate with, outside entities to use HPKP in a limited way, but using it flexibly so that new keys can be introduced in a reasonable way may require getting a signing cert, which is a much bigger deal.

- **incremental:** with a conforming client, any participating site is more secure (for the user), however, both client and server must participate.

- **minimal-impact:** clients and web-servers need small modifications.

- **no-break:** —

- **scal./maint./robust:** some small concern about how domain names change hands.

- **resource-friendly:** all the pin information needs to be stored, which could be problematic for memory-constrained clients. There are also concerns about attacks on memory resources via HPKP.

- **realistic:** HPKP only requires buy-in from browser vendors to get started. Given that this is a Google draft, that buy-in might be there.

6 Comparisons and Conclusions

We have surveyed three proposals for for improving the current condition of authentication for the secure web, each of which are fairly well-known.

- DANE offers the prospect of providing strong guarantees of domain name authentication. However, with "usage values" 2 and 3 it eliminates the defense in depth that the system of certificate authorities was supposed to bring. Moreover, DANE is built on top of DNSSEC, and DNSSEC adoption has not progressed very quickly.
- Certificate Transparency offers a mechanism by which domain/server owners can detect attackers that are impersonating their sites. However, it has a

number of pragmatic problems, as detailed above, and may increase the number of "false positive" warnings experienced by users.
- The HTTP Extension for Server-directed Pinning (HPKP) is designed to provide continuity authentication. HPKP suffer from the "poison pin" problem, namely that once the wrong pin gets in a client's pin store, the client will present the user with a "false positive" error message.

DANE (ignoring usage 2 and 3), Certificate Transparency and HPKP are pretty much orthogonal to one another, meaning that they could be used in combination without interfering with one another or overlapping in what they provide. In fact, used in conjunction we would have stronger domain name authentication (from DANE), continuity authentication (from HPKP) and improved impostor discoverability (from Certificate Transparency).

We finish up by looking beyond the proposals analyzed in this paper, and asking whether the analysis suggests new ideas to investigate or has any other interesting implications. The first thing we would like to point out is that instead of viewing proposals like these as trying to "fix" authentication for the TLS-protected web, we should evaluate a proposal by clearly understanding the authentication or forensics property it is trying to provide, and analyzing the extent to which it provides that property along with the positive and negative impacts on the other properties that would result from adopting the proposal.

A second point is that once we stop looking for a single, monolithic, universal fix to authentication for the TLS-protected web, the importance of "orthogonality" of proposals becomes quite clear, by which we mean that the adoption of a proposal wouldn't interfere with existing mechanisms or other proposed improvements. When proposals are orthogonal they can be composed, and that strengthens authentication. DANE's usages 2 and 3 are unfortunate precisely because they ruin orthogonality. Without usage 2 or 3, DANE and the current certificate infrastructure compose nicely.

A third point is that when we view these various mechanisms as providing evidence for one or more of the four authentication properties, we see each connection attempt as making a case for accepting the proposed identity of the server on the other end. It might be reasonable to pin the "shape" of that evidence — i.e. what kind of evidence was presented. So, for example, suppose a user has been using client-directed pinning, and DANE (without usage 2 or 3) is used in conjunction with the usual Certificate validation process. The user tries to go to https://example.com and there is an error — the public key in the Certificate presented by the server does not match what the browser has pinned for the name example.com. However, the DANE record is validated, and the certificate chains to a trust anchor. The decision about whether to trust this server despite the pin mismatch is unclear. Now suppose that on prior connections, example.com had sent the client an EV certificate, and suppose the client had pinned that fact. It would not be at all unreasonable to base the trust decision on whether or not the certificate presented by the server is an EV certificate. Note that pinning the "shape" of the authentication evidence provided by a site has the really nice property that it actually provides increased security to sites that choose to employ stronger authentication evidence. In the example above, the client would have "pinned" the facts that example.com employs DANE and uses an EV certificate. Thus, an attack will generate a warning to the user unless the attacker both subverts DNSSEC and gets a fraudulent EV certificate for example.com. Taking control of DNS records might be enough to get an ordinary certificate for example.com, but it shouldn't be enough to get an EV certificate. This kind of pinning could also make gradual adoption of some of these proposals easier. For example, if the client pinned the fact that a given site used Certificate Transparency (i.e. sent an SCT) in prior connections, then the client could be configured to *require* CT for that site from that point on, but not for other sites. This would eliminate the problem of clients being flooded with false positives. One of the interesting things about Perspectives [6] is that it explicitly presents itself as a mechanism for providing evidence about identity, not as a procedure that proscribes trust decisions. That's a powerful and flexible idea.

The fourth and final point is a suggestion that we feel falls out of this analysis. We start with the observation that if clients were to do client-directed pinning of end-entity certificates and servers would do OCSP stapling, then most of the time there would be strong authentication of the server on the basis of those two pieces of evidence alone, and and the connection could proceed[4]. The process would be quick and involve very little overhead provided that the certificates match, and the point is that they usually would. So the question really is how to deal with the infrequent situation in which the above is unable to confirm authentication. This can happen when a client connects with a given name for the first time, when a different end entity certificate is sent by the server, or when an end-entity certificate is revoked. Making the right decision in these cases, and doing it to the greatest extent possible without user intervention, is crucial. However, since these situations are infrequent (as well as important), when they do arise it would be acceptable to have the client take substantially longer to make a decision, or to gather information to present to the user in case it is necessary. We suggest research into mechanisms or the development of standards that allow the client to collect a lot of relevant data in order to make a strong case for or against trusting the server. As a very simple example, suppose there was a standard way for a client that was not able to authenticate the server using the pinning & stapling mechanism above, to fetch additional certificates for the server. A client could implement a policy requiring that, in this event, it is able to fetch an additional certificate that contains the same TLS key, the same domain name, and chains to a different trust anchor (without cross signing).

---

[4] Client-directed pinning would ensure that the certificate hadn't changed since the client's last connection to the site, and OCSP stapling would ensure that the certificate had not been revoked — at least as of some reasonably recent point in the past. Online Certificate Status Protocol (OCSP) stapling is a piece of the modern certificate infrastructure. It allows a server to send clients a message, signed by the relevant certificate authority, that asserts that as of a certain point in time, the server's certificate has not been revoked. This is a nice alternative to contacting OCSP servers to check for revocation, or either pushing or pulling blacklists.

This increases the difficulty of a man-in-the-middle attack significantly, since the attacker would have to obtain fraudulent certificates from two different trust anchors. This is orthogonal to other proposals, and it strengthens all of them. For example, HPKP has potential problems with unplanned key transitions. With a mechanism like this, an organization that is forced to deal with an unplanned key transition could have strong evidence (we've suggested multiple certificates as a possible form) that a client could fetch on that single connection for which HPKP broke. The client could be convinced with overwhelming evidence and accept the TLS connection — without user intervention. The delay caused by fetching and analyzing this extra evidence would only be incurred once, then HPKP would suffice for subsequent connections.

References

1. Dierks, T., Rescorla, E.: The transport layer security (TLS) protocol version 1.1. RFC 5246, RFC Editor (April 2006)
2. Housley, R., Santesson, S.: Update to directorystring processing in the internet X.509 public key infrastructure certificate and certificate revocation list (CRL) profile. RFC 5280, RFC Editor (August 2006)
3. Herley, C.: So long, and no thanks for the externalities: the rational rejection of security advice by users. In: Proceedings of the 2009 Workshop on New Security Paradigms Workshop, NSPW 2009, pp. 133–144. ACM, New York (2009)
4. Sunshine, J., Egelman, S., Almuhimedi, H., Atri, N., Cranor, L.F.: Crying wolf: an empirical study of ssl warning effectiveness. In: Proceedings of the 18th Conference on USENIX Security Symposium, SSYM 2009, pp. 399–416. USENIX Association, Berkeley (2009)
5. Marlinspike, M., Perrin, T.: Trust assertions for certificate keys. Internet-Draft draft-perrin-tls-tack-02, IETF Secretariat (January 2013)
6. Wendlandt, D., Andersen, D.G., Perrig, A.: Perspectives: improving ssh-style host authentication with multi-path probing. In: USENIX 2008 Annual Technical Conference on Annual Technical Conference, ATC 2008, pp. 321–334. USENIX Association, Berkeley (2008)
7. CA/Browser Forum. Guidelines for the issuance and management of extended validation certificates (March 2014),
   https://cabforum.org/wp-content/uploads/
   EV-SSL-Certificate-Guidelines-Version-1.4.6.pdf
8. Hoffman, P., Schlyter, J.: The DNS-based authentication of named entities (DANE) transport layer security (tls) protocol: TLSA. RFC 6698, RFC Editor (August 2012)
9. National Institute of Standards and Technology (NIST),
   http://www.dnsops.gov/dnssec-perform.html
10. Laurie, B., Langley, A., Kasper, E.: Certificate transparency. RFC 6962, RFC Editor (June 2013)
11. Evans, C., Palmer, C., Sleevi, R.: Public key pinning extension for http. Internet-Draft draft-ietf-websec-key-pinning-08, IETF Secretariat (July 2013)
12. Barker, E., Barker, W., Burr, W., Polk, W., Smid, M.: Recommendation for key management - part 1: General (revision 3). Technical Report NIST Special Publication 800-57, National Institute of Standards and Technology (March 2007)

Standardization Transparency

An Out of Body Experience

Phillip H. Griffin

Griffin Information Security Consulting, Raleigh, North Carolina, USA
phil@phillipgriffin.com

Abstract. This paper examines the issue of transparency in standards setting organization processes used to select security techniques for standardization. Analysis of data collected from interviews, electronic mail, and other documentation is presented as a narrative in two case studies. A Kaleidoscope conference case study illustrates the positive impacts of open participation on improving transparency through the reduction of bias in the selection process. These impacts include more timely inputs from researchers on emerging technology issues, and greater diversity in the sources of creative new ideas and solutions considered for standardization. Restrictions imposed on the selection process by government control of national body activities are described through a second case study of practice in the United States. Finally, recommendations are proposed on actions standards setting organizations can take to broaden participation in the selection of techniques for standardization and to strengthen communications between standards developers and the research community.

Keywords: openness, security, standardization, transparency.

1 Introduction

Transparency is an important issue in the development of security standards for information and communication technology (ICT). Standards development organizations (SDOs) implement processes aimed at enhancing the perception of transparency by standards users and other stakeholders. When SDO transparency is mentioned in research literature, discussion is often confined to processes within a standards development lifecycle that starts with the idea for a new standard. Though the names of lifecycle components vary by SDO, those under SDO control typically begin with creation of a standards project.

The initial selection of techniques for standardization occurs much earlier, before the project is created. A selected technique may stem from an idea, a novel approach, or a perceived need for a standard. Participation in the selection process may be limited to a small number of SDO members and not involve most stakeholders. The participants may be biased or serve narrow interests. The selection process may lack transparency, since it occurs on the edge of the standards development lifecycle, before formal SDO approval and project creation. Transparency in the initial selection process is the focus of this paper, and transparency is defined here as the perception

L. Chen and C. Mitchell (Eds.): SSR 2014, LNCS 8893, pp. 57–68, 2014.

of an objective observer that "the process of selecting security techniques for standardization" is "as scientific and unbiased as possible" [1]. Accounts are presented as case studies of the ITU Kaleidoscope conference and United States (US) security standardization practice. These reports are informed by personal experience as a participant in US SDO activities and analysis of data collected from interviews, electronic mail, and other documentation used to examine selection process transparency in global standards setting organizations.

The standardization branch of the International Telecommunication Union (ITU) is the ITU-T. Along with the International Organization for Standardization (ISO) and the International Electrotechnical Commission (IEC), these three are considered the global 'de-jure' standards setting bodies [2]. ISO, IEC, and ITU follow "the principle of territorial representation" with their voting members being "national SDO or other national representatives" [3]. There are committees within the global SDO that produce international information security standards (e.g., in ITU-T Study Group 17 Security, ISO/TC68/SC2 Financial Services Security, and ISO/IEC JTC 1/SC 27 IT Security techniques).

Together, ISO, IEC and ITU-T are the "most prominent official international SDO (with national membership)" [3] that produce information security standards. Through its Kaleidoscope conferences, ITU has opened up the standardization selection process to be more inclusive. These conferences attract members of the research and development community that may not participate in national SDO activities. Kaleidoscope provides 'an out-of-body experience' that does not depend on national body approval or processing of proposals considered for standardization.

Through Kaleidoscope conference papers, presentations, discussions, and social events, ITU has found a way to increase the transparency and openness of its standardization selection process. These conferences augment formal ITU standardization practices to include more academic and research institutions and to foster greater input from these important sources of standardization ideas. Section 2 of this paper presents a case study of a paper written by this author that was accepted for the 2013 Kaleidoscope conference. That section describes the impact of the author's paper on information security and data privacy standards.

Governments have an important role to play in national SDO by ensuring transparency and facilitating open participation, and "in stimulating innovation and standardization" in the area of ICT security techniques [4]. Since openness "appears feasible only in a local (maybe national) context", governments should ensure balanced representation in the minority of stakeholders that select technologies for standardization and that form the delegations attending international standardization committees [3]. From the viewpoint of standards as public goods, governments have a responsibility to promote collaboration among diverse participants having widely different perspectives, expertise, and resources. Governments can act to promote transparency and openness by requiring SDOs to provide "access to standardisation activities without obliging SMEs to become a member of a national standardisation body", "free access or special rates to participate in standardisation activities", and "free access to draft standards" [5].

In the United States, government agencies have taken managerial control over the SDOs that support international standardization of biometrics and IT security techniques. The scope of the primary US security SDO has been changed from one solely devoted to international standards to one whose members also focus on the development and promotion of government sourced national standards. SDO control has been used to ignore technical criticism of agency-sponsored work and to circumvent consensus agreement processes. Section 3 of this paper presents a case study of US standardization practice and its impact on information security standardization.

2   Kaleidoscope Conference Case Study

2.1   Background

The International Telecommunication Union (ITU) was founded in Paris in 1865 to become the first formal standards organization. The ITU was established by the European states as a public-private partnership to create "a forum for the negotiation of standards to ensure network interoperability" [6]. Since its founding, ITU has evolved into a worldwide organization with "a membership of 193 countries and over 700 private-sector entities", including 63 academic and research institutions [2].

The ITU began holding its Kaleidoscope conferences in 2008, with the goal of hosting one conference per year from a different location around the world. An important purpose of these events is to facilitate greater communication between the academic research community and standards developers. To ensure diverse global participation, ITU selects a new conference theme for each event and an academic institution or research organization to serve as conference host [2]. Through the presentation of original, multi-disciplinary papers around a central theme, Kaleidoscope conferences foster dialog between research and practitioner attendees.

Academia is viewed as an important external source of new concepts, technologies, and ideas for ITU-T standardization. Using conference presentations, technical papers, and discussions, the ITU seeks to identify new areas for standardization from outside of its standards development community to support the creation of new or improved ICT products and services. Through its Kaleidoscope conferences, ITU can respond to the increasing "globalization of technology development" and the need for greater transparency and broader participation in standardization by "opening up of the process to a wider range of actors" [6].

ITU organizes Kaleidoscope conferences in collaboration with the Institute of Electrical and Electronics Engineers (IEEE). ITU publishes accepted academic papers in conference proceedings and ensures their wide dissemination by making them freely available for download from its web site. IEEE lists each paper from the complete proceedings in the IEEE XPlore Digital Library. IEEE also publishes select conference papers that have a high potential impact on standardization in a special standards edition of the IEEE Communications Magazine.

2.2   Openness

The principle of openness in international SDOs is one that has traditionally "put more emphasis on input legitimacy" [3]. To achieve the perception of openness, SDOs establish processes to facilitate stakeholder input as standards are being created, extended, or modified to correct defects. SDO charters, directives, and processes attempt to ensure that "all affected individuals and organizations have the opportunity to get involved in the decision-making process" [3] as standards are being developed. The ITU Kaleidoscope program seeks to apply the principle of openness earlier in the standards development process, to the stage at which security techniques are considered and selected for standardization.

Open participation, inclusiveness, and freedom of discourse are key features of the Kaleidoscope paper acceptance process. Paper submissions may be submitted by anyone from any of the 193 ITU member countries, including students, academics, inventors, and researchers. No topic or content approval by the author's national standards body is required. Authors need not have any experience or prior involvement in standardization activities, and need not be a member of an SDO.

The scope of ITU-T standardization has evolved as telecommunication networks have matured to include standards for multimedia, cloud computing, telebiometrics, information security, and more. Kaleidoscope authors might address any of these areas. Original technical papers can be submitted at no cost, though accepted papers must be presented at the conference by one of the authors to be published. The cost of attending may present a financial obstacle to some authors, and no accommodation is made by ITU for teleconference presentation. There is no academic affiliation requirement, and papers are selected from both academia and industry submissions based strictly on merit. The nationality of the author, the ranking of the author institution or the company size or worth are not considered in the acceptance process.

Submissions are anonymous and peer reviewed against criteria listed in a published rubric using a double-blind process. Unlike many academic conferences, the Kaleidoscope conference uses a scoring rubric to guide authors and reviewers. This rubric is used to evaluate submitted papers based on five criteria: content that demonstrates excellent or novel research, originality, clarity of communication, relevance to the conference objectives, and standards. The standards criterion is weighted higher than the others are. Taken together the criteria call for authors to submit clearly articulated original research that is relevant to the conference theme and that could have an impact on future ICT standardization.

2.3   Impact on Standardization

From the first six Kaleidoscope conferences held between 2008 and 2014, ITU received important contributions that would influence the course of their standardization efforts. Authors submitted papers to all conferences that identified new areas for ITU-T standardization. Their contributions included a new network architecture, a standards-based service management implementation, and a business model for next generation networks. A recent paper led directly to a new information security standards project for secure messaging to support cryptographic key management techniques.

In the 2008 Innovations in NGN (next generation networks) conference, the best paper described an architecture and business model for an open heterogeneous mobile network. This paper by Murata, Hasegawa, Murakami, Harada, and Kato [7] inspired the creation of a new ITU-T Focus Group on Future Networks (FG FN). Subsequently, future networks became a key theme in ITU-T Study Group 13, underlying their standardization efforts in the areas of cloud computing, NGN, and mobile networks.

The 2009 Innovations for Digital Inclusion conference provided an important model and implementation methodology for quality of service (QoS) management for internet service providers. This award winning contribution was based on extensive review and study of the ITU-T Recommendation E.802 framework described in a paper by Ibarrola, Xiao, Liberal, and Ferro [8]. As participants of an ITU-T Academic Member institution, these authors regularly provide input to the protocols and test specifications standardization work of ITU-T Study Group 11.

The 2013 Building Sustainable Communities conference would provide the first new information security standards project to come from an ITU Kaleidoscope conference presentation. My paper on telebiometric information security and safety management would win an award and reveal security and other defects in several widely used international standards [9] and in the cryptographic messaging used to protect their information. Affected standards included the International Civil Aviation Organization (ICAO) standard for electronic passports, the ISO/IEC 24761 Authentication context for biometrics (ACBio), and the ISO/IEC 19785-4 Common Biometric Exchange Format Framework (CBEFF) – Security block format specifications. Example defects are shown in Fig. 1.

My Kaleidoscope paper noted that while there were several "widely deployed CMS standards", including the "RSA Public Key Cryptography Standard (PKCS) #7, the Secure Electronic Mail (S/MIME) CMS standard defined by the Internet

```
LDSSecurityObject {
    iso(1) identified-organization(3) icao(ccc) mrtd(1)
        security(1) ldsSecurityObject(1)
}                                                          Invalid

LDSSecurityObjectVersion ::= INTEGER { V0(0) }

AlgorithmIdentifier FROM PKIX1Explicit88

AlgorithmIdentifier ::= SEQUENCE {              Deprecated, Invalid
    algorithm   OBJECT IDENTIFIER,
    parameters  ANY DEFINED BY algorithm OPTIONAL
}
                                     Deprecated
```

Fig. 1. Invalid and deprecated ICAO schema definitions (Source: IEEE 52.1, p. 188)

Engineering Task Force (IETF), and the X9.73 Cryptographic Message Syntax: ASN.1 and XML standard, there was no "normative international CMS standard" that used valid ASN.1 syntax to define its messages [10]. All of the defective security standards identified in my paper relied on the IETF CMS SignedData type for data integrity and origin authenticity. They all relied on secure messaging that was "based on X.208, the deprecated 1988 version of ASN.1 that was withdrawn as a standard in 2002", and which contained known defects that "were never corrected before it was abandoned" by ISO/IEC and ITU-T [10].

Another defect identified in my paper involved the protection provided by "the optional signature block defined in the ISO/IEC 19785 CBEFF standard" [10]. A cryptographic message wrapper that encapsulates the entire template provides stronger protection of biometric templates in some environments than using the optional CBEFF security block. Using a message wrapper approach "allows a trivial attack on reference templates to be detected using signature verification" [10]. As described in Fig. 2, when used in "environments in which the optional signature block is not required to be present, it is possible for a low skill attacker to remove the entire signature block to thwart the signature safeguard's effectiveness" [10].

Common Biometric File Format (CBEFF)
Information Data Structure

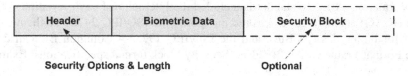

Editing the *Header* & deleting the *Security Block* removes the *Biometric Data* protection provided by a digital signature, an act that may not be detected in some environments.

Fig. 2. Security vulnerability (Source: IEEE 52.1, p. 189)

My Kaleidoscope paper proposed four areas for standardization. One information security proposal called for the creation of an international Cryptographic Message Syntax (CMS) standard containing correct messaging syntax and based on the current ASN.1 standards. A new CMS standard could be used to correct defects identified in the ICAO, ACBio, and CBEFF standards. A second proposal called for the standardization of a new SigncryptedData CMS message type. Two biometric information management and security proposals called for the creation of a standard event journal and security alert system for logging distributed biometric system security and safety events, and for the creation of a telebiometric system heartbeat message that could support optional inclusion of digitally signed or signcrypted ACBio system security verification reports [9].

2.4   Security Standards Proposals

During the 2013 conference in Kyoto, the Kaleidoscope Secretariat and her staff informed me that they had submitted my conference paper and presentation deck to the delegates of the ongoing ITU-T Study Group 17 Security meeting. A staff member arranged for me to make a presentation to the Question 9 (telebiometrics) standardization delegates meeting that week in Geneva. A member of the ITU Telecommunications Standards Bureau and the conference hosts arranged to provide an office next to the conference auditorium, a laptop, and a headset for the remote presentation of my proposals for new standardization.

The presentation was well received and followed by a brief question and answer session. However, neither of the biometric security management proposals described in the presentation would result in new standardization projects. The author would later discuss these proposals with a US delegate to SG17 who had not attended the Question 9 meeting in Geneva. After being thanked for presenting new standardization proposals to the committee, the delegate advised the author that the National Institute of Standards and Technology (NIST) was opposed to any biometrics standardization work being performed outside of the ISO/IEC JTC 1/SC 37 Biometrics group. The delegate noted that it would not be possible to propose new biometric standards development work from the US without NIST approval, and that the Kaleidoscope biometric security management proposals would not be put forward for consideration as new work items. No documentation was offered as evidence by the delegate that their remarks were an official US government position.

My 2013 Kaleidoscope paper and presentation slide deck were also provided to Study Group 17 Question 11 (Specification and Implementation Languages) for review. The two secure message proposals in the paper were viewed favorably, and the Rapporteur to the Study Group 17 plenary from France submitted a new work item proposal. The proposal was approved, and a project was started to create an international CMS standard that would correct and extend the defective Internet Engineering Task Force (IETF) CMS standard. The new CMS project would include a Signcrypted data type based on the ISO/IEC 29150 Signcryption security standard. No US approval was required to submit or approve the CMS new work item proposal.

3 US Practice Case Study

3.1 Government Roles

The American National Standards Institute (ANSI) is the US national body member of two of the three global SDOs responsible for information security standardization: ISO and IEC. The third SDO is ITU-T, which produces some security standards jointly with ISO and IEC, and whose US member is the US Department of State. ANSI establishes "the standards setting process that US national SDOs need to implement" to ensure their operations are based on "attributes like openness and transparency" [11]. There are over 200 ANSI-accredited SDOs in the US system, each typically associated with a single industry sector. NIST plays a special role in US standardization.

NIST "co-ordinates standards policy among the federal agencies", who are "encouraged to actively contribute to standards setting and to standards policy" as required by the National Technology Transfer and Advancement Act (NTTAA) [11].

Ideally, the role of NIST would be indirect, and as suggested by Sherif and Seo, limited to "promoting an environment in which firms can be innovative" and "promoting quality and performance standards with a global focus" [4]. However, NIST does not perform a limited, indirect role in US information security standardization. NIST contributes significant technical expertise to the development of international security standards. NIST also contributes the time of employees who have served as talented technical editors (i.e., ISO/IEC 29150, ISO/IEC 19790, etc.). However, a primary focus of NIST has been on national standards that serve government agency needs rather than on global standards that serve wider US interests [12].

Jakobs recognizes that the US government plays important coordination and promotional roles in standardization. He notes that with respect to technical details, "the US administration does not intervene in the process" of standardization, "nor does it mandate any standards" [11]. Rather, proposals developed by US SDOs result in voluntary, consensus standards. However, for information security standards there have been notable exceptions to this view.

These exceptions have resulted in defects remaining uncorrected for years in some standards. This was noted at Kaleidoscope 2013 in my presentation describing CBEFF defects and the defective information exchange syntax used in a widely referenced IETF cryptographic message standard. Government restrictions on the development of new standards are illustrated in the treatment of proposals for telebiometric security management standards described above. In the case of the NIST Role Based Access Control (RBAC) standard, there has been direct government intervention in the technical details of a security standard and in the consensus standards process.

3.2   Role Based Access Control

The NIST timeline for its RBAC standard starts in 2004. In that year, ANSI International Committee for Information Technology Standards (INCITS) adopted the "RBAC proposal as an industry consensus standard" [13]. Consensus agreement on the proposal was reached by the INCITS Executive Board, but it was never reached in the committee assigned by INCITS to review and comment on the technical details of the proposal. While under review by the T4 IT Security Techniques committee, the RBAC proposal failed to pass successive ballots. These failures were due to technical defects that subsequent research described as "limitations, design flaws, and technical errors" [14].

The authors of the RBAC proposal never addressed editorial and technical defects cited in the ballot comments of T4 members. Following a second failed ballot, T4 invited the authors to discuss how the concerns of the committee might be resolved, but received no response. Instead, NIST intervened directly in the standardization process. Using its standing as a member of the INCITS Executive Board, NIST appealed directly to the board for approval. Ignoring the sustained technical concerns raised by T4 members, the INCITS board approved the RBAC proposal as ANSI INCITS 359-2004, despite "a number of spelling and technical errors in the standard" [14].

Hoel argues that the role of government in standardization should be minimal. If government is to support technology innovation, direct government control is not needed, but instead, "initiatives to support consensus processes, in which governments do not have the final word regarding technical details" [15]. In the case of RBAC, NIST bypassed the consensus building process, and failed to reach agreement with fellow T4 members or to respond to their criticism of its work.

Following the negative committee response to its RBAC proposal, the role of NIST in the standardization of security techniques changed. NIST went from being an important technical contributor to assuming a more direct and controlling managerial role over the US SDO for information security. In 2005, NIST proposed that the INCITS Executive Board create a new security committee, one that could focus on the development of US national standards in its program of work [12], standards in which the T4 committee had expressed no interest. The INCITS board approved the proposal and created the CS1 Cyber Security committee. NIST has chaired CS1 since its inception, giving it greater control over US contributions to international security standardization.

In its recommendation to create INCITS CS1, NIST proposed that the T4 committee be stripped of most of its Technical Advisory Group (TAG) responsibilities in ISO/IEC JTC 1/SC27 IT Security techniques. Despite the growing global adoption of the ISO/IEC 17799 Code of practice for information security management [16], the NIST proposal asserted that there was not "wide consensus on best practices for information security management" and proposed development of a US national standard for "Risk Based Information Security Management" based on work "in progress at NIST" [12]. NIST proposed further development of national standards, such as extension of the RBAC standard and development of an IT security metrics standard with no indication given that these would be proposed for international standardization. This direct intervention in US security SDO practices by NIST changed the direction of US efforts from a focus on international information security standards to a focus on competing national alternatives.

3.3   Cryptographic Message Syntax

RSA Data Security first proposed the Secure/Multipurpose Internet Mail Extensions (S/MIME) for IETF standardization in the mid-1990s. The RSA proposal was based on their proprietary secure messaging standard, RSA Public Key Cryptography Standard 7 (PKCS #7) Cryptographic Message Syntax (CMS). NIST has a long history of involvement in the development of IETF CMS and in its promotion.

The 2002 NIST Special Publication (SP) 800-49 Federal S/MIME Client Profile recommends a version of IETF CMS whose message schema is based on the 1988 and 1990 Abstract Syntax Notation One (ASN.1) standards, X.208 and X.209 [17]. Both X.208 and X.209 were withdrawn as international standards that same year, 2002, due to well-known deficiencies and documented defects [18]. Their withdrawal followed several years in which they had been formally deprecated, and users in IETF and elsewhere encouraged to migrate to the current ASN.1 standards.

X.208 and X.209 were replaced and superseded by the X.680 and X.690 series of ASN.1 standards in 2002. These replacements corrected all known defects in the withdrawn versions, but these defects were never removed from X.208 and X.209. Efficient national language support in the current ASN.1 standards was never added to X.208. The XML Encoding Rules (XER), and the Distinguished Encoding Rules (DER) that provide the unambiguous data representations required by CMS and other cryptographic protocols were never defined for use with the X.208 syntax. Reliance on invalid and deprecated cryptographic message schema for data security in the AC-Bio, CBEFF, and ICAO standards does not enhance their ability to provide information assurance and security.

Recently, reference to IETF 3852 CMS began to appear in several important international biometrics and information security standards, despite its continued reliance on withdrawn standards to define its secure information exchange messages. Affected standards that depend on CMS for data security include ICAO electronic passports, ISO/IEC 24761 biometric security, ISO/IEC 19785-4 biometrics, and the ANSI/NIST-ITL biometric information exchange standard widely used by law enforcement and defense agencies. The design of the optional US National Security Agency (NSA) Type-98 information assurance record in the ITL standard is based on the same approach described in Fig. 2.

A paper presented at the 2013 ITU Kaleidoscope conference in Kyoto described these issues and recommended corrections [9]. The presentation proposed that ITU-T create a new international CMS standard that would correctly use the current ASN.1 schema definition standards. Once completed, a new corrected CMS standard could be referenced in other international security standards. Like most ITU-T standards it would be freely available to SDOs and implementers. Following approval of a new CMS project by Study Group 17, work began in January 2014 on a joint ISO/IEC 24824-4 standard and ITU-T 'x.CMS' recommendation.

The selection of CMS security techniques for standardization did not require the approval of the national body of the proposer, or on national body procedural support. The proposal did not depend on membership in a national body SDO. ITU Kaleidoscope conferences provide a venue that allows proposals to be presented, discussed, and evaluated strictly on their technical merits. These conferences establish an important new open and transparent process for the initial selection of security techniques for standardization that can circumvent restrictions that may be imposed by government controlled SDO.

4   Conclusion and Recommendations

From its roots as the first international standards organization, ITU has again shown the way forward for global SDO that strive to remain relevant in an increasingly interdependent world of rapidly changing technology. ITU Kaleidoscope conferences foster greater openness and transparency in the initial selection of security techniques for international standardization. Kaleidoscope conferences invite new proposals for standardization from the broad community of academic and research institutions.

Participation is not limited to national SDO members, but open to all standards stakeholders, to everyone who lives in an ITU member country.

Kaleidoscope conferences are hosted by academic and research institutions at different locations around the world. These events bring standards developers together with inventors and researchers who are closest to new and emerging technologies that may be appropriate for standardization. The Kaleidoscope proposal process shortens the gap between the discovery of innovative solutions and their standardization, and can circumvent restrictions imposed by government controlled national SDO.

ITU could improve their conferences by allowing authors to present their work remotely by teleconference or videoconference. Author presentations could also be recorded and made freely available by ITU in audio or video format, as is presently the case for USENIX security conferences. Recorded presentations could provide an additional educational benefit and enhance the ability of the conference to foster the free exchange of research information.

The US economy is diverse and immense. The broad interests of US standards stakeholders cannot be served by a single government agency alone. The government should not have the final say on the technical details of any standard, especially security standards such as RBAC or CMS. NIST should play only an indirect role in international standardization, a role that facilitates open, transparent processes that benefit all standards stakeholders.

NIST should not exert managerial control over SDO activities or use SDO management boards to quash technical criticism of their work. They should limit their involvement to promoting standards adoption and use, providing contributions of technical expertise, and promoting broad and inclusive participation. In an increasingly global economy, NIST would better serve the interests of US citizens and businesses by promoting the development and adoption of international standards, rather than promoting their own information security and information security management work as sources for national standards alternatives.

Recently, security techniques proposed at a Kaleidoscope academic conference resulted in creation of a new international standardization project in ITU. All of the global security SDOs could benefit from the closer ties to researchers afforded by the ITU model. The SC 27 IT Security techniques and TC68/SC2 Financial Services Security groups should provide a means for academic papers and conference presentations to inform their information security standardization selection processes. They should encourage greater openness through wider participation in the development of new standards by academic and research institutions.

The ITU Kaleidoscope conference serves as a model to be replicated by others. Conferences with a focus on information security techniques that can provide new inputs for international security standardization should be sponsored by security standards setting bodies. Sponsored conferences should be attended by SDO representatives who can transform proposals into actions, such as the creation of study groups and new work item proposals. Examples of promising conferences to be considered include the new 2014 Security Standardization Research (SSR) conference in the United Kingdom and the 2014 International Conference on Smart Computing in Hong Kong.

References

1. SSR 2014: Security Standardisation Research, http://www.ssr2014.com/
2. International Telecommunication Union, http://www.itu.int
3. Werle, R., Iversen, E.: Promoting legitimacy in technical standardization. Science, Technology & Innovation Studies 2 (2006)
4. Sherif, M.H., Seo, D.: Government role in information and communications technology innovations. In: Innovations for Digital Inclusions. ITU-T Kaleidoscope, pp. 1–5. IEEE (2009)
5. Regulation of the European Parliament and Council, (EU) No 1025/2012, http://eur-lex.europa.eu/legal-content/EN/TXT/?uri=CELEX:32012R1025
6. Graham, I.: Reflexive Standardization of Network Technology. In: Proceedings of the ITU Kaleidoscope Academic Conference, pp. 83–88 (2011)
7. Murata, Y., Hasegawa, M., Murakami, H., Harada, H., Kato, S.: The architecture and a business model for the open heterogeneous mobile network. In: Proceedings of the 2008 ITU Kaleidoscope Academic Conference: Innovations in NGN, pp. 143–150 (2008), http://www.itu.int/pub/T-PROC-KALEI-2008/en
8. Ibarrola, E., Xiao, J., Liberal, F., Ferro, A.: Quality of Service management for ISP: A model and implementation methodology based on ITU-T Rec. E.802 framework. In: Proceedings of the 2009 ITU Kaleidoscope Academic Conference: Innovations for Digital Inclusion, pp. 35–42 (2009), http://www.itu.int/pub/T-PROC-KALEI-2009
9. Griffin, P.: Telebiometric Security and Safety Management. In: Proceeding of the 2013 ITU Kaleidoscope Academic Conference: Building Sustainable Communities, pp. 127–134 (2013), http://www.itu.int/pub/T-PROC-KALEI-2013
10. Griffin, P.: Telebiometric Security and Safety Management. IEEE Communications Magazine 52(1), 186–192 (2014)
11. Jakobs, K.: ICT Standardisation in China, the EU, and the US. In: Innovations for Digital Inclusions, K-IDI 2009. ITU-T Kaleidoscope, pp. 1–6. IEEE (2009)
12. INCITS Proposal to create new security technical committee CS1, in050057 (2005), http://csrc.nist.gov/groups/SNS/rbac/documents/in050057.pdf
13. National Institute of Standards and Technology - Computer Security Resource Center, http://csrc.nist.gov/groups/SNS/rbac/faq.html#timeline
14. Li, N., Byun, J., Bertino, E.: A Critique of the ANSI Standard on Role Based Access Control. IEEE Security & Privacy 5(6), 41–49 (2007)
15. Hoel, T.: Paradoxes in LET standardisation–towards an improved process. In: Proceedings of the 21st International Conference on Computers in Education. Asia-Pacific Society for Computers in Education, Indonesia (2013)
16. Backhouse, J., Hsu, C., Silva, L.: Circuits of power in creating de jure standards: shaping an international information systems security standard. MIS Quarterly 30, 413–438 (2006)
17. National Institute of Standards and Technology SP 800-49 Federal S/MIME V3 Client Profile, http://csrc.nist.gov/publications/nistpubs/
18. ITU-T Recommendation X.208, https://www.itu.int/rec/T-REC-X.208/en

Size-Efficient Digital Signatures with Appendix by Truncating Unnecessarily Long Hashcode*

Jinwoo Lee and Pil Joong Lee

Department of Electrical Engineering, POSTECH, Republic of Korea
{woojung3,pjl}@postech.ac.kr

Abstract. Digital signature mechanism with appendix(DSwA) is a type of digital signature in which, after the message, a signature Σ is appended. When DSwA is constructed based on the discrete logarithm problem, Σ is composed of a pair (R, S). When R is a hashcode with bit length γ and S is an element of subgroup of order q with bit length β, it is recommended to adjust γ and β to be similar because the security strength depends on the smaller value between γ and β. However in some circumstances only hash functions with longer output could be available. Then γ becomes unnecessarily longer than β, and hence the longer Σ is appended. For the above case, we propose a generalized method for reducing the size of Σ by truncating R by β without loss of any security strength. Our proposed method can be applied to mechanisms like KCDSA, SDSA, EC-KCDSA, and EC-SDSA in ISO/IEC 14888-3: Digital signatures with appendix — Part 3: Discrete logarithm based mechanisms.

Keywords: Digital Signature, KCDSA, SDSA, ISO/IEC 14888-3.

1 Introduction

The notion of digital signature is an alternative for handwritten signatures in digital world [11]. The concept of a digital signature first appeared in [9]. In their scheme they proposed that each user can publish a public key, which is used for verifying signatures, while keeping secret a private key, which is used for producing signatures.

There are two types of digital signature mechanisms [3]. If the whole message has to be stored and/or transmitted along with the signature, the mechanism is called a "digital signature mechanism with appendix (DSwA)". If the verification process reveals all or part of the message, the mechanism is called a "digital signature mechanism giving message recovery". In this work, we consider the

* This research was supported in part by the MSIP (Ministry of Science, ICT and Future Planning), Korea, under the ITRC (Information Technology Research Center) support program (NIPA-2014-H0301-14-1004) supervised by the NIPA (National IT Industry Promotion Agency) and in part by the MSIP (Ministry of Science, ICT and Future Planning), Korea, under the "IT Consilience Creative Program" (NIPA-2014-H0201-14-1001) supervised by the NIPA (National IT Industry Promotion Agency).

L. Chen and C. Mitchell (Eds.): SSR 2014, LNCS 8893, pp. 69–78, 2014.

DSwA which is constructed based on the Discrete Logarithm Problem(DLP). DSwA based on DLP is specified in a International Organization for Standardization/International Electrotechnical Commission Joint Technical Committee 1 (ISO/IEC JTC 1) 14888-3 "Information technology — Security techniques — Digital signatures with appendix — Part 3: Discrete logarithm based mechanisms" [4].

DSwA based on DLP has a digital signature in which, after the message, a signature Σ of pair (R, S) is appended. When R is a hashcode with bit length γ and S is an element of subgroup of order q with bit length β, it is recommended to adjust γ and β to be similar. This is because the security strength depends on the smaller value between γ and β.

To prevent unnecessarily long hashcode R from being used, some standards made a restriction. For example, in [7] it is recommended to use SHA-224 with q of length 224-bits and SHA-256 with q of length 256-bits, but it is allowed to use SHA-256 with q of length 224-bits when only SHA-256 is available, while SHA-224 is not.

There are many DSwA based on DLP in which R is a hashcode. Among twelve mechanisms in the standard document [4], four mechanisms, Korean Certificate-based Digital Signature Algorithm (KCDSA), Schnorr Digital Signature Algorithm (SDSA), Elliptic Curve KCDSA (EC-KCDSA), and Elliptic Curve SDSA (EC-SDSA), are such cases. In this paper we propose a generalized method of truncating R by β for mechanisms with R as a hashcode. By doing this, when R is longer than β, Σ with shorter length can be made without loss of any security strength.

We note here that the idea of truncating R by β was first given by Eric Peeters (personal communication, February 23, 2011), although he only mentioned about EC-KCDSA.

2   Preliminaries

2.1   General Model for Digital Signatures with Appendix Based on Discrete Logarithm Problem

Generation of Domain Parameters. For mechanisms of DSwA based on DLP, the set of domain parameters includes the following parameters:

- $\mathcal{E}$, a finite commutative group;
- q, a prime divisor of $\#\mathcal{E}$;
- G, a generator of the subgroup of order q

A private signature key of a signer is a randomly or pseudo-randomly generated secret integer X such that $0 < X < q$. The corresponding public verification key Y is an element of $\mathcal{E}$ and is computed as $Y = G^{X^D}$. The value of D is one of two values, -1 and 1.

Signature Process. Signature process makes use of a randomizing value K, which is used to produce a witness R. The signature Σ for the message M is the pair (R, S).

1. Producing the randomizer: For each signature, the signer freshly generates a secret randomizer K with $0 < K < q$. The output of this stage is K, which shall be kept secret and destroyed safely after use.
2. Producing the pre-signature: The inputs to this stage are the randomizer K and signature key X, with which the signer computes the pre-signature, Π.
3. Computing the witness: The inputs to this stage are Π and optionally M. If M is used here, then M is not used in the next "Computing the assignment" stage. The output of the witness function is the witness R.
4. Computing the assignment: The inputs to the assignment function are R, and optionally Y. If M is not used in the above "Computing the witness" stage, then M is used as input too in this stage. The output of the assignment function is assignment V.
5. Computing the second part of the signature: The inputs to this stage are K, X, V, domain parameter q, and D. The signer solves the signature equation for S.
6. Constructing the signed message: The signed message is obtained by the concatenation of a message and the signature, so $M\|\Sigma$. Here, the signature Σ is the pair (R, S).

Verification Process

1. Retrieving the witness: From the signed message $M\|\Sigma$, the verifier retrieves the signature Σ and divides it into R and S. Also, the verifier checks the range or the bit length of the signature elements, R and S, according to the rule specified by each signature process. If the predefined rule is violated, the signature shall be rejected.
2. Retrieving the assignment: The inputs to the assignment function consist of R, and optionally Y and M. If M is used here, then M is not used in the next "Recomputing the witness" stage. The assignment V is recomputed as the output from the assignment function.
3. Recomputing the pre-signature: The inputs to this stage are the set of domain parameters, Y, V, R, and S. The verifier computes the element Π'.
4. Recomputing the witness: The computations at this stage are the same as in computing the witness of signature process. The input is Π'. If M is not used in the above "Retrieving the assignment" stage, then M is used as input too in this stage. The output is the recomputed R'.
5. Verifying the witness: The signature is verified if the recomputed R' is equal to R.

2.2   Conversion Functions

– Conversion from a bit sequence to an integer ($BS2I$): A g-long sequence of bits $\{x_1, ..., x_g\}$ is converted to an integer by the rule

$$\{x_1, ..., x_g\} \rightarrow x_1 \cdot 2^{g-1} + x_2 \cdot 2^{g-2} + ... + x_{g-1} \cdot 2 + x_g.$$

– Conversion from an integer to a bit sequence ($I2BS$): An integer x in the range $0 \le x < 2^g$ may be converted to a g-long sequence of bits by using its binary expansion as shown below:

$$x = x_1 \cdot 2^{g-1} + x_2 \cdot 2^{g-2} + ... + x_{g-1} \cdot 2 + x_g \to \{x_1, ..., x_g\}.$$

2.3   Hash Functions

In this work, we use hash functions from [2]. Note that SHA-224 is identical to SHA-256, except that the IV is different and the output is truncated to 224 bits [6]. This truncation method is also used with SHA-384, Snefru, and Tiger [10].

3   Applicable Mechanisms of Digital Signatures with Appendix Based on Discrete Logarithm Problem

Our proposed method can be applied when R is a hashcode and S is an integer such that $0 < S < q$. In this section, we present KCDSA and SDSA from ISO/IEC 14888-3 2nd edition [4].

– KCDSA
  - $R = h(I2BS(\Pi))$
  - $V = BS2I(R \oplus h(I2BS(Y \bmod 2^l \| M))) \bmod q$
  - $S = X(K - V) \bmod q$
– SDSA
  - $R = h(I2BS(\Pi) \| M)$
  - $V = -BS2I(R) \bmod q$
  - $S = (K - VX) \bmod q$

As shown above, both KCDSA and SDSA are mechanisms in which R is a hashcode and S is an integer which satisfies $0 < S < q$.

Followings are illustrations of KCDSA and SDSA. The detailed explanations which are same with the general model is excluded.

KCDSA: Korean Certificate-based Digital Signature Algorithm

– Domain parameters: A finite commutative group $\mathcal{E}$ is Z_p^*, D is -1.
– Generation of signature key and verification key: The signature key of a signer is randomly or pseudo-randomly generated random integer X such that $0 < X < q$. The corresponding public verification key Y is $G^{X^{-1}} \bmod p$.
– Signature process:
  1. Producing the randomizer: The signer randomly or pseudo-randomly generates an integer K such that $0 < K < q$.
  2. Producing the pre-signature: The input to this stage is K. The signer computes $\Pi = G^K \bmod p$.
  3. Computing the witness: The signer computes $R = h(I2BS(\Pi))$.

4. Computing the assignment: The signer computes $V = BS2I(R \oplus H) \bmod q$, where $H = h(Y'\|M)$ is the hashcode of the concatenation of $Y' = I2BS(Y \bmod 2^l)$ and message M. Here, mod 2^l is used for truncating Y by least significant l bits.

5. Computing the second part of the signature: The signature is (R, S) where $S = X(K - V) \bmod q$.

6. Constructing the signed message: A signed message is the concatenation of a message and the signature, so $M\|(R, S)$.

- Verification process:

1. Retrieving the witness: The verifier retrieves R and S from the signature. The verifier then checks whether the following conditions hold or not:
   - $0 < S < q$;
   - The bit length of R is equal to the output bit length of the employed hash-function h.

   If any of the above conditions does not hold the signature shall be rejected.

2. Retrieving the assignment: This stage is identical to computing the assignment. The inputs to the assignment function consist of R and M. V' is recomputed as output from the computing the assignment.

3. Recomputing the pre-signature: The inputs to this stage are Y, S, V', and domain parameters. The verifier obtains a recomputed value Π' using the formula $\Pi' = Y^{(S \bmod q)} G^{(V' \bmod q)} \bmod p$.

4. Recomputing the witness: The computations at this stage are the same as in computing the witness. The verifier executes the witness function. The input is Π'. The output is the recomputed R'.

5. Verifying the witness: The verifier compares the recomputed R' to the value of R. If $R' = R$, then the signature is valid.

SDSA: Schnorr Digital Signature Algorithm

- Domain parameters: A finite commutative group $\mathcal{E}$ is Z_p^*, D is 1.
- Generation of signature key and verification key: The signature key of a signer is a randomly or pseudo-randomly generated secret integer X such that $0 < X < q$. The corresponding public verification key Y is $G^X \bmod p$.
- Signature process:

1. Producing the randomizer: The signer randomly or pseudo-randomly generates an integer K such that $0 < K < q$.

2. Producing the pre-signature: The input to this stage is K, and the signer computes $\Pi = G^K \bmod p$.

3. Computing the witness: The signer computes the witness R as the hashcode of the pre-signature Π and the message M so $R = h(I2BS(\Pi)\|M)$.

4. Computing the assignment: The input to the assignment function is R. $V = -BS2I(R) \bmod q$.

5. Computing the second part of the signature: The signature is (R, S) where $S = (K - VX) \bmod q$.

6. Constructing the signed message: A signed message is the concatenation of a message and the signature, so $M||(R, S)$.
- Verification process:
  1. Retrieving the witness: The verifier retrieves R and S from the signature. The verifier checks to see that R is within the range of the hash function and that $0 < S < q$.
  2. Retrieving the assignment: This stage is identical to computing the assignment. The input to the assignment function is R. V' is recomputed as output from the computing the assignment.
  3. Recomputing the pre-signature: The inputs to this stage are Y, S, V', and domain parameters. The verifier obtains a recomputed value Π' by computing it using the formula $\Pi' = Y^{(V' \bmod q)} G^{(S \bmod q)} \bmod p$.
  4. Recomputing the witness: The computations at this stage are the same as in computing the witness. The verifier executes the witness function. The inputs are Π' and M. The output is the recomputed R'.
  5. Verifying the witness: The verifier compares the recomputed R' to the retrieved version of R. If $R' = R$, then the signature is verified.

4   The Proposed Method

4.1   Construction

The following construction can be applied to the mechanisms of the standard document [4] in which R is a hashcode and S is an integer of length β such that $0 < S < q$. This is a generalized method of truncating R by β when γ, length of R, is longer than β. By doing this, signature Σ with shorter length can be made without loss of any security strength. A detailed discussion on the security strength of the proposed method is in section 5. For each mechanism, all the other stages are same with the general model except for the following stages:

- Signature process:
  - Computing the witness: The signer computes R by witness function, $f(\cdot)$. If γ is longer than β, then the witness function $f(\cdot)$ is replaced by $f'(\cdot) = I2BS(BS2I(f(\cdot)) \bmod 2^\beta)$. Here, $\bmod\ 2^\beta$ is used for truncating R by least significant β bits.
  - Computing the assignment: If γ is longer than β, then all the operands which is used with R should be truncated by β.
- Verification process:
  - Retrieving the witness: The verifier retrieves R and S from the signature. The verifier then checks whether the following conditions hold or not:
    * $0 < S < q$;
    * If the length of the value γ is not longer than β, the bit length of R is equal to the output bit length of the employed hash-function h;
    * If the length of the value γ is longer than β, the bit length of R is equal to β.

If any of the above conditions does not hold the signature shall be rejected.

In short, we truncate R by β when γ is longer than β. By doing this the size of the signature Σ decreases without loss of security strength. The security of the proposed method has no problem if the hash function used does not lose its security when truncated.

Our construction can be applied to KCDSA, SDSA, EC-KCDSA, and EC-SDSA, which are in the standard document [4]. Here, we present detailed stages of modified KCDSA in 4.2, which is recently included in N13975 Text of ISO/IEC 3$^{\rm rd}$ WD 14888-3 [5]. Korean domestic standard was also modified in [7], accordingly. We also present a detailed stages of modified SDSA in 4.3. The other mechanisms, EC-KCDSA and EC-SDSA, can also be improved by our method, but is not illustrated here. We note that EC-KCDSA is also modified in [5] and [8] as we proposed.

4.2   Modified KCDSA

- Signature process:
    - Computing the witness: The signer computes $R = h(I2BS(\Pi))$. If γ is longer than β, then the computation of R is replaced by $R = I2BS(BS2I$ $(h(I2BS(\Pi)))$ mod $2^{\beta})$. Here, mod 2^{β} is used for truncating R by least significant β bits.
    - Computing the assignment: The signer computes $V = BS2I(R \oplus H)$ mod q, where $H = h(Y'\|M)$ is the hashcode of the concatenation of $Y' = I2BS(Y$ mod $2^{l})$ and message M. The value of Y' is a bit string of length l. If γ is longer than β, then the computation of H is replaced by $H = I2BS(BS2I$ $(h(Y'\|M))$ mod $2^{\beta})$.
- Verification process:
    - Retrieving the witness: The verifier retrieves R and S from the signature. The verifier then checks whether the following conditions hold or not:
        * $0 < S < q$;
        * If the length of the value γ is not longer than β, the bit length of R is equal to the output bit length of the employed hash-function h;
        * If the length of the value γ is longer than β, the bit length of R is equal to β.

    If any of the above conditions does not hold the signature shall be rejected.

4.3   Modified SDSA

- Signature process:
    - Computing the witness: The signer computes $R = h(I2BS(\Pi)\|M)$. If γ is longer than β, then the computation of witness is replaced by $R = I2BS(BS2I(h(I2BS(\Pi)\|M))$ mod $2^{\beta})$. Here, mod 2^{β} is used for truncating R by least significant β bits.

    – Verification process:
        • Retrieving the witness: The verifier retrieves R and S from the signature.
          The verifier then checks whether the following conditions hold or not:
              ∗ $0 < S < q$;
              ∗ If the length of the value γ is not longer than β, the bit length of R
                is equal to the output bit length of the employed hash-function h;
              ∗ If the length of the value γ is longer than β, the bit length of R is
                equal to β.
          If any of the above conditions does not hold the signature shall be re-
          jected.

5   Security

Figure 1 represents the original DSwA scheme and a modified DSwA scheme to which our proposed method is applied. There, DSwA scheme with hash function is Λ and modified DSwA scheme with truncated hash function is Λ'. Both Λ and Λ' have a signature Σ of a pair (R, S). R is a hashcode for Λ, and a truncated hashcode for Λ'. S is an element of subgroup of order q. In Λ, bit length of R is γ, bit length of S is β, and $\gamma > \beta$. In Λ', bit length of R is γ', bit length of S is β, and $\gamma' = \beta$.

The security strength of Λ solely depends on the smaller value between γ and β. Therefore, Λ has security strength of β. If the security strength of Λ' is also β then modified scheme applying our method is secure. To satisfy this, truncated hashcode should not have security strength less than β. Thus, for the modified scheme to be secure, truncation of the hashcode should not have any side effects rather than reducing its length.

However, not every hash function is free from the side effects of truncation. For example, SHA-0 is easily broken through near-collsions when truncated [12]. Truncating SHA-0 by the first 142 bits makes unsafe variant. On the other hand, the hash functions that we consider in this paper are safe from near-collisions.

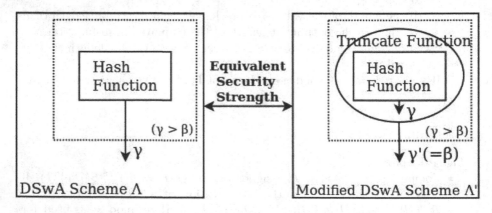

Fig. 1. DSwA Scheme with hash function (left) and truncated hash function (right)

Moreover, in [2], the output for the hash functions is the hashcode which is derived by taking the leftmode L_H(length of a hashcode) bits of the final output string from the last round function. This means that a truncated hash funtion is allowed in the ISO/IEC specification.

As shown above, extra assumptions about hash functions are needed to perform a formal security proof of truncated hash functions, since not every hash function is free from the side effects of truncation. To the best of our knowledge, there is no such proof for now. In this paper we remain the formal security proof as an open question and assume that truncation of the proper hashcode does not have any side effects rather than reducing its length.

6 Conclusion

We proposed a general method for reducing the size of $\Sigma(= (R, S))$ by truncating R when R is a hashcode of length γ and S is an integer of length β such that $0 < S < q$. For such cases, it is preferred to adjust γ and β to be similar, because the security strength depends on the smaller value between γ and β. If γ becomes longer than β, it makes longer Σ without increasing any security strength. However in some cases only hash functions with longer output can be available. In that case our method can be useful to reduce the size of Σ without loss of any security strength (We assume that the truncation of the proper hashcode does not have any side effects rather than reducing its length as discussed in the section 5. For the moment, we leave formal security proof as an open question). If our method is applied, length of Σ is reduced from $\gamma+\beta$ to 2β. Applying our method to a mechanism in which R is a hashcode of SHA-256 and S is an integer of 224-bit results in a visible reduction of the length of the Σ from 480-bit to 448-bit. Among twelve mechanisms in the standard document [4], our proposed method can be applied to four mechanisms, Korean Certificate-based Digital Signature Algorithm (KCDSA), Schnorr Digital Signature Algorithm (SDSA), Elliptic Curve KCDSA (EC-KCDSA), and Elliptic Curve SDSA (EC-SDSA).

References

1. ISO/IEC 10118-1 (3rd edn.): Information technology — Security techniques — Hash-functions — Part 1: General (2000)
2. ISO/IEC 10118-3 (3rd edn.): Information technology — Security techniques — Hash-functions — Part 3: Dedicated hash-functions (2004)
3. ISO/IEC 14888-1 (2nd edn.): Information technology — Security techniques — Digital signatures with appendix — Part 1: General (2008)
4. ISO/IEC 14888-3 (2nd edn.): Information technology — Security techniques — Digital signatures with appendix — Part 3: Discrete logarithm based mechanisms (2006)
5. N13975 Text of ISO/IEC 3rd WD for 3rd edition of 14888-3: Information technology — Security techniques — Digital signatures with appendix — Part 3: Discrete logarithm based mechanisms (2014)

6. FIPS 180-4: Secure Hash Standard (SHS) (2012)
7. TTAK.KO-12.0001/R3: Digital Signature Mechanism with Appendix — Part 2: Korean Certificate-based Digital Signature Algorithm KCDSA (2014)
8. TTAK.KO-12.0015/R2: Digital Signature Mechanism with Appendix — Part 3: Korean Certificate-based Digital Signature Algorithm using Elliptic Curves EC-KCDSA (2014)
9. Diffie, W., Hellman, M.E.: New directions in cryptography. IEEE Trans. Inform. Theory, 644–654 (1976)
10. Kelsey, J.: SHA-160: A Truncation Mode for SHA256 (and most other hashes). Cryptographic Hash Workshop, NIST (2005)
11. Matyas, S.: Digital signatures — an overview. Computer Networks, 87–94 (1979)
12. Biham, E., Chen, R.: Near-collisions of SHA-0. In: Franklin, M. (ed.) CRYPTO 2004. LNCS, vol. 3152, pp. 290–305. Springer, Heidelberg (2004)

Blinded Diffie-Hellman

Preventing Eavesdroppers from Tracking Payments

Duncan Garrett and Michael Ward

EMVCo Security Working Group
www.emvco.com

Abstract. In this paper we present a novel form of ECC Diffie-Hellman key agreement that provides privacy and anti-tracking for contactless payments. The payer's device can be authenticated by a payment terminal using a static public key with associated certificates belonging to the payer's device; however, a passive eavesdropper is unable to determine the static data and keys that might otherwise be used to identify and track the payer. The new protocol has better performance than alternative protocols; it avoids the payer's device having to support signature algorithms with dedicated hashes and it has a security proof given in [3]. The new protocol does not appear in any standards known to the authors.

Keywords: Elliptic curve cryptography, Diffie-Hellman, Key agreement, Privacy, Payments, Standards.

1   Introduction

The purpose of this paper is to describe a new protocol, called 'Blinded Diffie-Hellman', for establishing and using a secure channel between a payment card and a merchant terminal.

EMVCo (www.emvco.com) has stated [7] that it needs such a protocol as the basis of future payment technology which is intended to use elliptic curve cryptography (ECC) rather than RSA.   In the absence of suitable standardised techniques, EMVCo has developed new key agreement technology.   Two candidate mechanisms were published in November 2012 [7], and subsequently a security proof for the more efficient mechanism, the Blinded Diffie-Hellman protocol, was published by Brzuska, Smart, Warinschi and Watson [3].   This paper provides a detailed description of this protocol and of its security and performance benefits.

The rest of the paper is organized as follows. We first describe the EMV security requirements for the EMV secure channel and then we review the landscape of currently available standards and mechanisms. In Section 3 we describe the only protocol in the literature (Station-to-Station protocol) that meets the EMV security objectives, and then describe the Blinded Diffie-Hellman protocol that also meets the EMV security objectives but is more efficient. In Section 4 we compare the performance of the new protocol with the Station-to-Station protocol and briefly describe its security

L. Chen and C. Mitchell (Eds.): SSR 2014, LNCS 8893, pp. 79–92, 2014.

properties and associated proof of security (more details of which are provided in [3]). Section 5 contains concluding remarks.

2   EMV Security Requirements and the Standards Landscape

2.1   EMV Security Requirements

The EMV specifications form the basis for debit and credit card transactions in many parts of the world. The current specifications enable the card to locally authenticate itself to the merchant's terminal using an RSA based PKI and also to remotely authenticate itself (and the payment details) to the card-issuing bank using symmetric cryptography. Local card authentication is achieved by the card creating a digital signature in response to a random challenge from the terminal and the terminal verifying this signature using public key certificates provided by the card and a root public key installed in the terminal. However a consequence of choosing this method of local card authentication is that a passive bugging device located near the terminal might eavesdrop on a contactless transaction and discover the card account number and the public key of the card. Even if the card account number could be protected, the public key of the card is still on its own sufficient to track card use at this terminal and at other terminals whose transactions are similarly eavesdropped.

It would clearly be desirable if, when making the changes to introduce elliptic curve cryptography, the part of the EMV protocol that provides local card authentication could be enhanced to provide protection against such passive eavesdropping. In [7] EMVCo states the security objectives for a new secure channel protocol as follows:

The protocol takes place between a payment card and a merchant terminal and establishes a secure channel between the card and terminal.

The security objectives of the secure channel are

— to provide authentication of the card to the terminal,
— to detect modifications to the communications, and
— to protect against eavesdropping of transactions and card tracking.

It is given that the card contains

— a private key unique to the card,
— the corresponding card public key certified by the card issuer, and
— the corresponding issuer public key certified by a Payment System,

and that the terminal contains the corresponding Payment System public key.

The card and terminal achieve a secure channel by using a key agreement protocol based on Diffie-Hellman, a session key derivation function and a block-cipher based authenticated encryption algorithm (see [29]). The payments may take place in situations that require fast contactless transactions and where the contactless card is only briefly in range of the terminal's contactless field, so good performance of the key agreement protocol is critical.

2.2   Standards Landscape

Key management standards within ISO/IEC [22], ANSI [21], IETF [14,15], NIST [18], IEEE [13] and SECG [16] do not appear to have methods for hiding the authenticated party's public key. For example ISO/IEC 11770-3 [23] defines 12 key agreement mechanisms, classifying them in terms of number of passes, implicit key authentication, key confirmation, entity authentication, public key operations, forward secrecy and key freshness. However none of the mechanisms offer secrecy of the public key and only two of them (mechanisms 2 and 11) support one-sided key agreement where only one party is authenticated. The Handbook of Applied Cryptography [12] discusses variants of the Diffie-Hellman algorithm that provide anonymity and this includes the Station-to-Station protocol which does afford anonymity for the participants; however it is described as a scheme where both parties have public keys which can be authenticated and this is not the situation for the EMV system.

One might also consider key transport mechanisms using public key encryption methods such as El Gamal and ECIES as specified in ISO/IEC 18033-2 [27]. However, although such key transport mechanisms provide confidentiality for the payload/message, they do not in themselves protect the confidentiality of the public key used nor offer joint key control.

Finally, there do exist standardised methods for achieving privacy and anonymity objectives (e.g. ISO/IEC 20008 Anonymous Digital Signatures [30], ISO/IEC 20009 Anonymous Authentication [31], ISO/IEC 18370 Blind Signatures [28]); however the privacy issues that these methods aim to solve are different to those considered in this paper (e.g. they provide anonymity via group signatures).

The only 'standard' method in the literature that might meet the EMV privacy objectives would appear to be a one-sided version of the Station-to-Station protocol (see [12]) where both card and terminal conduct a Diffie-Hellman key agreement using ephemeral ECC keys to establish a secure channel and then the card digitally signs its ephemeral key using its static and certified ECC keys. However, the performance of this protocol is not very good (see Section 4) and this therefore led to the design of a new alternative protocol [7] referred to as "Blinded Diffie-Hellman" that is described in detail in the next section.

3   Description of Protocols

In this section we describe the Station-to-Station protocol using elliptic curve cryptography. We also describe a variant that meets the one-sided constraints of EMV described in the previous section and the new Blinded Diffie-Hellman protocol that also meets the one-sided constraints of EMV.

In the descriptions below, the following conventions are used:

— Conversion between integers and byte strings and between elliptic curve points and byte strings is not considered. Rules for this are defined in ISO/IEC 15946 [26].

- For simplicity the descriptions do not include important aspects such as bounds checking, parsing and checking that points are on curves.
- Optimisations such as point compression are not considered.
- The derivation of a single symmetric key for authenticated encryption is shown, whereas in practice it is expected that two uni-directional keys would be derived according to standard methods (see Section 4.3).

3.1   Station-to-Station Protocol

The original Station-to-Station protocol described in [12] is an extension of the Diffie-Hellman protocol that allows the establishment of a shared secret key between two parties A and B with mutual entity authentication and mutual explicit key authentication. The Station-to-Station protocol also facilitates anonymity whereby the identities of A and B may be protected from eavesdroppers.

The Station-to-Station protocol also employs digital signatures. In our description we assume the use of elliptic curve cryptography for the Diffie-Hellman aspects, so the signature mechanism could, for example, be ECDSA or a variant thereof (see [25]). In the description below we introduce a key derivation function KDF which is not mentioned in [12].

Summary:
▪ Parties A and B exchange 3 messages
Result:
▪ key agreement, mutual entity authentication, explicit key authentication, anonymity.
Notation:
▪ G is a generator of a group of points on an elliptic curve whose order is n, and $d \cdot G$ is scalar multiplication of G by an integer d (i.e. the point on the curve resulting from adding G to itself d times).   ▪ $E(k)[m]$ denotes encryption of m under key k using an encryption algorithm E.   ▪ $S_A(m)$ denotes A's signature (e.g. using ECDSA) on m.   ▪ Cert(A) denotes A's public key and associated public key certificates.   ▪ KDF is a key derivation function that generates, from a point on the elliptic curve, a symmetric key suitable for the encryption algorithm E.
Protocol steps: 1. A generates an ephemeral private key d_a and generates A's ephemeral public key as $Q_a = d_a \cdot G$ 2. A → B : $Q_a = d_a \cdot G$

3. B generates an ephemeral private key d_b and generates B's ephemeral public key $Q_b = d_b \cdot G$

4. B computes key $K_b = \text{KDF}(d_b \cdot Q_a)$ and signs $Q_b \parallel Q_a$

5. $B \rightarrow A : Q_b \parallel E(K_b)[\text{ Cert(B) } \parallel S_B(Q_b \parallel Q_a)]$

6. A computes key $K_a = \text{KDF}(d_a \cdot Q_b)$. A authenticates B's public key using B's certificates and verifies B's signature on $Q_b \parallel Q_a$. A only accepts the validity of key K_a if the signature verifies successfully. A signs $Q_a \parallel Q_b$

7. $A \rightarrow B : E(K_a)[\text{ Cert(A) } \parallel S_A(Q_a \parallel Q_b)]$

8. B authenticates A's public key using A's certificates and verifies A's signature on $Q_a \parallel Q_b$. B only accepts the validity of key K_b if the signature verifies successfully.

3.2   One-Sided Station-to-Station Protocol

We now describe the variant of the Station-to-Station protocol, which we refer to as 'one-sided Station-to-Station'. This variant would suit the requirements of EMV, namely where only one party, the card, has a static certified public key. Note that according to [7] the EMV secure channel will use authenticated encryption instead of plain encryption, so this is included in the description below.

In our description below we denote the participants as C and T (rather than A and B) to better reflect that the participants are a payment card and a payment terminal, respectively.

Summary:

- Parties C and T exchange 3 messages

In the description below, party C is the card and party T is the terminal. Subscripts c and t are used for their associated data.

Result:

- key agreement, entity authentication of party C to party T, explicit key authentication to party T, anonymity.

Notation:

- G is a generator of a group of points on an elliptic curve whose order is n, and $d \cdot G$ is scalar multiplication of G by an integer d (i.e. the point on the curve resulting from adding G to itself d times).

- $Æ(k)[m]$ denotes encryption of m under key k using an authenticated encryption algorithm $Æ$.
- $S_C(m)$ denotes C's signature (e.g. using ECDSA) on m.
- Cert(C) denotes C's public key and associated public key certificates.
- KDF is a key derivation function that generates, from a point on the elliptic curve, a symmetric key suitable for the encryption algorithm $Æ$.

Protocol steps:

1. C generates an ephemeral private key d_c and generates C's ephemeral public key as $Q_c = d_c \cdot G$

2. $C \rightarrow T : Q_c = d_c \cdot G$

3. T generates an ephemeral private key d_t and generates T's ephemeral public key $Q_t = d_t \cdot G$

4. T computes key $K_t = \text{KDF}(d_t \cdot Q_c)$

5. $T \rightarrow C : Q_t \parallel Æ(K_t)[\, Q_t \parallel Q_c]$

6. C uses Q_t to compute key $K_c = \text{KDF}(d_c \cdot Q_t)$ and uses K_c to authenticate and decrypt the payload $Q_t \parallel Q_c$. (With correct operation of the protocol $K_c = K_t$; if authentication fails then as soon as this happens the protocol is terminated). C signs $Q_c \parallel Q_t$

7. $C \rightarrow T : Æ(K_c)[\, \text{Cert}(C) \parallel S_C(Q_c \parallel Q_t)]$

8. T uses K_t to authenticate and decrypt the payload. (With correct operation of the protocol $K_c = K_t$; if authentication fails then as soon as this happens the protocol is terminated) and authenticates C's public key using the certificates and verifies C's signature on $Q_c \parallel Q_t$. T only accepts the validity of key K_t if the signature verifies successfully.

3.3   Blinded Diffie-Hellman Protocol

We now describe the Blinded Diffie-Hellman protocol, a protocol that also suits the requirements of EMV, namely where only party C, the card, has a static certified public key, but which also has advantages compared to the one-sided Station-to-Station protocol (see Section 4.1).

Summary:

- Parties C and T exchange 3 messages

In the description below, party C is the card and party T is the terminal. Subscripts c and t are used for their associated data.

Note that in this description C has a static private key d_c and corresponding public key Q_c with associated certificates $\text{Cert}(Q_c)$.

Result:

- key agreement, entity authentication of party C to party T, explicit key authentication to party T, anonymity.

Notation:

- G is a generator of a group of points on an elliptic curve whose order is n, and $d \cdot G$ is scalar multiplication of G by an integer d (i.e. the point on the curve resulting from adding G to itself d times).
- $\text{Æ}(k)[m]$ denotes encryption of m under key k using an authenticated encryption algorithm Æ.
- $\text{Cert}(Q_c)$ denotes C's public key and associated public key certificates.
- KDF is a key derivation function that generates, from a point on the elliptic curve, a symmetric key suitable for the encryption algorithm Æ.

Protocol steps:

1. C generates a random integer r $(1 \leq r < n)$ and generates C's blinded public key $R = r \cdot Q_c$.

2. $C \rightarrow T : R$

3. T generates an ephemeral private key d_t and generates T's ephemeral public key $Q_t = d_t \cdot G$

4. T computes key $K_t = \text{KDF}(d_t \cdot R)$

5. $T \rightarrow C : Q_t \parallel \text{Æ}(K_t)[\,D_1\,]$, where D_1 can be an optional application-level message

6. C uses Q_t to compute key $K_c = \text{KDF}(rd_c \cdot Q_t)$ and optionally uses K_c to authenticate and decrypt the payload D_1. (With correct operation of the protocol $K_c = K_t$; if authentication fails then as soon as this happens the protocol is terminated.)

7. $C \rightarrow T : \mathbb{E}(K_c)[\ r \parallel \text{Cert}(Q_c) \parallel D_2\ ]$ where D_2 is optional data that the card may wish to send protected to the terminal.

8. T uses K_t to authenticate and decrypt the payload. (With correct operation of the protocol $K_c = K_t$; if authentication fails then as soon as this happens the protocol is terminated) and authenticates C's public key using the certificates and verifies that R received in Step 2 is equal to $r \cdot Q_c$.

The diagram below illustrates the Blinded Diffie-Hellman key agreement using the notation defined above and where the dotted arrow represents a secure channel created using the derived AES keys.

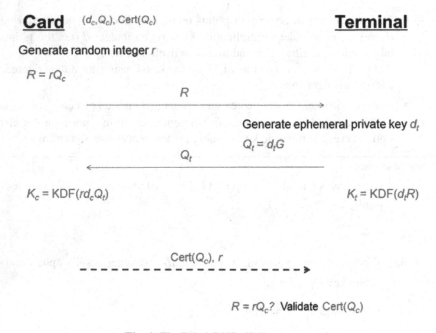

Fig. 1. The Blind-Diffie Hellman protocol

4   Performance and Security

This section addresses the performance and security of the Blinded Diffie-Hellman protocol.

4.1   Performance Comparison with Alternative Protocol

This section makes a performance comparison between the one-sided Station-to-Station protocol and the Blinded Diffie-Hellman protocol. As can be seen from the extract below, the message flows for the two protocols are similar.

Step	Station-to-Station	Blinded Diffie-Hellman
2. $C \rightarrow T$	$Q_c = d_c \cdot G$	$R = r \cdot Q_c$
5. $T \rightarrow C$	$Q_t \parallel \mathcal{E}(K_t)[\ Q_t \parallel Q_c]$	$Q_t \parallel \mathcal{E}(K_t)[\ D_1]$
7. $C \rightarrow T$	$\mathcal{E}(K_c)[\ \mathrm{Cert}(C) \parallel S_C(Q_c \parallel Q_t\ )]$	$\mathcal{E}(K_c)[\ r \parallel \mathrm{Cert}(Q_c) \parallel D_2\ ]$

Although the authenticated encryption payload is significantly larger for Station-to-Station compared to Blinded Diffie-Hellman, the authenticated encryption is likely to consume significantly less time than the elliptic curve computations. However, for completeness, we outline the possible impact.

Suppose, for example, that the protocols use an elliptic curve defined over a 256-bit prime field, the public keys (represented as (x,y)-coordinates on the curve) and signatures occupy 64 bytes and the certificates occupy over 256 bytes and the random blinding factor r is 32 bytes. Note that here the certificates are assumed to include at least the public key of the card, a signature of the issuing bank on that public key, the public key of the issuing bank, and a signature of the payment system on that public key.

Then, for the authenticated encryption payloads in steps 5 and 7, the size of the payload for the Station-to-Station is roughly 384 bytes and for Blinded Diffie-Hellman is roughly 288 bytes. If processing of the authenticated encryption by the card would take about 50ms for 100 bytes then this in itself might make Station-to-Station as much as 50ms slower than Blinded Diffie-Hellman.

However the main performance penalty introduced by Station-to-Station is the cost of the ECC operations. With contactless card transactions performance can be critical and the main performance concern is the time taken for ECC operations on the card-side, especially the ECC scalar multiplications. We see that the Blinded Diffie-Hellman protocol requires two scalar multiplications:

- Blinding the card public key in Step 1, and
- Calculating $rd_c \cdot Q_t$ in Step 6

whereas the one-sided Station-to-Station protocol requires two scalar multiplications:

- Generating the ephemeral public key in Step 1, and
- Calculating $d_c \cdot Q_t$ in Step 6

and a signature generation in Step 6.

Thus if we allocate 100ms for a card scalar multiplication or ECDSA calculation (although in reality a signature calculation involves more than a simple scalar multiplication depending on the algorithm) then Blinded Diffie-Hellman would require 200ms (plus time for the authenticated encryption and communications processing); whereas the one-sided Station-to-Station would require 300ms (plus time for the authenticated encryption and communications processing). This 100ms difference between the two protocols is very significant when one considers requirements for special environments such as transit ticketing.

Note that these performance considerations focus on the card computations required while the card is in the contactless field and although pre-computations may be possible it is likely that both protocols could benefit equally from this (e.g. step 1 for both protocols might be pre-computed).

So clearly Blinded Diffie-Hellman is more efficient and has the added benefit that the card does not need to implement ECC signature algorithms (e.g. ECDSA) and any associated hash function (e.g. SHA256).

The Blinded Diffie-Hellman mechanism would also provide improved performance if a smaller than full-size blinding factor would be used. However in this case the security proof no longer holds (see next section).

4.2   Security of Blinded Diffie-Hellman

As stated in Section 2.1, the EMV security objectives for the secure channel using the Blinded Diffie-Hellman protocol are

- to provide authentication of the card to the terminal (entity authentication),
- to detect modifications to the communications (message authentication), and
- to protect against eavesdropping of transactions and card tracking (privacy and unlinkability).

The first and second of these together ensure bilaterally that messages transmitted over the resulting secure channel are guaranteed to come from the other party engaged in the protocol and that for the terminal this other party is guaranteed to be the card identified by the card public key certificate that was verified. This does not, and because the card does not authenticate the terminal it clearly cannot, provide assurance to the card that it is corresponding with a legitimate terminal.

Similarly for the third objective, if the card is engaged in a legitimate transaction with a legitimate terminal then privacy and unlinkability are achieved even in the presence of active adversaries capable of intercepting, modifying, or injecting messages. However this does not rule out the possibility that a rogue device could perform the whole protocol with a card. Thus, in particular, communications confidentiality is assured for the corresponding parties against eavesdroppers (passive and active), but active adversaries can still engage with a card by pretending to be a terminal for the whole protocol run.

Brzuska, Smart, Warinschi, and Watson [3] have proved the security of the Blinded Diffie-Hellman key agreement protocol on the basis that the secure channel uses secure authenticated encryption and the PKI ensures secure authentication of the card public key. Their proof uses reductions in the Random Oracle Model and assumes the intractability of hard problems on the elliptic curve: the Gap Diffie-Hellman problem, the Diffie-Hellman problem and the Discrete Logarithm problem. The proof uses a modular security technique wherein the reduction takes a component of a hard problem and embeds this into the card's blinded public key. See [9] and [3] for more details of this technique.

[3] provides full details of the security proof for the Blinded Diffie-Hellman protocol (an earlier version can also be found at http://www.iacr.org/2013/031).

Although performance would certainly be improved by using a smaller blinding factor (e.g. 128 bits), the security proof from [3] would no longer hold. Indeed this possibility has also been considered in [4] where it is shown that bounded height discrete logarithm attacks may become feasible when using shorter blinding factors.

Note that because the card initiates the protocol rather than the terminal, it is possible that a terminal may be able to partially control the value of the uni-directional keys. As described in Section 8 of ISO/IEC 11770-3 [23], the terminal could control the value of s bits in the established key at the cost of generating 2^s candidate values for their ephemeral key in the time interval between discovering the card's blinded public key and choosing their ephemeral key. However with no terminal authentication and with the requirement for very fast transactions, this type of attack is not so relevant. If it would be considered relevant then the terminal could instead initiate the protocol (and it is understood that the security proof of [3] would still hold in this case) and/or the terminal could be required to make an initial commitment of its ephemeral key.

The Blinded Diffie-Hellman key agreement protocol requires the use of a secured communications channel after the key has been agreed and derived. This secure channel is constructed using the derived keys and enables the card public key certificates and the blinding factor to be sent protected to the terminal so that the terminal can authenticate the card public key and validate the blinded public key received previously. The secure channel should use a standard authenticated encryption algorithm (e.g. mechanism from ISO/IEC 19772 [29]) and a standard technique for deriving keys (see next section).

4.3   Uni-directional Keys and Key Derivation Function KDF()

According to the security proof in [3], the terminal and card should use message counters to ensure statefulness and should use uni-directional keys. Such uni-directional keys can be easily derived as $(K_1, K_2) = \text{KDF}(Q)$, where

- for the terminal $Q = d_t \cdot R$
- for the card $Q = r\, d_c \cdot Q_t$

where Q is a point on the elliptic curve and is the same for both terminal and card (assuming the protocol is operating correctly).

The key derivation function KDF() can be any standard key derivation function (see for example ISO/IEC 11770-6 [24] and NIST SP-800-56A [18], NIST SP-800-56C [19], NIST SP-800-108 [20]) so long as it is secure and known to both card and terminal.

For completeness this section concludes by providing an example where the key derivation function KDF() generates two AES keys K_1 and K_2 by encrypting two fixed strings using a key derivation key *KDK* that has been generated by MACing the x-coordinate of Q with an all-zero key (per CMAC in ISO/IEC 9797-1 [22] and NIST SP800-38B [17]). As with the performance example, the example below assumes that the elliptic curve is defined over a 256-bit prime field.

1. Key Extraction

Convert the x-coordinate of Q to be the 256-bit string $Z = Z_0 \| Z_1$ where Z_i is a 128-bit string for i=0,1.

2. Randomness Extraction

Compute a 128-bit *KDK* as an AES128 CBC MAC of Z using the all zero key (no padding)

- $KDK = \text{AES128}[0](\text{AES128}[0](Z_0) \text{ xor } Z_1)$

3. Key Expansion

Derive uni-directional keys as

- $K_1 = \text{AES128}[KDK] (0x00 \| S)$
- $K_2 = \text{AES128}[KDK] (0x01 \| S)$

where the value S could be any 15 byte field and could be chosen to represent a KDF version number, card/terminal identifiers, authenticated encryption algorithm identifier, an extra card nonce, and/or the bit-length of the total keying material produced.

Note that one could alternatively take Z as a function of the x and y coordinates, however, the use of the x-coordinate only is consistent with NIST SP-800-56A [18].

5   Conclusions

This paper has described a new key agreement protocol that has important privacy and performance properties that are not available in existing standards. The protocol has a security proof and a potential application in card payments.

Acknowledgements. We would like to thank the anonymous referees and Chris Mitchell for their helpful comments.

References

1. Blake-Wilson, S., Menezes, A.: Authenticated Diffie-Hellman key agreement protocols. In: Tavares, S., Meijer, H. (eds.) SAC 1998. LNCS, vol. 1556, pp. 339–361. Springer, Heidelberg (1999)
2. Blake-Wilson, S., Johnson, D., Menezes, A.: Key agreement protocols and their security analysis. In: Darnell, M. (ed.) Cryptography and Coding 1997. LNCS, vol. 1355, pp. 30–45. Springer, Heidelberg (1997)
3. Brzuska, C., Smart, N.P., Warinschi, B., Watson, G.: An Analysis of the EMV Channel Establishment Protocol. In: ACM CCS 2013, pp. 373–386. ACM (2013)
4. Blackburn, S., Scott, S.: The discrete logarithm problem for exponents of bounded height. J. Computation and Mathematics 17(Special Issue A), 148–156 (2014)
5. Canetti, R., Krawczyk, H.: Analysis of key-exchange protocols and their use for building secure channels. In: Pfitzmann, B. (ed.) EUROCRYPT 2001. LNCS, vol. 2045, pp. 453–474. Springer, Heidelberg (2001)
6. Dagdelen, O., Fischlin, M., Gagliardoni, T., Marson, G.A., Mittelbach, A., Onete, C.: A Cryptographic Analysis of OPACITY (2013), http://www.iacr.org/2013/234
7. EMVCo: EMV ECC Key Establishment Protocols. Draft, 1st edn. (2012), http://www.emvco.com/specifications.aspx?id=243
8. Goldberg, G., Stebila, S., Ustaoglu, B.: Anonymity and one-way authentication. In: Key Exchange Protocols. Designs, Codes and Cryptography, vol. 67(2), pp. 245–269 (May 2013)
9. Kudla, C., Paterson, K.G.: Modular security proofs for key agreement protocols. In: Roy, B. (ed.) ASIACRYPT 2005. LNCS, vol. 3788, pp. 549–565. Springer, Heidelberg (2005)
10. Hankerson, D., Menezes, A., Vanstone, S.: Guide to Elliptic Curve Cryptography. Springer (2004) ISBN 0-387-95273-X
11. Law, L., Menezes, A., Qu, M., Solinas, J., Vanstone, S.: An efficient protocol for authenticated key agreement, Dept. C & Q, Univ. of Waterloo, CORR 98-05 (1998)
12. Menezes, A., Oorschot, P., Vanstone, S.: Handbook of Applied Cryptography. CRC Press (1997)
13. IEEE P1363: A standard for RSA, Diffie-Hellman, and Elliptic-Curve cryptography (1999)
14. IETF RFC 2631, Diffie-Hellman Key Agreement Method (June 1999)
15. IETF RFC 4492, Elliptic Curve Cryptography (ECC) Cipher Suites for Transport Layer Security (TLS) (2006)
16. Certicom Research, Standards for Efficient Cryptography (2000)
17. NIST Special Publication 800-38B, Recommendation for Block Cipher Modes of Operation: the CMAC Mode for Authentication (May 2005)
18. NIST Special Publication 800-56A, Recommendation for Pair-Wise Key Establishment Schemes Using Discrete Logarithm Cryptography (Revised) (March 2007)
19. NIST Special Publication 800-56C, Recommendation for Key Derivation through Extraction-then-Expansion (November 2011)
20. NIST Special Publication 800-108, Recommendation for Key Derivation using Pseudorandom Functions (Revised) (October 2009)
21. ANSI X9.63, Public Key Cryptography for the Financial Services Industry Key Agreement and Key Transport Using Elliptic Curve Cryptography (2011)
22. ISO/IEC 9797-1: Information technology — Security techniques — Message authentication codes — Part 1: Mechanisms using a block cipher (2011)
23. ISO/IEC 11770-3: Information technology — Security techniques — Key management — Part 3: Mechanisms using asymmetric techniques (2008)

24. ISO/IEC CD 11770-6. Information technology — Security techniques — Key management — Part 6: Key derivation

25. ISO/IEC 14888-3: Information technology — Security techniques — Digital signatures with appendix — Part 3: Discrete logarithm based mechanisms (2006)

26. ISO/IEC 15946-1: Information technology — Security techniques — Cryptographic techniques based on elliptic curves (2008)

27. ISO/IEC 18033-2: Information technology — Security techniques — Encryption algorithms — Part 2: Asymmetric ciphers (2006)

28. ISO/IEC CD 18370-1. Information technology — Security techniques — Blind digital signatures — Part 1: General

29. ISO/IEC 19772: Information technology — Security techniques — Authenticated Encryption (2009)

30. ISO/IEC 20008-1: Information technology — Security techniques — Anonymous digital signatures — Part 1: General (2013)

31. ISO/IEC 20009-1: Information technology — Security techniques — Anonymous entity authentication — Part 1: General (2013)

Security Goals and Evolving Standards

Joshua D. Guttman, Moses D. Liskov, and Paul D. Rowe

The MITRE Corporation, Bedford MA, USA
{guttman,mliskov,prowe}@mitre.org

Abstract. With security standards, as with software, we cannot expect to eliminate all security flaws prior to publication. Protocol standards are often updated because flaws are discovered after deployment. The constraints of the deployments, and variety of independent stakeholders, mean that different ways to mitigate a flaw may be proposed and debated.

In this paper, we propose a criterion for one mitigation to be at least as good as another from the point of view of security. This criterion is supported by rigorous protocol analysis tools. We also show that the same idea is applicable even when some approaches to mitigating the flaw require cooperation between the protocol and its application-level caller.

1   Introduction

Security standards, which often contain errors, evolve over time as people correct them. Often, their flaws may be discovered after considerable product deployment, which places great pressure on the choice of mitigation. The constraints of the operational deployments and the need to satisfy the various stakeholders involved is crucial, but so is a precise understanding of the attack and what it enables.

How are we to choose among various proposed alternatives for mitigating a flaw? A security flaw is a failure of the protocol to meet a goal—possibly one not well understood until after the flaw becomes apparent—and a revised understanding of the goals of the protocol is necessary to ensure that the mitigation is secure. Obviously, satisfying a goal that was not previously met is essential to any mitigation. However, any two alternatives both of which eliminate a particular insecure scenario may not have equivalent security implications.

In this paper, we describe a formal language for expressing protocol security goals that supports "enrich-by-need" analysis tools such as the Cryptographic Protocol Shapes Analyzer (CPSA) [18]. We will use it to guide our description, even though there are other tools that are using other versions of enrich-by-need [5,13].

Each "enrich-by-need" analysis process starts from a scenario containing some protocol behavior, and also some assumptions about freshness and uncompromised keys. The analysis returns a set of result scenarios that show all of the minimal, essentially different ways that the starting scenario could happen. When

L. Chen and C. Mitchell (Eds.): SSR 2014, LNCS 8893, pp. 93–110, 2014.

the starting scenario is undesirable (e.g. a confidentiality failure), we would like this to be the empty set. When the starting scenario is the behavior of one principal, then the results indicate what authentication guarantees the protocol ensures.

For each run of the analyzer, there is a formula expressing the security goal that the analysis justifies. These security goals are independent of the particular protocol variant described.

Our Contributions. We make two main contributions in this paper. First, we show how to compare security consequences by analyzing alternatives in a formal framework. The protocol analyses determine a partial order on protocols. When one protocol Π_1 is above another protocol Π_2 in this order, that means that Π_1 achieves at least as much security as Π_2 with regard to the starting scenario of the analysis.

Second, we show that our techniques are flexible enough to accommodate a variety of viewpoints on the protocol and its goals during the remediation discussion. Whether an attack represents a flaw in the protocol, or a flaw in the interface between protocol and application, or perhaps some mixture of the two, is often debatable. Our techniques may apply to a model of the protocol at any level of detail, and we describe how models that include interfaces relate to models that do not. A model of this sort may help to clarify what goals are important for a protocol to offer to the applications that use it.

Structure of This Paper. In Section 2, we consider the Kerberos public key extension PKINIT. The initial version contained a flaw allowing a man-in-the-middle attack. Two alternatives emerged to fix this. We describe how to evaluate them with CPSA and the security goal language it suggests, giving a clear result: From the security point of view, the choices are equally good. In Section 3, we explain the core ideas of CPSA, which motivates our choice of protocol goal languages in Section 4.

In Section 5, we turn to cross level issues, illustrated with the TLS vulnerabilities arising from renegotiation. One natural diagnosis of this flaw is that the higher-level application is not aware enough of how the API it uses engages in the TLS protocol. In Section 6, we show that our techniques support discussion at this level also. Higher-level goals can thus be described purely in terms of observable application behavior.

2   Example: Kerberos PKINIT

PKINIT [22] is an extension to Kerberos [17] that allows a client to authenticate to the Kerberos authentication server (KAS) and obtain a ticket-granting ticket using public-key cryptography. This is intended to ease the management burden of establishing shared secrets (specifically passwords) and maintaining them, which the standard Kerberos exchange requires.

Cervesato et al. found a flaw in PKINIT version 25 [4], which was already widely deployed. The flaw was eventually fixed in version 27. Figure 1 shows

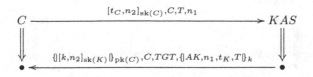

Fig. 1. PKINIT version 25, where $TGT = \{\!|AK, C, t_K|\!\}_{k_T}$

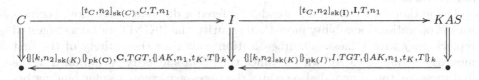

Fig. 2. Attack on flawed PKINIT, where $TGT = \{\!|AK, I, t_K|\!\}_{k_T}$

the expected message flow between the client and the KAS in v. 25. The client provides the KAS with its identity C, the identity T of the server it would like to access, and a nonce n_1. It also includes a signature over a timestamp t_C and a second nonce n_2 using the client's private key sk(C). The KAS replies by creating a fresh session key k, signing it together with the nonce n_2 and encrypting the signature using the client's public key pk(C). It uses the session key k to protect another session key AK to be used between the client and the subsequent server T, together with the nonce n_1 and an expiration time t_K for the ticket. The ticket TGT is an opaque blob from the client's perspective because it is an encryption using a key shared between K and T. It contains AK, the client's identity C and the expiration time t_K of the ticket.

In Cervesato et al.'s attack [4] (Fig. 2), an adversary I has obtained a private key to talk with the KAS. I uses it to forward any client C's initial request, passing it off as a request from I. I simply replaces C's identity with I's own, re-signing the timestamp and nonce n_2. When the KAS responds, I re-encrypts the response for C, this time replacing the identity I with C. In the process, the adversary learns the session key k, and thus can also learn the subsequent session key AK. This allows the attacker to read any subsequent communication between the client and the next server T. Moreover, the adversary may impersonate the ticket granting server T to C, because C believes the only other entity with knowledge of AK is T.

The attack arises from a lack of cryptographic binding between the session key k, and the client's identity C [4]. When C completes the two-message exchange, although she knows the KAS must have recently produced the keying material (due to the binding between k and n_2), it would be incorrect to conclude that the KAS intended the key to be used by C. Identifying this as the root cause of the attack suggests a natural fix, namely including the client's

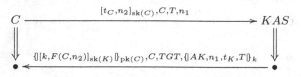

Fig. 3. Generic fix for PKINIT

identity C in the signed portion of the second message. Indeed this is the first suggestion in [4].

The authors of the PKINIT standard offered a different suggestion. For reasons of operational feasibility more than security, the PKINIT authors suggested replacing n_2 with a message authentication code over the entirety of the first message, keying the MAC with k. Since the client's identity is contained in the first message, this proposal also creates the necessary cryptographic binding between k and C.

Cervesato et al., working with a manual proof method, opted to verify a generic scheme for mitigating the attack, ensuring that the two proposals were instances of the scheme. This allowed them to avoid the time-consuming process of writing proofs for any other proposals that might also fit this scheme. They verified that the attack is prevented if n_2 is replaced with any expression $F(C, n_2)$ that is injective on those values (i.e. $F(C, n_2) = F(C', n_2')$ implies $C = C'$ and $n_2 = n_2'$).

We obtain the first proposal by instantiating F as the identity: $F(C, n_2) = (C, n_2)$. The second proposal results by instantiating F as the MAC of the client's request: $F(C, n_2) = H_k([t_C, n_2]_{sk(C)}, C, T, n_1)$. Since the MAC provides second preimage resistance, the injectivity requirement holds with overwhelming probability (Fig. 3).

The PKINIT parable illustrates recurring themes in protocol standard development and maintenance. Frequently, attacks show us that we care about previously unstated, unrecognized security goals. PKINIT does achieve some level of authentication, but it fails a more stringent type of authentication. In Lowe's terms [12], PKINIT achieves recent aliveness for both the client and the KAS because each party signs time-dependent data. However, PKINIT fails weak agreement which requires each side to know the other party was engaged in the protocol *with them*. When we see the attack, it forces us to identify explicitly the goal that the flawed protocol does not meet.

But an attack itself does not uniquely identify a security goal. We learned that it is important for the client to be guaranteed that it agrees with the KAS on the client's identity, but what about other values such as the expiration time of the ticket? Operational difficulties might arise if the client is unaware of this expiration time, but are there any security consequences? Indeed a key contribution of [4] is to state carefully what security goal the repair provides.

This goal can be achieved by different mitigations. Issues of efficiency, ease of deployment, or robustness to future protocol modifications may influence various stakeholders to prefer different mitigations. In our PKINIT example, the researchers opted for a change that was minimally invasive to their formal representation, thereby highlighting the root cause of the problem. The protocol designers had more operational context to constrain the types of solutions they deemed feasible.

While a pair of choices might both manage to satisfy some stated security goal, one of them may actually satisfy strictly stronger goals than another. We propose a goal language (Section 4) to express when a protocol mitigation is at least as good as a competitor—or strictly better than it—from the security point of view.

3   Enrich-by-Need Protocol Analysis

Our approach to protocol analysis is based on what we call the "point-of-view principle." Most of the security goals we care about in protocol design and analysis concern the point of view of a particular participant P. P *knows* that it has sent and received certain messages in a particular order. P may be willing to *assume* that certain keys are uncompromised, which for us means that they will be used only in accordance with the protocol in question. And P may also be willing to *assume* that certain randomly chosen values will not also be independently chosen by another participant, whether a regular (compliant) participant including P itself on another occasion, or an adversary.

The protocol analysis question is, given these facts and assumptions, what follows about what may happen on the network? These conclusions are of two main kinds. Positive conclusions assert that some regular participant Q has taken protocol actions. These are authentication goals. They say that P's message transmissions and receptions authenticate Q as having taken some corresponding actions, subject to the assumptions. Negative conclusions are generally non-disclosure assertions. They say that a value cannot be found available on the network in a particular form; often, that a key k cannot be observed unprotected by encryption on the network.

Skeletons and Cohorts. The enrich-by-need process starts with a representation of the hypothesis. We will refer to these representations of behavior and assumptions as *skeletons* $\mathbb{A}$. The skeleton $\mathbb{A}_0$ we start from includes some behavior of P, together with the stipulated assumptions. At any point in the enrich-by-need process, we have a set $\mathcal{S}$ of skeletons to work with. Initially, $\mathcal{S} = \{\mathbb{A}_0\}$.

At each step, we select one of these skeletons $\mathbb{A} \in \mathcal{S}$, and ask if the behavior of the participants recorded in it is possible. When a participant receives a message, then the adversary should be able to generate that message, using messages that have been sent earlier, without violating the assumptions. In this case, we regard that reception as "explained," since we know how the adversary can arrange to deliver the expected message. We say that that particular reception is *realized*. When every reception in a skeleton $\mathbb{A}$ is realized, we call $\mathbb{A}$ itself realized. It then represents—together with behavior that the adversary can supply—a possible complete execution. We collect the realized skeletons in a set $\mathcal{R}$.

If the skeleton $\mathbb{A} \in \mathcal{S}$ we select is not realized, then we use a small number of rules to generate an enrichment step. An enrichment step takes one unrealized reception and considers how to add some or all of the information that the adversary would need to generate its message. It returns a *cohort* of skeletons, meaning a finite set $\{\mathbb{A}_1, \ldots, \mathbb{A}_i\}$ of skeletons which together supply this information to the adversary in all of the ways that the regular participants could supply it. We update $\mathcal{S}$ by removing $\mathbb{A}$ and adding the cohort members: $\mathcal{S}' = (\mathcal{S} \setminus \{\mathbb{A}\}) \cup \{\mathbb{A}_1, \ldots, \mathbb{A}_i\}$.

As a special case, a cohort may be the empty set, i.e. $i = 0$, and in this case $\mathbb{A}$ is discarded and nothing replaces it. This occurs when there are no possible behaviors of the regular participants that would explain the required reception. Then the skeleton $\mathbb{A}$ cannot contribute any executions (realized skeletons).

This process may not terminate, and in fact the underlying class of problems is undecidable [9]. However, when it does terminate, it yields a finite set $\mathcal{R}$ of realized skeletons with a crucial property: For a class of security goals, if they have no counterexample in the set $\mathcal{R}$, then the protocol really achieves that goal [11]. Moreover, we can inspect the members of $\mathcal{R}$ and determine whether any of them is a counterexample. We call the members of $\mathcal{R}$ *shapes*, and they represent the *minimal, essentially different* executions consistent with the starting point.

Enrich-by-need protocol analysis originates with Meadows's NPA [13]. Dawn Song's Athena [20] applied the idea to strand spaces [21]. Two systems in use currently that use the enrich-by-need idea in a form close to what we describe here are Scyther [5] and CPSA [18]. See [10,11] for a comprehensive discussion, and for more information about our terminology here.

In particular, we will use the term *regular strand* to mean a local run of a particular principal in a single compliant local session of a protocol. A regular strand (or often, we will just say strand) contains a sequence of transmission and reception actions. We will refer to any one of these actions as a *node*.

Example 1: Initiator's Authentication Guarantee in PKINIT. Suppose that the client C has executed a strand of the client role in the fixed PKINIT, where for now we will instantiate $F(C, n_2) = (C, n_2)$. Suppose also that we are willing to assume that the authentication server K has an uncompromised signature key $\mathrm{sk}(K)$. We annotate this assumption as $\mathrm{sk}(K) \in \mathsf{non}$, meaning that $\mathrm{sk}(K)$ is non-compromised.

$$\tag{1}$$

$$\mathsf{Client}[C, K, T, n_1, n_2, t_C, t_K, k, AK] \quad \mathrm{sk}(K) \in \mathsf{non}$$

This is our starting point $\mathbb{A}_0$. C receives a message that contains the digital signature $[k, (C, n_2)]_{\mathrm{sk}(K)}$, and we know that the adversary cannot produce this because $\mathrm{sk}(K)$ is uncompromised. Thus, this second node of the local run is unrealized.

To explain this reception, we look at the protocol to see what ways a regular participant might create a message of the form $[k, (C, n_2)]_{\mathrm{sk}(K)}$. In fact, there is only one. Namely, the second step of the KAS role does so. Knowing the KAS sends this signature means it will agree on the parameters used: K, k, C, n_2. However, we do not yet know anything about the other parameters used in K's strand. They could be different values $t'_C, T', n'_1, TGT', AK', t'_K$. Thus, we obtain a cohort containing a single skeleton $\mathbb{A}_1$ that includes an additional KAS strand with the specified parameters.

$$\mathrm{sk}(K) \in \mathsf{non} \tag{2}$$

Client$[C, K, T, n_1, n_2, t_C, t_K, k, AK]$ KAS$[C, K, T', n'_1, n_2, t'_C, t'_K, k, AK']$

This skeleton is now already realized, because, with this weak assumption, the adversary may be able to use C's private decryption key to obtain k and modify the authenticator $\{\!|AK, n_1, t_K, T|\!\}_k$ as desired. The adversary might also be able to guess k, e.g. if K uses a badly skewed random number generator. Similarly, the components that are not cryptographically protected are under the power of the adversary.

We can we now re-start the analysis with two additional assumptions to eliminate these objections. First, we add $\mathrm{sk}(C)$ to non. Second, we assume that K randomly generates k, and we write $k \in \mathsf{unique}$, meaning that k is chosen at a unique position. We are uninterested in the negligible probability of a collision between k and a value chosen by another principal, even one chosen by the adversary.

$$\mathrm{sk}(K), \mathrm{sk}(C) \in \mathsf{non} \qquad\qquad k \in \mathsf{unique} \tag{3}$$

Client$[C, K, T, n_1, n_2, t_C, t_K, k, AK]$ KAS$[C, K, T', n'_1, n_2, t'_C, t'_K, k, AK']$

This skeleton is not realized, because with the assumption on $\mathrm{sk}(C)$, the adversary cannot create the signed unit $[t'_C, n_2]_{\mathrm{sk}(C)}$; it must come from a compliant principal. Examining the protocol, this can only be a client strand with matching parameters C, n_2, t'_C, i.e. a local run Client$[C, K'', T'', n''_1, n_2, t'_C, t''_K, k'', AK'']$. Curiously, there are two possibilities now. This strand could be identical with the one already in the diagram, in which case the doubly-primed parameters are identical with the unprimed ones. Or alternatively, it might be another client strand that has also by chance selected the same n_2, since we have not assumed $n_2 \in \mathsf{unique}$. In this case, the doubly-primed variables are not constrained. If we further add the assumption that $n_2 \in \mathsf{unique}$, then the second client strand must coincide with the first.

In all of these cases, C and K do not have to agree on TGT, since this item is encrypted with a key shared between K and T, and C cannot decrypt it or check any properties about what he receives.

This summary of enrich-by-need protocol analysis illustrates several important points. Authentication properties are built up by successive inferences of regular behavior, driven by some message component the adversary cannot build. When two inferences are possible the method branches, potentially resulting in a set of outputs. Various levels of authentication may be achieved according to which parameters principals agree on, and which parameters may vary. Secrecy properties are met when we can infer that no execution is compatible with the disclosure of the secret.

More formally, there is a notion of homomorphism between skeletons [10]. Given a starting point $\mathbb{A}_0$, with shapes $\mathbb{C}_1, \ldots, \mathbb{C}_i$, for each $\mathbb{C}_j$, there is a homomorphism H_j from $\mathbb{A}_0$ to $\mathbb{C}_j$. Moreover, every homomorphism $K \colon \mathbb{A}_0 \to \mathbb{D}$ from $\mathbb{A}_0$ to a realized skeleton $\mathbb{D}$ agrees with at least one of the H_j. Specifically, we can regard K as the result of adding more information after one of the H_j. We mean that we can always find some $J \colon \mathbb{C}_j \to \mathbb{D}$ such that K is the composition $K = J \circ H_j$.

4   A Language of Protocol Goals

Shape Analysis Formulas. The pattern of enrich-by-need protocol analysis suggests how to express the security properties of protocols. These security properties are essentially implications, that say that if the circumstances described in the starting point $\mathbb{A}_0$ hold, then some further information must hold. Given a skeleton, we can summarize all of the information in it in the form of a conjunction of atomic formulas. We call this formula the *characteristic formula* for the skeleton, and write $\mathsf{cf}(\mathbb{A})$. Thus, a CPSA run with starting point $\mathbb{A}_0$ is essentially exploring the security consequences of $\mathsf{cf}(\mathbb{A}_0)$.

When CPSA reports that $\mathbb{A}_0$ leads to the shapes $\mathbb{C}_1, \ldots, \mathbb{C}_i$, it is telling us that any formula that is true in all of these skeletons, and is preserved by homomorphisms, is true in all realized skeletons $\mathbb{D}$ accessible from $\mathbb{A}_0$. The set of formulas preserved by homomorphism are called *positive existential*, and are those formulas built from atomic formulas, $\wedge, \vee$, and $\exists$. By contrast, formulas using negation $\neg\phi$, implication $\phi \implies \psi$, or universal quantification $\forall y \,.\, \phi$ are not always preserved by homomorphisms.

Thus, the disjunction of the characteristic formulas of the shapes $\mathbb{C}_1, \ldots, \mathbb{C}_i$ tell us just what security goals $\mathbb{A}_0$ leads to. However, we can be somewhat more precise. The skeleton $\mathbb{C}_j$ may have nodes that are not in the image of $\mathbb{A}_0$, and it may involve parameters that were not relevant in $\mathbb{A}_0$. Thus, $\mathbb{A}_0$ will not determine exactly which values these new items take in $\mathbb{C}_j$, e.g. which session key is chosen on some local run not present in $\mathbb{A}_0$. Thus, these new values should be existentially quantified. Effectively, these are all the variables that do not appear in $\mathsf{cf}(\mathbb{A}_0)$. Thus, for each $\mathbb{C}_j$, let $\overline{y_j}$ list all the variables in $\mathsf{cf}(\mathbb{C}_j)$ that are not in $\mathsf{cf}(\mathbb{A}_0)$. Let $\overline{x}$ list all the variables in $\mathsf{cf}(\mathbb{A}_0)$. Then this run of CPSA has validated the formula:

$$\forall \overline{x} \,.\, (\mathsf{cf}(\mathbb{A}_0) \implies \bigvee_{1 \leq j \leq i} \exists \overline{y_j} \,.\, \mathsf{cf}(\mathbb{C}_j)) \tag{4}$$

The conclusion $\bigvee_{1 \le j \le i} \exists \overline{y_j} \cdot \mathrm{cf}(\mathbb{C}_j)$ is the strongest formula that is true in all of the $\mathbb{C}_j$.

We call the formula (4) the *shape analysis formula* for this run of CPSA. In the special case where $i = 0$, so that the conclusion of the implication is the empty disjunction, (4) is $\forall \overline{x} \cdot \mathrm{cf}(\mathbb{A}_0) \Longrightarrow \mathsf{false}$, or equivalently $\forall \overline{x} \cdot \neg \mathrm{cf}(\mathbb{A}_0)$, since the empty disjunction is the constantly false formula.

The Goal Language $\mathcal{GL}(\Pi)$. So far, we have discussed characteristic formulas without concern for the vocabulary we use to build them. We choose a vocabulary that is motivated by the kinds of analysis CPSA does. In particular, it is adapted to expressing which instances of roles have occurred, and how far each has progressed. It also allows us to say what value each parameter takes; we have already seen that a prime category of flaw occurs when local runs that should agree on a parameter do not. The language also expresses the orderings among events, and assumptions on uncompromised keys and fresh values.

However, it is also designed to have the minimum possible expressiveness. It contains no arithmetic; it contains no inductively defined data-types such as terms; and it has no ability to describe the syntax of messages. As a consequence, its formulas are preserved under a class of "security preserving" transformations between protocols [11]. Also, for interesting restricted classes of protocols, the set of security goals they achieve is decidable [8]. These properties require careful control over expressiveness.

In this language, we may summarize Eqn. (1) by the formula:

$$\mathtt{ClientDone}(n) \wedge \mathsf{Peer}(n, K) \wedge \mathsf{Non}(\mathsf{sk}(K)) \tag{5}$$

This asserts of a node n that it completes a client run, i.e. it is the second event on that local run. It also asserts that the peer parameter of this node n is a name K such that the signature key of K is non-compromised. The letters n, K here are free variables, and this formula is satisfied under an assignment of values to the free variables if those values have the properties we mentioned. A precise semantics is given in [11].

Observe that we don't have to say $\mathtt{ClientStart}(m)$, referring to the first node of the run. The presence of a second node ensures that the previous step occurred, and we don't need to say anything in particular about it.

Turning to Eqn. (2), the conjuncts of Eqn. (5) still hold. There is however also another strand, which is a complete KAS run. We also know that several of its parameters agree with those of C:

$$\begin{aligned} (5) &\wedge \mathsf{Self}(n, c) \wedge \mathsf{AuthNonce}(n, n_2) \wedge \\ \mathtt{KASDone}(m) &\wedge \mathsf{Self}(m, K) \wedge \mathsf{AuthNonce}(m, n_2) \wedge \\ &\mathsf{Peer}(m, c) \wedge \quad \mathsf{Preceq}(m, n) \end{aligned} \tag{6}$$

The shape analysis formula that results has a single disjunct in the conclusion:

$$\forall n, K \cdot (5) \Longrightarrow \exists m, c, n_2 \cdot (6) \tag{7}$$

More generally, suppose that we are given a protocol Π. It has a number of roles, and each of its roles has a number of nodes. For each of these nodes, the goal language $\mathcal{GL}(\Pi)$ has a *role position predicate*. The two predicates $\texttt{ClientDone}(n)$ and $\texttt{KASDone}(m)$ used above are examples. Each one is a one-place predicate that says what kind of node its argument n, m refers to.

On each node, there are parameters. The *parameter predicates* are two place predicates. Each one associates a node with one of the values that has been selected when that node occurs. For instance, $\texttt{Self}(n, c)$ asserts that the *self* parameter of n is c. This allows us to assert agreement between different strands. $\texttt{Peer}(m, c)$ asserts that m appears to be partnered with the same principal who is in fact the *self* parameter of n.

The role position predicates and parameter predicates vary from protocol to protocol, depending on how many nodes the protocol has, and how many parameters. The predicate names may be chosen as convenient. For instance, we may choose to use the same predicates for two different protocols, using this to emphasize structural similarities between them.

All protocols also have some shared common vocabulary (summarized in Table 1) that helps to express the structural properties of bundles. $\texttt{Preceq}(m, n)$ asserts that one node occurs before another; $\texttt{Coll}(m, n)$ says that they lie on the same strand. $\texttt{Non}(v)$ and $\texttt{Unq}(v)$ express non-compromise and freshness (unique origination). $\texttt{pk}(a)$ and $\texttt{sk}(a)$ relate a principal to its keys, $\texttt{ltk}(a, b)$ represents the long-term key of two principals, and $\texttt{inv}(k)$ is the inverse of a key.

Table 1. Protocol-independent vocabulary of the languages $\mathcal{GL}(\Pi)$

Functions:	$\texttt{pk}(a)$	$\texttt{sk}(a)$	$\texttt{inv}(k)$
	$\texttt{ltk}(a, b)$		
Relations:	$\texttt{Preceq}(m, n)$	$\texttt{Coll}(m, n)$	$=$
	$\texttt{Unq}(v)$	$\texttt{UnqAt}(n, v)$	$\texttt{Non}(v)$

Using Shape Analysis Formulas to Evaluate Alternatives. We now return to the intuitive notions of "good enough" and "as good as" and describe how these notions can be rigorously reflected through a combined understanding of shape analysis formulas and goal languages.

The notion of "good enough", naturally, is to be defined relative to a goal or set of goals. A protocol Π is good enough if all the required goals are satisfied in all executions. Each goal is of the form

$$\forall \overline{x} . (\Phi \implies \bigvee_{1 \leq j \leq i} \exists \overline{y_j} . \Psi_j) \tag{8}$$

where Φ and Ψ are conjunctions of atomic formulas of the goal language $\mathcal{GL}(\Pi)$. We can evaluate whether Π achieves the goal by running CPSA starting from a suitable skeleton, the "characteristic skeleton" of Φ. If some Ψ_j is satisfied in each of the resulting shapes, the goal is achieved.

"At Least as Good as." We relativize our definition of one protocol being at least as good as another to a particular hypothesis Φ. This hypothesis should be a formula of both $\mathcal{GL}(\Pi_1)$ and $\mathcal{GL}(\Pi_2)$. In that case, Π_2 is *at least as good as Π_1 relative to* the hypothesis Φ if, for every goal of the form $\forall \bar{x} \,.\, \Phi \Longrightarrow \bigvee \Psi_j$ with this hypothesis Φ, if Π_1 achieves this goal, then so does Π_2.

We can write $\Pi_1 \trianglelefteq_\Phi \Pi_2$ to mean that Π_2 is at least as good as Π_1 relative to Φ.

Two protocols are *equally good relative to* Φ if each is at least as good as the other relative to it.

We can establish $\Pi_1 \trianglelefteq_\Phi \Pi_2$ from shape analysis formulas. Φ determines a skeleton $\mathbb{A}_0$ in protocol Π_1 and a skeleton $\mathbb{B}_0$ in protocol Π_2. Suppose the set of shapes for $\mathbb{A}_0$ are $\mathbb{C}_1, \ldots, \mathbb{C}_n$, and the set of shapes for $\mathbb{B}_0$ are $\mathbb{D}_1, \ldots, \mathbb{D}_m$. If the disjunction of the characteristic formulas of the $\mathbb{D}_i$ entails the disjunction of the characteristic formulas of the $\mathbb{C}_j$, then $\Pi_1 \trianglelefteq_\Phi \Pi_2$:

$$\bigvee_{i \leq m} \exists \overline{y_i} \,.\, \mathsf{cf}(\mathbb{D}_i) \implies \bigvee_{j \leq n} \exists \overline{z_j} \,.\, \mathsf{cf}(\mathbb{C}_j). \tag{9}$$

If Π_1 achieves a goal $\forall \bar{x} \,.\, \Phi \Rightarrow \Psi$, then $\bigvee_{j \leq n} \exists \overline{z_j} \,.\, \mathsf{cf}(\mathbb{C}_j) \implies \Psi$. Hence, Π_2 achieves the goal also, because in protocol Π_2,

$$\forall \bar{x} \,.\, \Phi \Rightarrow \bigvee_{1 \leq i \leq n} \exists \overline{y_i} \,.\, \mathsf{cf}(\mathbb{D}_i)$$

$$\Rightarrow \bigvee_{1 \leq j \leq m} \exists \overline{z_j} \,.\, \mathsf{cf}(\mathbb{C}_j) \Rightarrow \Psi$$

5 Example: TLS Renegotiation

Our method gives a good criterion for deciding that two protocols are equally good—or one is at least as good as the other—relative to a hypothesis Φ. But as described, this method applies if the fix consists of internal protocol modifications only. In the remainder of this paper, we would like to consider the possibility that the fix involves both modifying the protocol and also the information that it passes up to the application on behalf of which it is acting. Thus, part of the resolution is for the application to use this additional information correctly, so as to achieve its security goals. The underlying protocol is obligated to provide it with accurate information, and signaling when relevant events occur. We start with an example.

Transport Layer Security (TLS) [7] is a globally deployed protocol designed to add confidentiality, authentication and data integrity between two communicating applications. It is secure, scalable, and robust enough to protect e-commerce transactions performed over HTTP. Despite the success of TLS it has been forced to evolve over time, in part due to the discovery of various flaws in the design logic.

One such flaw, discovered in 2009 by Marsh Ray, concerns renegotiating TLS parameters. It works on the boundary between the TLS layer and the application

layer it supports. [19] contains a good description of the flaw; we give a brief
summary.

```
Client                        Attacker                         Server
------                        --------                         ------
                        <----------- Handshake ----------->
                        <======= Initial Traffic ========>
<--------------------- Handshake ============================>
<=================== Client Traffic ==========================>
```

Fig. 4. TLS renegotiation attack

Fig. 4 (borrowed from [19]) is a high-level picture of the attack. The attacker first creates a *unilaterally* authenticated session with the server in the
first handshake. Thus, the server authenticates itself to the attacker, but not
vice versa. The attacker and server then exchange initial traffic protected by
this TLS session. Later, a renegotiation occurs, possibly when the application
at the server requires *mutual* authentication for some action. The attacker then
allows the client to complete a handshake with the server, adding and removing
TLS protections. The client's handshake occurs in the clear (depicted by <-->
in Fig. 4), while the server's handshake is protected by the current TLS session.
The attacker has no access to this newly negotiated session, but the server may
retroactively attribute data sent in the previous session to the authenticated
client. The server may then perform a sensitive action in response to a request
sent by the attacker, but based on the credentials subsequently provided by the
client. Which level is to blame for this attack?

– Does TLS fail to achieve a security goal that it should achieve?
– Or should the application take responsibility? It accepts some data out of
  a stream that is not bilaterally authenticated, and lumps it with the future
  data which will be bilaterally authenticated.
– Or is there shared responsibility? Perhaps TLS should provide clearer indications to the application when a change in the TLS properties takes place,
  and then the application should heed these indications.

In fact TLS was updated with a renegotiation extension [19]. TLS renegotiation now cryptographically binds the new session to the existing session. If a
server completes a mutually authenticated renegotiation with a client, then the
current session was also negotiated with the same client. However, the authors
of [19] also note:

> While this extension mitigates the man-in-the-middle attack described
> in the overview, it does not resolve all possible problems an application
> may face if it is unaware of renegotiation.

As Bhargavan et al. [2]'s recent attacks showed, the practically important issue
was not in fact resolved by this.

However, for applications to take partial responsibility, some signals and commands must be shared between TLS and the application. Enrich-by-need protocol analysis—coupled with our goal language—fits in naturally here. With a little effort the goal language can be updated to address the multilayer nature of flaws such as this.

6   Goals for Protocol Interfaces

We now describe how to express protocol goals to make cross-level choices explicit.

The job of TLS, acting in either direction, is to take a stream of data from a transmitting application, and to deliver as much as possible of this stream to the receiving application. When the sender is authenticated to the receiver, TLS guarantees that the portion delivered is an initial segment of what the authenticated sender transmitted. When the mode offers confidentiality, no other principal should learn about the content (as opposed to the length).

Naturally, these goals are subject to the usual assumptions, such as that the certificate authorities are trustworthy, that the private keys are uncompromised, and that randomness was freshly chosen.

When a renegotiation occurs, this affects what the application should rely on. If a handshake authenticates a client identity C, then the authentication guarantee should apply to the data starting when the cipher spec changes. We will call the period starting from a cipher spec change, and lasting until the next one (if any) an *epoch*, and part of the work of a handshake is to agree on an epoch ID between the endpoints. Thus, any guarantee should apply throughout an epoch. Authentication guarantees for the Client Traffic should definitely not apply to the Initial Traffic of Fig. 4, which lies on the other side of an epoch boundary from the authenticated traffic.

The interface between the protocol and its application, then, is an essential part of expressing the guarantees that the application should rely on. We can enlarge our notion of "protocol" to include signals across the API as well as the message roles. This enlarged protocol can itself be used to define a goal language. Some goal formulas refer only to API events and their parameters, not to the nuts-and-bolts events of the core protocol itself. These formulas express API-level goals. They are of course true only if the lower level protocol behaves properly. However, their content speaks explicitly only about the events of interest to the upper level.

Such a language allows us to apply our notions of "good enough" and "at least as good as", showing that they are relevant to our enlarged, API-aware protocols.

Representing APIs. An API is a set of signals and commands that may occur in a certain pattern, between an application and the service that implements the API. In our case, the service directly controls actual protocol interactions. The API thus consists of the signals (from service to application) and commands (from application to service), and how the reception of those signals and commands line up with the implemented protocol behavior.

The communication between service and application is of a different nature than the communication that takes place in protocol execution: in particular, this communication is not observable or controllable by a typical network adversary.

We enlarge our notion of the protocol to one that includes the signals and commands as well as how they interact with protocol messages. This enlarged notion of the protocol allows us to describe a goal language. In that language, an *application goal* is a goal expressible in terms directly referring to application-level roles.

Enrich-By-Need Analysis for APIs. In order to evaluate "good enough" or "as good as" for application goals, we need an enrich-by-need analysis that respects the distinct nature of communication between the API and the protocol service. There is some recent research on analyzing protocols with state that could be relevant. But no special machinery is required beyond protocol messages: all we need to do is emulate the information passed between application and service as a secure channel independent from all others involved in the protocol.

Let Π be a protocol. An *API-enhanced version* of Π is a protocol Π' that has a set of new nodes api such that the result of omitting the nodes api from Π' yields Π. A goal formula Φ is an *API goal* if it refers only to nodes in api, and their parameters.

If Φ is an API goal then its truth or falsehood can be established by an appropriate enrich-by-need analysis of Π'. In other words, "good enough" can be established through enrich-by-need analysis just as was the case for protocol-level goals.

Furthermore, for any particular API goal, its antecedent references a certain subset of API role events and variables. Thus, it can be meaningful to compare two APIs (even for exactly the same underlying protocol) in terms of goals that are API goals for both APIs. In such circumstances, our notion of "as good as" applies here.

TLS Renegotiation, Revisited. Now we state an example of an application-level goal for TLS that addresses the interface concerns specific to the renegotiation flaw. In particular, the authors of [19] point out the dangers of an application being unaware of renegotiation. The flaw that arises from the renegotiation attacks is most easily understood from the application level. Here, we describe that goal in a formal manner.

The application is aware of data being exchanged over a TLS connection, and may also query for the status of the connection. Consider the following set of predicates:

- $\mathtt{DataSend}(n)$: a command was issued at node n to send data over TLS.
- $\mathtt{DataRecv}(n)$: a signal was received at node n that data was received from TLS.
- $\mathtt{DataVal}(n, d)$: d is the data involved in the $\mathtt{DataSend}$ or $\mathtt{DataRecv}$ event occurring at node n.
- $\mathtt{Self}(n, ID)$: ID is the identifier of the actor at node n.

- $\mathtt{Status}(n)$: a status signal was received at node n about a TLS connection.
- $\mathtt{EID}(n, eid)$: eid is the epoch ID involved in the $\mathtt{DataSend}$, $\mathtt{DataRecv}$, or $\mathtt{Status}$ event occurring at node n.
- $\mathtt{Client}(n, cID)$: cID is either the ID of the client reported as the authenticated client in a status event, or "anon" otherwise.
- $\mathtt{Server}(n, sID)$: sID is the ID of the server reported in a status event.

Informally, the goal we will describe is that if the server receives d over some TLS connection, and also gets a report about the status of that same connection, then either the status report identifies the client as anonymous, or the identified client actually sent the data d. Formally,

$$\mathtt{DataRecv}(n) \wedge \mathtt{DataVal}(n, d) \wedge \mathtt{Self}(n, s) \wedge \mathtt{EID}(n, eid),$$
$$\wedge \quad \mathtt{Status}(m) \wedge \mathtt{Server}(m, s) \wedge \mathtt{Client}(m, c) \wedge \mathtt{EID}(m, eid)$$
$$\Rightarrow \quad\quad\quad c = \text{"anon"} \vee$$
$$\exists n' : \mathtt{DataSend}(n') \wedge \mathtt{Self}(n', c) \wedge \mathtt{DataVal}(n', d) \wedge \mathtt{EID}(n', eid)$$

for all values of the free variables.

Here the first line of the hypothesis assumes that a data reception signal occurred for s at node n, involving data d, epoch identifier eid. The second line assumes that a status was checked for epoch identifier eid, and the signal was received at node m and indicates that the client is c and the server is s. If both of these conditions are met for a common eid, the goal states that either c is "anon" (indicating that the client is not authenticated), or otherwise that the client c actually sent d. This last claim is the mirror-image of the first line: namely, that a data transmission command occurred for c at node n, involving data d and epoch identifier eid.

Note that the goal does not specify that $\mathtt{Preceq}(m, n)$: in other words, the status may be reported even after the data is received and we still expect the status to reflect an accurate assessment of the identity of the client if the client is not anonymous. This is precisely what goes wrong in the renegotiation attack: the adversary initiates an anonymous session, causes data to be received, and then convinces an honest client to renegotiate the session so that it is later reported as authenticated.

Goals and Mandates. Describing the goal for TLS at this level is natural, but the discussion ultimately must match the mandate of the standard itself: to specify the actual protocol messages and to advise about how applications are to be informed on its use. Delving into details of the interface, in this case, is not appropriate (but if it was, the notions of "good enough" and "as good as" apply just as well to the more complex interface-inclusive protocol model). However, stakeholders present in the discussion will be able to comment on the constraints they are under.

One particular constraint is that a TLS API will ultimately aim to set up a simple type of data stream functionality in which status issues are separated from data signals. In other words, the $\mathtt{Status}$ and $\mathtt{DataRecv}$ events cannot occur at the same node. Another important constraint is that status reports be limited to a current status, so that the API is not responsible for maintaining an exhaustive status history.

7   Related Work and Conclusion

The full literature on the use of formal methods for analyzing cryptographic protocols is too vast to summarize here, although we would direct the reader to [15] for a (now classic) survey. As methods and tools have become more developed, they have been effectively applied to the analysis of published standards, demonstrating their maturity and applicability [14,16,1,2,4]. Many of these efforts have explicitly engaged with the the relevant standards body to ensure their input was reflected in the standard.

We are not the only ones to propose a logical language of security goals. Numerous efforts attempt to use a formal logic in which to reason directly about cryptographic protocols [3,6]. While they do provide formal statements of security goals, the proof methods do not lend themselves to natural comparisons of the goals that various protocols might achieve. [12] contains a hierarchy of authentication goals demonstrating the relationship between the goals themselves, and [5,1] integrate the hierarchy with an enrich-by-need analysis method.

The process of standardizing cryptographic protocols is both difficult and important. Getting a variety of stakeholders to converge on a single point of view requires careful consideration of all proposed options and a clear way of comparing them. Although many unchangeable constraints exist pertaining to issues such as efficiency, computational limitations, or backwards compatibility, the security of the designed protocol is typically of paramount importance. Unfortunately, the term "security" can mean different things in different contexts. A clear and precise formulation of the security requirements of a protocol can help focus group discussions on the precise outcomes that it is most important for a protocol to achieve. It can also help to distinguish those constraints that pertain to security from operational constraints, allowing committees to better understand the space of trade-offs for design decisions.

In this paper we demonstrated how automated tools based on formal methods can assist in this complicated decision-making process. We presented a formal language in which to express security goals. We focused on how the enrich-by-need method of protocol analysis integrates with this goal language and demonstrated its applicability to the historical case of mitigating a flaw in the PKINIT protocol. We then demonstrated how the goal language might be adapted to accommodate goals that lie at the intersection of security protocols and the applications they support, reinforced by the example of a previously discovered flaw in TLS renegotiation.

While we believe the goal language paired with enrich-by-need protocol analysis is particularly interesting, we also believe other formal languages and tools could be used in a quite similar way. The precision and clarity that comes from the abstraction of formal methods must be balanced against the practical considerations of potential implementations. We present here a vision for how the results of formal analyses can be incorporated with real-world decision making processes to focus discussion and strengthen the security of the resulting standards.

References

1. Basin, D.A., Cremers, C., Meier, S.: Provably repairing the ISO/IEC 9798 standard for entity authentication. Journal of Computer Security 21(6), 817–846 (2013)
2. Bhargavan, K., Delignat-Lavaud, A., Fournet, C., Pironti, A., Strub, P.-Y.: Triple handshakes and cookie cutters: Breaking and fixing authentication over TLS. In: IEEE Symposium on Security and Privacy (2014)
3. Burrows, M., Abadi, M., Needham, R.: A logic of authentication. ACM Transactions on Computer Systems 8, 18–36 (1990)
4. Cervesato, I., Jaggard, A.D., Scedrov, A., Tsay, J.-K., Walstad, C.: Breaking and fixing public-key Kerberos. Inf. Comput. 206(2-4), 402–424 (2008)
5. Cremers, C., Mauw, S.: Operational Semantics and Verification of Security Protocols. Springer (2012)
6. Datta, A., Derek, A., Mitchell, J.C., Roy, A.: Protocol composition logic (PCL). Electr. Notes Theor. Comput. Sci. 172, 311–358 (2007)
7. Dierks, T., Rescorla, E.: The Transport Layer Security (TLS) Protocol Version 1.2. RFC 5246 (Proposed Standard), Updated by RFCs 5746, 5878, 6176 (August 2008)
8. Dougherty, D.J., Guttman, J.D.: Decidability for lightweight Diffie-Hellman protocols. In: IEEE Symposium on Computer Security Foundations (2014)
9. Durgin, N., Lincoln, P., Mitchell, J., Scedrov, A.: Multiset rewriting and the complexity of bounded security protocols. Journal of Computer Security 12(2), 247–311 (1999), Initial version appeared Workshop on Formal Methods and Security Protocols (1999)
10. Guttman, J.D.: Shapes: Surveying crypto protocol runs. In: Cortier, V., Kremer, S. (eds.) Formal Models and Techniques for Analyzing Security Protocols. Cryptology and Information Security Series. IOS Press (2011)
11. Guttman, J.D.: Establishing and preserving protocol security goals. Journal of Computer Security 22(2), 201–267 (2014)
12. Lowe, G.: A hierarchy of authentication specification. In: CSFW, pp. 31–44 (1997)
13. Meadows, C.: The NRL protocol analyzer: An overview. The Journal of Logic Programming 26(2), 113–131 (1996)
14. Meadows, C.: Analysis of the Internet Key Exchange Protocol using the NRL Protocol Analyzer. In: IEEE Symposium on Security and Privacy, pp. 216–231 (1999)
15. Meadows, C.: Formal methods for cryptographic protocol analysis: Emerging issues and trends. IEEE Journal on Selected Areas in Communications 21(1), 44–54 (2003)
16. Mitchell, J.C., Roy, A., Rowe, P., Scedrov, A.: Analysis of EAP-GPSK authentication protocol. In: Bellovin, S.M., Gennaro, R., Keromytis, A.D., Yung, M. (eds.) ACNS 2008. LNCS, vol. 5037, pp. 309–327. Springer, Heidelberg (2008)
17. Neuman, C., Yu, T., Hartman, S., Raeburn, K.: The Kerberos Network Authentication Service (V5). RFC 4120 (Proposed Standard), Updated by RFCs 4537, 5021, 5896, 6111, 6112, 6113, 6649, 6806 (July 2005)

18. Ramsdell, J.D., Guttman, J.D.: CPSA: A cryptographic protocol shapes analyzer (2009), http://hackage.haskell.org/package/cpsa
19. Rescorla, E., Ray, M., Dispensa, S., Oskov, N.: Transport Layer Security (TLS) Renegotiation Indication Extension. RFC 5746 (Proposed Standard) (February 2010)
20. Song, D.X.: Athena: A new efficient automated checker for security protocol analysis. In: Proceedings of the 12th IEEE Computer Security Foundations Workshop. IEEE CS Press (June 1999)
21. Thayer, F.J., Herzog, J.C., Guttman, J.D.: Strand spaces: Proving security protocols correct. Journal of Computer Security 7(2/3), 191–230 (1999)
22. Zhu, L., Tung, B.: Public Key Cryptography for Initial Authentication in Kerberos (PKINIT). RFC 4556 (Proposed Standard), Updated by RFC 6112 (June 2006)

Analysis of the IBM CCA Security API Protocols in Maude-NPA[*]

Antonio González-Burgueño[1], Sonia Santiago[1], Santiago Escobar[1],
Catherine Meadows[2], and José Meseguer[3]

[1] DSIC-ELP, Universitat Politècnica de València, Spain
{agonzalez,ssantiago,sescobar}@dsic.upv.es
[2] Naval Research Laboratory, Washington DC, USA
meadows@itd.nrl.navy.mil
[3] University of Illinois at Urbana-Champaign, USA
meseguer@illinois.edu

Abstract. Standards for cryptographic protocols have long been attractive candidates for formal verification. It is important that such standards be correct, and cryptographic protocols are tricky to design and subject to non-intuitive attacks even when the underlying cryptosystems are secure. Thus a number of general-purpose cryptographic protocol analysis tools have been developed and applied to protocol standards. However, there is one class of standards, security application programming interfaces (security APIs), to which few of these tools have been applied. Instead, most work has concentrated on developing special-purpose tools and algorithms for specific classes of security APIs. However, there can be much advantage gained from having general-purpose tools that could be applied to a wide class of problems, including security APIs.

One particular class of APIs that has proven difficult to analyze using general-purpose tools is that involving exclusive-or. In this paper we analyze the IBM 4758 Common Cryptographic Architecture (CCA) protocol using an advanced automated protocol verification tool with full exclusive-or capabilities, the Maude-NPA tool. This is the first time that API protocols have been satisfactorily specified and analyzed in the Maude-NPA, and the first time XOR-based APIs have been specified and analyzed using a general-purpose unbounded session cryptographic protocol verification tool that provides direct support for AC theories. We describe our results and indicate what further research needs to be done to make such protocol analysis generally effective.

Keywords: IBM 4758 Common Cryptographic Architecture, security Application Programming Interfaces (security APIs), symbolic cryptographic protocol analysis, automatic reasoning modulo XOR theory.

[*] Antonio González-Burgueño, Sonia Santiago and Santiago Escobar have been partially supported by the EU (FEDER) and the Spanish MINECO under grants TIN 2010-21062-C02-02 and TIN 2013-45732-C4-1-P, and by Generalitat Valenciana PROMETEO2011/052. José Meseguer has been partially supported by NSF Grant CNS 13-10109.

L. Chen and C. Mitchell (Eds.): SSR 2014, LNCS 8893, pp. 111–130, 2014.

1   Introduction

Standards for cryptographic protocols have long been attractive candidates for formal verification. Cryptographic protocols are tricky to design and subject to non-intuitive attacks even when the underlying cryptosystems are secure. Furthermore, when protocols that are known to be secure are implemented as standards, the modifications that are made during the standardization process may introduce new security flaws. Thus a considerable amount of work has been done in the application of formal methods to cryptographic protocol standards [26,4,25,1]. In this work the protocols are treated *symbolically*, with the cryptosystems treated as black-box function symbols. The formal methods tool attempts to show that there is no way an attacker, by interacting with the protocol and applying the cryptographic functions symbols in any order, can break the security of the protocol. Such tools can be used both to search for attacks and to prove security with respect to the symbolic model.

Such symbolic formal analyses can be of great benefit to standards development. The environment in which these standards must be developed makes it difficult to maintain security. Standards often must compromise between different and conflicting requirements. The main focus is often interoperability instead of security. Moreover, standards are often evolving documents; they do often must be updated as new requirements arise. However, standards are chiefly intended as guides to implementation, and often contain little information about the security decisions that were made in the design of previous versions of the protocol.[1] All of this means that security flaws often creep into a standard even when it is based on a protocol that was originally secure. Symbolic formal analysis can provide a rapid means of evaluating and re-evaluating the security of a standard and the security requirements it must satisfy as it evolves.

Most symbolic analysis work has concentrated on standards for key generation and secure communication, as these are the types of protocols that are most widely standardized. However, recently another type of application has begun to attract interest: secure Application Programming Interface (API) protocols. This is the functionality a secure device provides for use by applications that run on it. The API allows the application to authenticate itself to the device and perform the functions it is authorized to do. However, it must also be constructed so that the application can not use it to perform any actions that it is *not* authorized to do. For example, it should not be able to obtain cryptographic keys in the clear. It is clearly more economical, both from the point of view of guaranteeing security and producing applications, if APIs are standard across different platforms, and as a result such standards as the IETF's GSSAPI [21] have appeared. But even when an API is not standardized across different platforms, but is created by a single company or other entity to guide application implementers in the use of the devices it creates, it still has many of the properties of a standard. The focus of the documentation is more on implementation than explaining security decisions,

---

[1] The IETFs insistence on a Security Considerations section in every document is an attempt to address this last problem.

and the APIs often evolve as the hardware and the requirements it must satisfy evolve. Moreover, they are widely distributed and available for formal analysis. Thus lessons learned by analysis of APIs that are not official standards can still be useful to the designers of such standards.

APIs face many of the same issues as key distribution protocols. However, although some of the earliest formal cryptographic protocol analysis work was applied to security APIs [19,22,23], it was a long time before there was any work following up on that. Indeed, it was not until more recent work uncovered security problems in a number of well-known APIs, such as Bond's discovery of flaws in the IBM 4758 Common Cryptographic Architecture API (CCA-API) [3] that this again became an active area of investigation. Even so, application of symbolic formal methods tools for cryptographic protocol analysis to this problem have not been that common until recently. Even now, work has mostly concentrated on developing special-purpose algorithms and tools fine-tuned for specific classes of APIs, rather than expanding general-purpose cryptographic protocol analysis tools to deal with this kind of problem. Indeed, even though Bond's attacks on CCA were discovered almost fifteen years ago, and they have become one of the benchmarks for symbolic protocol analysis, general-purpose tools often still struggle with them.

One of the reasons we believe that general-purpose symbolic cryptographic protocol analysis tools have not been applied yet that widely to security APIs is that the analysis of some API protocols involves features that are not usually considered in the analysis of cryptographic protocols. An illustrative example of this case is the work of Mukhamedov *et. al.* [28]. In this work the authors analyze a fragment of the API for a Trusted Platform Module in ProVerif [2], but encountered problems in encoding state information and in handling such information during the analysis. However, we note that the model of APIs and their desired behavior is possible to formalize and verify by hand, as in [5,10]; the issue here is implementing the appropriate functionality into cryptographic protocol analysis tools.

Another reason that is perhaps harder to address is that many of the APIs rely on properties of the cryptoalgorithm that are not supported by many of the tools, or are supported only partially. Many of these properties can be expressed as equations describing the behavior of the crypto system. For example, CCA-API makes extensive use of exclusive-or, which satisfies the following equations, where $*$ denotes the exclusive-or symbol:

$$x * (y * z) = (x * y) * z \qquad \text{(associativity)}$$
$$x * y = y * x \qquad \text{(commutativity)}$$
$$x * 0 = x \qquad \text{(neutral element)}$$
$$x * x = 0 \qquad \text{(self-cancellation)}$$

Although there are a number of tools, e.g. ProVerif, that can deal with equations that can be expressed as rewrite rules (that is, that can be given an orientation), tools that can deal with equations that involve both associativity and commutativity (AC) are rarer. Even those tools that do support AC theories do not always support exclusive-or. For example, the Tamarin tool is optimized [27] for modular exponentiation and bilinear pairing, but has not been applied to or optimized for exclusive-or. However, the problem is not necessarily completely intractable. Hand proofs have been developed for some APIs, a number of decision procedures have been developed for the bounded session model, in which the attacker can interact with the protocol only a finite number of times [6,7], and others have been developed for the unbounded session model for certain subclasses of protocols [8,32,9]. In particular, the class of algorithms addressed by [9] is focused on IBM-CCA-like protocols, and has been applied to several versions of IBM-CCA, including the ones analyzed in this paper. Steel [31] has also proposed the use of *XOR constraints* and applied them to the analysis of the IBM CCA protocols, as well as some key exchange protocols using XOR, such as a modified version of Needham-Schroeder. However, this also assumes a bounded session model, e.g. a bounded number of executions of the API operators.

There have, however, been some notable exceptions to this rule, in which general cryptographic protocol analysis tools that allow search in the unbounded session model have been applied to protocols using exclusive-or. One is the Maude-NPA protocol analysis tool [12], which supports equational theories having *finite variant decompositions*, which includes exclusive-or. It has been used successfully to analyze a number of protocols that use exclusive-or, e.g. in [13,15], but had not been applied to cryptographic APIs until now. The other is the work of Küsters and Truderung [20], who give an algorithm for compiling a class of xor-based protocols called *XOR-linear* to protocols that can be analyzed via ProVerif, a tool that does not in itself support AC theories. Not all XOR-based protocols are XOR-linear, but in some cases it is possible to transform a protocol to an XOR-linear protocol that is equivalent to the original with respect to secrecy properties. Küsters and Truderung perform such a transformation for the IBM CCA, and then use their algorithm to analyze it in ProVerif.

In this paper we apply Maude-NPA to the analysis of IBM CCA. We analyze not only the original protocol, but the different fixes provided by IBM in [16], and the different XOR-linear versions provided by Küesters and Truderung in [20]. In particular, we seek to reproduce Bond's attack on the different versions. We demonstrate that it is indeed possible to perform analyses of APIs using XOR, and in some cases to achieve termination. We also discuss what needs to be done to improve Maude-NPA's performance.

In addition we demonstrate the use of *never patterns* to refine and guide our search. Maude-NPA finds attacks by searching backwards from an *attack pattern*. A never pattern is a state pattern that can be added to the attack pattern to reduce the size of the search space; if Maude-NPA creates a state in the search tree that contains an instantiation of a never pattern, then it does not look for any children of that state. We show how never patterns can be used in a way that reduces the size of the search space but is *complete* with respect to reachability of the original attack pattern. In some cases it may not be possible to maintain completeness; but we show how never patterns can be used in a way that maintains completeness *with respect to the existence of a particular attack trace or class of traces.* Both types of never patterns were used in the IBM-CCA analyses.

2 Maude-NPA

In this section we give a high-level summary of Maude-NPA, with particular interest paid to the use of never patterns. For further information, please see [12].

2.1 Preliminaries on Unification and Narrowing

We assume an order-sorted signature $\Sigma = (S, \leq, \Sigma)$ with a poset of sorts $(S, \leq)$ and an S-sorted family $\mathcal{X} = \{\mathcal{X}_s\}_{s \in S}$ of disjoint variable sets with each $\mathcal{X}_s$ countably infinite. $\mathcal{T}_\Sigma(\mathcal{X})_s$ is the set of terms of sort s, and $\mathcal{T}_{\Sigma,s}$ is the set of ground terms of sort s. We write $\mathcal{T}_\Sigma(\mathcal{X})$ and $\mathcal{T}_\Sigma$ for the corresponding order-sorted term algebras. For a term t, $\mathcal{V}ar(t)$ denotes the set of variables in t.

Positions are represented by sequences of natural numbers denoting an access path in the term when viewed as a tree. The top or root position is denoted by the empty sequence ϵ. The subterm of t at position p is $t|_p$ and $t[u]_p$ is the term t where $t|_p$ is replaced by u.

A *substitution* $\sigma \in \mathcal{S}ubst(\Sigma, \mathcal{X})$ is a sorted mapping from a finite subset of $\mathcal{X}$ to $\mathcal{T}_\Sigma(\mathcal{X})$. Substitutions are written as $\sigma = \{X_1 \mapsto t_1, \ldots, X_n \mapsto t_n\}$ where the domain of σ is $Dom(\sigma) = \{X_1, \ldots, X_n\}$ and the set of variables introduced by terms $t_1, \ldots, t_n$ is written $Ran(\sigma)$. The identity substitution is id. Substitutions are homomorphically extended to $\mathcal{T}_\Sigma(\mathcal{X})$. The application of a substitution σ to a term t is denoted by $t\sigma$.

A Σ-*equation* is an unoriented pair $t = t'$, where $t, t' \in \mathcal{T}_\Sigma(\mathcal{X})_s$ for some sort $s \in S$. Σ and a set E of Σ-equations, The E-equivalence class of a term t is denoted by $[t]_E$ and $\mathcal{T}_{\Sigma/E}(\mathcal{X})$ and $\mathcal{T}_{\Sigma/E}$ denote the corresponding order-sorted term algebras modulo E. An *equational theory* (Σ, E) is a pair with Σ an order-sorted signature and E a set of Σ-equations.

An E-*unifier* for a Σ-equation $t = t'$ is a substitution σ such that $t\sigma =_E t'\sigma$. For $\mathcal{V}ar(t) \cup \mathcal{V}ar(t') \subseteq W$, a set of substitutions $CSU_E^W(t = t')$ is said to be a *complete* set of unifiers for the equality $t = t'$ modulo E away from W iff: (i) each $\sigma \in CSU_E^W(t = t')$ is an E-unifier of $t = t'$; (ii) for any E-unifier ρ of $t = t'$ there is a $\sigma \in CSU_E^W(t = t')$ such that $\sigma|_W \sqsupseteq_E \rho|_W$ (i.e., there is a substitution η such that $(\sigma \circ \eta)|_W =_E \rho|_W)$; and (iii) for all $\sigma \in CSU_E^W(t = t')$, $Dom(\sigma) \subseteq (\mathcal{V}ar(t) \cup \mathcal{V}ar(t'))$ and $Ran(\sigma) \cap W = \emptyset$.

A *rewrite rule* is an oriented pair $l \to r$, where $l \notin \mathcal{X}$ and $l, r \in \mathcal{T}_\Sigma(\mathcal{X})_s$ for some sort $s \in S$. An *(unconditional) order-sorted rewrite theory* is a triple (Σ, E, R) with Σ an order-sorted signature, E a set of Σ-equations, and R a set of rewrite rules. The (R, E) rewriting relation $\to_{R,E}$ on $\mathcal{T}_\Sigma(\mathcal{X})$ is defined as: $t \to_{p,R,E} t'$ iff there exist $p \in \mathcal{P}os_\Sigma(t)$, a rule $l \to r$ in R, and a substitution σ such that $t|_p =_E l\sigma$ and $t' = t[r\sigma]_p$.

Let t be a term and W be a set of variables such that $Var(t) \subseteq W$, the R, E-*narrowing* relation on $\mathcal{T}_\Sigma(\mathcal{X})$ is defined as $t \leadsto_{p,\sigma,R,E} t'$ if there is a non-variable position $p \in Pos_\Sigma(t)$, a rule $l \to r \in R$ properly renamed s.t. $(Var(l) \cup Var(r)) \cap W = \emptyset$, and a unifier $\sigma \in CSU_E^{W'}(t|_p = l)$ for $W' = W \cup Var(l)$, such that $t' = (t[r]_p)\sigma$.

2.2 Maude-NPA Syntax and Semantics

Given a protocol $\mathcal{P}$, states are modeled as elements of an initial algebra $\mathcal{T}_{\Sigma_\mathcal{P}/E_\mathcal{P}}$, where $\Sigma_\mathcal{P}$ is the signature defining the sorts and function symbols (for the cryptographic functions and for all the state constructor symbols) and $E_\mathcal{P}$ is a set of equations specifying the *algebraic properties* of the cryptographic functions and the state constructors. Therefore, a state is an $E_\mathcal{P}$-equivalence class $[t]_E \in \mathcal{T}_{\Sigma_\mathcal{P}/E_\mathcal{P}}$ with t a ground $\Sigma_\mathcal{P}$-term.

In Maude-NPA a *state pattern* for a protocol P is a term t of sort State (i.e., $t \in \mathcal{T}_{\Sigma_\mathcal{P}/E_\mathcal{P}}(\mathcal{X})_{\mathsf{State}}$) which has the form $\{S_1 \& \cdots \& S_n \& \{IK\}\}$ where $\&$ is an associative-commutative union operator with identity symbol $\emptyset$. Each element in the set is either a *strand* S_i or the *intruder knowledge* $\{IK\}$ at that state.

The *intruder knowledge* $\{IK\}$ also belongs to the state and is represented as a set of facts using the comma as an associative-commutative union operator with identity element *empty*. There are two kinds of intruder facts: *positive* knowledge facts (the intruder knows m, i.e., $m \in \mathcal{I}$), and *negative* knowledge facts (the intruder *does not yet know* m but *will know it in a future state*, i.e., $m \notin \mathcal{I}$), where m is a message expression.

A *strand* [14] specifies the sequence of messages sent and received by a principal executing the protocol and is represented as a sequence of messages $[msg_1^-, msg_2^+, msg_3^-, \ldots, msg_{k-1}^-, msg_k^+]$ such that msg_i^- (also written $-msg_i$) represents an input message, msg_i^+ (also written $+msg_i$) represents an output message, and each msg_i is a term of sort Msg (i.e., $msg_i \in \mathcal{T}_{\Sigma_\mathcal{P}/E_\mathcal{P}}(\mathcal{X})_{\mathsf{Msg}}$).

Strands are used to represent both the actions of honest principals (with a strand specified for each protocol role) and the actions of an intruder (with a strand for each action an intruder is able to perform on messages). In Maude-NPA strands evolve over time; the symbol | is used to divide past and future. That is, given a strand $[m_1^\pm, \ldots, m_i^\pm \mid m_{i+1}^\pm, \ldots, m_k^\pm]$, messages $m_1^\pm, \ldots, m_i^\pm$ are the *past messages*, and messages $m_{i+1}^\pm, \ldots, m_k^\pm$ are the *future messages* ($m_{i+1}^\pm$ is the immediate future message). A strand $[msg_1^\pm, \ldots, msg_k^\pm]$ is shorthand for $[nil \mid msg_1^\pm, \ldots, msg_k^\pm, nil]$. An *initial state* is a state where the bar is at the beginning for all strands in the state, and the intruder knowledge is empty. A *final state* is a state where the bar is at the end for all strands in the state and there is no intruder fact of the form $m \notin \mathcal{I}$.

Since the number of states $\mathcal{T}_{\Sigma_\mathcal{P}/E_\mathcal{P}}$ is in general infinite, rather than exploring concrete protocol states $[t]_E \in \mathcal{T}_{\Sigma_\mathcal{P}/E_\mathcal{P}}$ Maude-NPA explores *symbolic state patterns* $[t(x_1, \ldots, x_n)]_E \in \mathcal{T}_{\Sigma_\mathcal{P}/E_\mathcal{P}}(\mathcal{X})$ on the free $(\Sigma_\mathcal{P}, E_\mathcal{P})$-algebra over a set of variables $\mathcal{X}$. In this way, a state pattern $[t(x_1, \ldots, x_n)]_E$ represents not a single concrete state but a possibly infinite set of such states, namely all the *in-*

stances of the pattern $[t(x_1, \ldots, x_n)]_E$ where the variables $x_1, \ldots, x_n$ have been instantiated by concrete ground terms.

The semantics of Maude-NPA is expressed in terms of *rewrite rules* that describe how a protocol moves from one state to another via the intruder's interaction with it. One uses Maude-NPA to find an attack by specifying an insecure state pattern called an *attack pattern*. Maude-NPA attempts to find a path from an initial state to the attack pattern via backwards narrowing (narrowing using the rewrite rules with the orientation reversed). Such a backwards narrowing sequence is called a *backwards path* from to the attack state. Maude-NPA attempts to find paths until it can no longer form any backwards narrowing steps, at which point it terminates. If it at that point it has not found an initial state, the attack pattern is judged *unreachable*. Note that Maude-NPA puts no bound on the number of sessions, so reachability is undecidable in general. Note also that Maude-NPA does not perform any data abstraction such as bound number of nonces. However, the tool makes use of a number of sound and complete state space reduction techniques that help to identify unreachable and redundant states, and thus make termination more likely.

2.3   Never Patterns in Maude-NPA

It is often desirable to exclude certain patterns from transition paths leading to an attack state. For example, one may want to determine whether or not authentication properties have been violated, e.g., whether it is possible for a responder strand to appear without the corresponding initiator strand. For this there is an optional additional field in the attack state containing the *never patterns*. Each never pattern is itself a state pattern. When we provide an attack state A and some never patterns $NP_1, \ldots, NP_k$ to Maude-NPA, every time the tool produces a state S via backwards narrowing from A, it checks whether there is a substitution θ such that $NP_i\theta =_{E_P} S$. If that is the case, the state is discarded. [2] We will write A with the never patterns $NP_1, \ldots, NP_k$ as $A \parallel \text{never}(NP_1) \ldots \parallel \text{never}(NP_k)$.

Although never patterns were introduced as a means for specifying authentication properties, they can also be used to reduce the search space. However, we want to preserve completeness as much as possible. Hence we make use of the following results.

Proposition 1. *Let M be a never pattern containing terms of the form $m \in \mathcal{I}$. Suppose that the state M is unreachable in Maude-NPA, Then for any state pattern S, if $S \parallel \text{never}(M)$ is unreachable, then so is S.*

Proof. (Sketch) Suppose that there is a backwards narrowing sequence from S to an initial state. Then it must pass through a state containing $M\theta$ for some substitution θ. But since M is unreachable, so is any state containing $M\theta$.

---

[2] Maude-NPA also checks whether $NP_i\theta$ satisfies *irreducibility constraints*, as described in [11].

We refer to never patterns that satisfy the conditions of Proposition 1 as *completeness-preserving never patterns*. Given a completeness-preserving never pattern, we can add it to any attack state without affecting its reachability.

In Proposition 2 below, we say that T is a *substate* of S, where T and S are state patterns, if every strand or intruder knowledge statement that appears in T also appears in S.

Proposition 2. *Let $S_0 \leadsto_{\sigma_1, R_{\mathcal{P}}, E_{\mathcal{P}}^{-1}} S_1 \ldots S_{k-1} \leadsto_{\sigma_k, R_{\mathcal{P}}, E_{\mathcal{P}}^{-1}} S_k = S$ be a backwards narrowing from an attack pattern S to an initial state, and M a never pattern containing terms of the form $m \in \mathcal{I}$, such that for each S_i there is no θ and T such that $M\theta =_{E_{\mathcal{P}}} T$, where T is a substate of S_i. Then the sequence $S_0 \leadsto_{\sigma_1, R_{\mathcal{P}}, E_{\mathcal{P}}^{-1}} S_1 \ldots S_{k-1} \leadsto_{\sigma_k, R_{\mathcal{P}}, E_{\mathcal{P}}^{-1}} S_k = S$ is a backwards narrowing sequence from $S \parallel never(M)$ to an initial state.*

Proof. (Sketch.) The result follows straightforwardly from the definition of never pattern.

We refer to never patterns that satisfy the conditions of Proposition 2 for a given attack trace as *attack-preserving never patterns*. We can use attack-preserving never patterns to help show that a new version of a protocol is immune to a known attack on the old one. Suppose that we have found an attack on a protocol, and we want to see whether a modified version of the protocol is immune to that attack. Suppose that the search space is intractibly large, even after adding completeness-preserving never patterns. We may be able to reduce the search space by adding attack-preserving never patterns. In that case, unreachability of the attack state with the attack-preserving never patterns does not necessarily imply unreachability of the attack state without these never patterns. But it *does* imply that a specific *class* of attacks, including the original attack we were concerned about, is no longer possible.[3]

We make use of both completeness-preserving and attack-preserving never patterns in our analysis of the IBM-CCA protocols. This is described in more detail in Section 5.

3 IBM CCA API

CCA stands for the Common Cryptographic Architecture API [17] as implemented on the hardware security module IBM 4758, which is an IBM cryptographic coprocessor widely used in security critical systems such as electronic payment and automated teller machine (ATM) networks.

The CCA API contains several protocols, namely the CCA-0 protocol, which is subject to an attack presented by Bond in [3], and other versions of this protocol

---

[3] Note that if the attack state with the attack-preserving never patterns is reachable, but the original attack is not found, that does not mean that the original attack is not subsumed by any of the found attacks. This is a result of Maude-NPA's state space reduction techniques, which make Maude-NPA produce only some of the possible attacks (but always at least one), when an attack state is reachable.

(CCA-1A, CCA-1B, CCA-2B, CCA-2C, and CCA-2E), designed to avoid this attack.

As explained in [18,9], the CCA is a key management system, which provides commands that use encrypted keys to achieve desired functions. A 168-bit triple-DES key, known as the *master key*, is stored in the security module's tamper-proof memory and is used to encrypt all other keys, which are then kept on the host computer. These other keys, known as *working keys*, are used to perform the various functions provided by the CCA API. There are several types of working keys, depending on the type of action they will be involved in. The CCA API supports the following functions and features:

- Encryption and decryption of data, using the DES algorithm [29].
- Message authentication code (MAC) generation, and data hashing functions.
- Generation and validation of digital signatures.
- Generation, encryption, translation and verification of Personal Identification Number (PIN) and transaction validation messages.
- General key management facilities.
- Administrative services for controlling the initialization and operation of the security module.

The CCA API uses four main types for classifying DES working keys, each of which is further sub-divided into more specific and restrictive types. A working key is stored outside of the security module, encrypted under the exclusive-or of the device's master key and the control vector representing the type of the key. The main key types, and their uses, are as follows:

- **Data Keys**: used for cryptographic operations on arbitrary data.
- **PIN Keys**: used for cryptographic operations on PINs.
- **Key Encryption Keys (KEK)**: used to encrypt and decrypt other working keys during transfer between security modules, and divided into import and export types.
- **Key Generation Keys**: used as input to a key generation algorithm.

The typing mechanism restricts the working keys that can be used for a particular command. For example, the PIN derivation key (PDK) used in the verification of a customer's PIN cannot be used to encrypt arbitrary data.

The following constants and variables are used throughout this section to denote the various control vectors, cryptographic keys and other exchanged data:

- constant DATA, IMP, EXP: control vectors for data, import-type key encryption, and export-type key encryption keys, respectively
- constant KP: a part of a key, and not a complete key
- constant KM: the security module's master key
- constants Km1, Km2, and Km3: Those are used as a simplification of the CCA protocol where it is assumed that the environment produces the term e(IMP * KP * KM, Km1 * Km2).
- variable *ekek*: an arbitrary key encryption key

- variable eK: a key generation key to encrypt messages
- variable T: an unknown, randomly generated, new cryptographic key or an arbitrary key type control vector. This variable is restricted to constants DATA, IMP, EXP and PIN.
- variables $km1$, $km2$, $km3$: i'th key part (used to build an arbitrary key)
- variable X: arbitrary (plain) data

In the following, we provide an informal description of the CCA APIs commands. Table 1 summarizes the exchange of messages performed for each command: messages in the left hand side of the "rule" denote the messages that need to be received; messages in the right hand side denotes messages that are sent as a result of the left hand messages being received. Note that **PKA Symmetric Key Import** is a later addition that converted a public key encryption of **eK** to a symmetric key encryption; it did not appear in the original CCA.

Table 1. CCA API commands and description

API command	Description
Encipher	$X, \{eK\}_{\{KM*DATA\}} \to \{X\}_{eK}$
Decipher	$\{X\}_{eK}, \{eK\}_{\{KM*DATA\}} \to X$
Key Export	$\{eK\}_{(KM*T)}, T, \{ekek\}_{\{KM*EXP\}} \to \{eK\}_{(ekek*T)}$
Key Import	$\{eK\}_{(kek*T)}, T, \{ekek\}_{\{KM*IMP\}} \to \{eK\}_{\{KM*T\}}$
Key Part Import First	$km1, T \to \{km1\}_{\{KM*KP*T\}}$
Key Part Import Middle	$km2, km1_{\{KM*KP*T\}}, T \to (km1 * km2)_{\{KM*KP*T\}}$
Key Part Import Last	$km3, km2_{\{KM*KP*T\}}, T \to (km2 * km3)_{\{KM*KP*T\}}$
Key Translate	$\{eK\}_{ekek1*T}, T, \{ekek1\}_{KM*IMP}, \{ekek2\}_{KM*EXP} \to \{eK\}_{(ekek2*T)}$
PKA Symmetric Key Import	$\{eK\ ;\ T\}_{PKA} \to \{eK\ \}_{KM*T}$

These commands are explained in more detail below.

- **Encipher** encrypts the given plaintext with the supplied data key. The data key can be either of the general **Key** type, or of one of the subtypes that allow data ciphering.
- **Decipher** decrypts ciphertext which has been encrypted under the supplied data key eK. The data key can be either of the general type, or of one of the subtypes that allow data deciphering.
- **Key Export** converts a working key eK encrypted under the local master key to one encrypted under the supplied export-type key encryption key $ekek$.

- Key Import converts a complete key eK encrypted by the supplied import-type *key encryption key* kek to one encrypted by the local master key KM.
- The Key Part Import commands can be used one after the other, by three different security officers, each in possession of one key part, to create the complete working import key. Note that $km1$, $km2$ and $km3$ are variables.
- Key Translate translates a key eK from encryption by an import key to encryption by an export key.
- PKA Symmetric Key Import converts a complete key eK encrypted by the a public key PKA to one encrypted by the local master key KM.

The exact steps that the security module performs for each command have not been included, since the process is virtually the same in all cases. The master key and all control vectors are known to the security module, and any additional information required is either passed on as a plaintext parameter, or is encrypted under a known key.

Converting these rules to Maude-NPA is straightforward; terms before the arrow are negative terms, and the term after the arrow is positive.

For further details about the specification of the CCA-API commands in Maude-NPA, we refer the reader to [15]. Complete specifications and the analyses outputs may be found in http://www.dsic.upv.es/šescobar/Maude-NPA_Protocols/API_Protocols.html

For example, the Maude-NPA version of Key Import is as follows

$$:: nil :: [(e(kek * T, eK))^-, (T))^-, (e(KM * IMP, kek))^-, (e(KM * T, eK))^+]$$

In [3] Bond points out that it is possible to obtain PDK in the clear by combining the commands in an unexpected way. This is described in Table 2, where we have reproduced the attack using the Maude-NPA tool. The terms preceded by a minus sign describe those the attacker needs to know to in order to perform an operation, while the terms preceded by a plus sign describes the term output by an operation.

3.1   IBM's Recommendations to Avoid CCA-0's Attack

In order to prevent the attack of the CCA-0 protocol described above, IBM suggested two recommendations in [16]. In the first one, they recommended the use of a public key version of Key Import, the PKA Symmetric Key Import described above. This version was broken by Cortier et al. in [9]. They then recommended that access control be used, and that no principal be allowed to execute both PKA Symmetric Key Import and Key Import. Following [9] we refer to this as CCA-1. We specify two versions of this in Maude-NPA: CCA-1A in which the attacker has access to Key Import, and CCA-2A, in which the attacker has access to PKA Symmetric Key Import.

In the second recommendation IBM proposed the use of a more elaborate form of *role-based* access control. Principals are assigned to roles determining which commands they are allowed to execute. The goal is to prevent one single

Table 2. Bond's Attack on CCA-0

Exchanged messages	Explanation
+(e(IMP * KP * KM, Km1 * Km2)),	The intruder receives e(IMP * KP * KM, km1 * km2) from the environment.
-(PIN * Km3), -(IMP), -(e(IMP * KP * KM, Km1 * Km2)), +(e(IMP * KM, PIN * Km1 * Km2 * Km3)),	It executes command "Key Part Import Last" where variable km3 is instantiated with Km3 * PIN. In this way he obtains e(IMP * KM, PIN * km1 *km2 * km3).
-(PIN * EXP * Km3), -(IMP), -(e(IMP * KP * KM, Km1 * Km2)), +(e(IMP * KM, PIN * EXP * Km1 * Km2 * Km3)),	The intruder uses the same command again, this time with variable km3 instantiated with PIN * EXP * Km3, obtaining e(IMP * KM, PIN * EXP * km1 * km2 * km3).
+(e(PIN * Km1 * Km2 * Km3, PDK)),	The intruder receives e(PIN * Km1 * Km2 * Km3, PDK)).
-(e(PIN * Km1 * Km2 * Km3, PDK)), -(null), -(e(IMP * KM, PIN * Km1 * Km2 * Km3)), +(e(KM, PDK)),	When PDK is imported, the intruder uses "Key Import" twice: The first time with inputs e(IMP * KM, PIN * Km1 * Km2 * Km3) and e(PIN * Km1 * Km2 * Km3, PDK) generating the message e(KM, PDK).
-(e(PIN * Km1 * Km2 * Km3, PDK)), -(EXP), -(e(IMP * KM, PIN * EXP * Km1 * Km2 * Km3)), +(e(EXP * KM, PDK)),	The second time "Key Import" is used with inputs e(IMP * KM, PIN * EXP * Km1 * Km2 * Km3), and e(PIN * Km1 * Km2 * Km3, PDK), which gives the message e(EXP * KM, PDK).
-(e(KM, PDK)), -(null), -(e(EXP * KM, PDK)), +(e(PDK, PDK))	Finally, using the "Key Export" command, the intruder gets e(PDK,PDK).

Table 3. CCA-API commands for IBM's recommendations

API command	CCA-1A	CCA-1B	CCA-2B	CCA-2C	CCA-2E
Encipher	✓	✓	✓	✓	✓
Decipher	✓	✓	✓	✓	✓
Key Export	✓	✓	✓	✓	✓
Key Import	✓		✓		✓
Key Part Import First			✓	✓	
Key Part Import Last				✓	
Key Test	✓	✓			
PKA Sym. Key Import		✓			
Key Translate		✓			

individual from having access to all the commands required to mount Bond's attack. IBM provided an example of the KEK transfer process involving five roles (A-E) such that no single role is able to mount the attack. Following [9], we refer to this as CCA-2. We specify three versions in Maude-NPA depending on which role the attacker is playing; CCA-2B, CCA-2C, and CCA-2E respectively. Since roles A and D do not have access to any of the operations, we do not supply specifications of them.

Table 3 summarizes the commands that each protocol can perform. In the original attack the intruder played the roles C and E together. Note that CCA-XY describes the actions prescribed for Role Y participating in protocol CCA-X.

4   Küsters' and Truderung's XOR-Linear Versions of CCA-Protocols

In [20], Küesters and Truderung analyzed the CCA API protocols in ProVerif via a protocol transformation technique. However, this work does not support full exclusive-or capabilities and requires restricting the analysis to *XOR-linear* protocols; see [20] for details on the XOR-linear property. Because of this, they needed to modify and transform by hand some of the CCA API commands, to produce an XOR-linear protocol equivalent to the original with respect to secrecy properties. We will refer to protocols using these manually-modified commands as "XOR-linear versions" of that protocols.

Table 4. Original specification of the protocol

API command	Description
Key Part Import First	$km1, T \rightarrow \{km1\}_{\{KM*KP*T\}}$
Key Part Import Middle	$km2, km1_{\{KM*KP*T\}}, T$ $\rightarrow (km1 * km2)_{\{KM*KP*T\}}$
Key Part Import Last	$km3, km2_{\{KM*KP*T\}}, T$ $\rightarrow (km2 * km3)_{\{KM*KP*T\}}$
Key Translate	$\{eK\}_{ekek1*T}, T, \{ekek1\}_{KM*IMP}, \{ekek2\}_{KM*EXP}$ $\rightarrow \{eK\}_{(ekek2*T)}$

Table 5. Küesters and Truderung version

API command	Description
KPI-First + KPI-Add/Middle	$km12, T \rightarrow \{KM * KP * IMP\}$
Key Part Import Last	$x, T, KM * KP * T \rightarrow (x)_{\{KM*T\}}$ x, IMP $\rightarrow (X * km12)_{\{KM*IMP\}}$
Key Translate	$\{eK\}_{ekek1*T}, T, \{ekek1\}_{KM*IMP}$ $\rightarrow transf(eK,T)$ $transf(eK,T), \{ekek2\}_{\{KM*EXP\}}$ $\rightarrow \{eK\}_{(ekek2*T)}$

As we can see in Tables 4 and 5, the XOR-linear versions of the CCA operators are as follows. The "KPI-First + KPI-Add/Middle" and "Key Part Import Last" API commands are the XOR-linear equivalent to the original "Key Part Import First", "Key Part Import Middle" and "Key Part Import Last" commands. Note that now the "Key Translate" command requires two steps instead of one in the original version. All the other commands remain the same without any transformation.

5   Maude-NPA's CCA Analysis

In this section we describe the results of Maude-NPA's analysis of both the original CCA protocols as proposed by IBM and the XOR-linear versions of Küesters and Truderung. We did not analyze the versions for ProVerif that were produced automatically from the XOR-linear versions, since these did not use XOR nor AC. In each case we asked if the attacker could learned $e(PDK, PDK)$, since this is the information that Bond's attacker needs to compute PDK. In the following we give the results of our Maude-NPA analyses, comparing the results for the original protocol with those for its XOR-linear transformation when one exists. The results are given in Table 7, where the number of states at depth N is the number of different N-length backwards narrowing sequences produced after N backwards narrowing steps. Complete specifications and the analyses outputs may be found in http://www.dsic.upv.es/ sescobar/Maude-NPA_Protocols/ API_Protocols.html.

In some cases, we were not able to obtain termination unaided, and were required to use never patterns as follows. We use two completeness-preserving never patterns: (i) e(Key, KM * Msg) inI and (ii) e(IMP * KM, Type * Key) inI. These have been proved unreachable in Maude-NPA. We also use several attack-preserving never patterns. One of these is (iii) PDK inI. This is motivated by the fact that in Bond's attack the intruder is trying to find $e(PDK, PDK)$ so that it can learn PDK, so we would not expect it to have learned PDK already. The others are different terms of the form $(X * Y) \in \mathcal{I}$ not used in Bond's attack. These are: (iv) $(Km1 * Y) \in \mathcal{I}$, (v) $(Km2 * Y) \in \mathcal{I}$, (vi) $(PDK * Y) \in \mathcal{I}$, (vii) $(KM * Y) \in \mathcal{I}$, and (viii) $(Y * e(K, Y)) \in \mathcal{I}$ where K and Y are variables. In Table 7 we give the cases in which we use and do not use never patterns. Since the protocols are similar, we use the same never patterns for all cases.

CCA-0. The CCA-0 protocol is insecure, since it is subject to the attack found by Bond in [3]. In this attack the intruder obtains a PIN derivation key in the clear, as in the IBM attack and, thus, can compute PINs from bank account numbers. This attack is the same found by Küesters and Truderung in [20].

Using the same assumptions as in [3] in terms of the role played by the intruder and its knowledge, Maude-NPA finds the attack of the CCA-0 protocol after 7 steps of protocol analysis. Table 6 shows the numbers of states generated at each depth of the backwards reachability analysis from an attack state in which the intruder has learned the expression e(PDK,PDK).

As we can see from rows 1 and 2 of Table 7, Maude-NPA finds the initial state for both protocols, the original CCA-0 and XOR-linear version, at the same depth of the backwards search tree. If the analysis is continued, the XOR-linear version produces a finite search space containing 2495 states in total. The use of the never patterns was required to guarantee termination for the more complex original protocol, but not the XOR-linear version. For both protocols Maude-NPA terminated at Step 7; that is, it produced no new states at Step 8.

CCA-1A. This protocol is XOR-linear and Küesters and Truderung do not transform it. Row 3 of Table 7 summarizes the result of the analysis of the pro-

Table 6. CCA-0 Analysis Output

Level	1	2	3	4	5	6	7
States	1	7	27	79	89	44	1
Solutions	0	0	0	0	0	0	1

Table 7. Experimental results

	Protocol	States	Depth	Terminates
1	CCA-0	291*	7	Yes
2	CCA-0-XOR-linear	2495	7	Yes
3	CCA-1A	21*	5	Yes
4	CCA-1B	48*	6	Yes
5	CCA-1B-XOR-linear	1	2	Yes
6	CCA-2B	324*	11	Yes
7	CCA-2C	131*	6	No
8	CCA-2C-XOR-linear	105	4	No
9	CCA-2E	385*	7	No

*This protocol analysis uses never patterns

tocol specified in Maude-NPA using never patterns. The search space terminates at step 5 (that is there are no states produced at step 6).

CCA-1B. This protocol is not XOR-linear and Küesters and Truderung manually transformed it. In Table 8 we can see the differences between the two CCA-1B protocols, the original and the XOR-linear versions. Rows 4 and 5 of Table 7 summarize the results of the analysis of both versions. As we can see, the XOR-linear version is extremely simple and the analysis is almost immediate in Maude-NPA, requiring no never patterns. The search space terminates at depth 6, finding no initial state.

CCA-2B. Row 6 of Table 7 summarizes the result of the analysis of CCA-2B. Note that this protocol is XOR-linear and Küesters and Truderung do not transform it. The search, using never patterns, terminates at depth 11, finding no initial state.

CCA-2C. Table 8 shows the original protocol and the XOR-linear version provided by Küesters and Truderung in [20]. Rows 7 and 8 of Table 7 summarize the results of the analysis of both versions. In these two cases we were not able to run Maude-NPA to termination. For the XOR-linear version we were able to run Maude-NPA to depth 4, and depth 6 in the original version.

CCA-2E. Row 9 of Table 7 summarizes the result of the analysis of the protocol specified in Maude-NPA. Note that this protocol is XOR-linear and Küesters and Truderung do not transform it. In this case we were able to run Maude-NPA to depth 7, but were not able to achieve termination.

In all the cases in which we were unable to achieve termination, the issue does not seem to be so much state explosion as the time to produce all the states at a given depth. Indeed, in the case of CCA-2E, Maude-NPA found 76 states at step 6 and 54 states at step 7, suggesting that it might have terminated if run to a greater depth.

Table 8. Original and Küesters-Truderung versions of CCA-1B and CCA-2C

API command	Description
CCA 1-B Original Key Translate	$\{eK\}_{ekek1*T}$, T, $\{ekek1\}_{KM*IMP}$, $\{ekek2\}_{KM*EXP}$ $\rightarrow \{eK\}_{(ekek2*T)}$
CCA-1B-XOR-linear Key Translate	$\{eK\}_{ekek1*T}$, T, $\{ekek1\}_{KM*IMP} \rightarrow$ transf(eK,T) transf(eK,T), $\{ekek2\}_{\{KM*EXP\}} \rightarrow \{eK\}_{(ekek2*T)}$
CCA-2C Original Key Part Import Last	km3, $(km2)_{\{KM*KP*T\}}$, T $\rightarrow (km2 * km3)_{\{KM*KP*T\}}$
CCA-2C-XOR-linear Key Part Import Last	x, T, KM * KP * T $\rightarrow (x)_{\{KM*T\}}$ x, IMP $\rightarrow (X * km12)_{\{KM*IMP\}}$

6   Discussion

We have demonstrated that in certain cases Maude-NPA is indeed able to prove properties of XOR-based cryptographic APIs. This is to the best of our knowledge the first application of a general-purpose unbounded session cryptographic protocol analysis tool that directly models the properties of XOR to XOR-based cryptographic APIs. However, there were a number of performance issues that affected termination. We discuss these in more detail below, and what can be done to address them.

We do not provide a detailed comparison of the performance of the different tools, since the way protocols are modeled and security properties proved vary from case to case. For example, Cortier et al. analyze a slightly different version of CCA-2 in which principals are given greater privileges. Also, each of the analyses of Cortier et al., Küsters and Truderung, and ourselves makes different assumptions about the initial knowledge available to the attacker. On the other hand, we can make some general comparisons. Küsters and Truderung are able to achieve termination in all cases, although this comes at a cost, since it is not clear that all protocols can be converted to XOR-linear versions, and it is unknown whether the conversion process can be made sound and complete with respect to authentication as well as secrecy properties. Cortier et. al. are able to obtain termination for CCA-1, but for CCA-2 they ran their algorithm only up to a certain bound, and then verified informally that the attacker could not gain any useful terms by interacting further with the protocol.[4] Maude-NPA terminated for CCA-0, CCA-1, and CCA-2B with never patterns, and for the XOR-linear versions without. However, it had more problems with CCA-2C and CCA-2E, even the XOR-linear version. We note that state explosion could be controlled with never patterns; the main problem we were experiencing was that Maude-NPA took longer and longer to complete its search at a given depth, even though it might not be producing that many states.

The chief cause of state explosion seems to be the failure of Maude-NPA state space reduction techniques, in particular the Maude-NPA *grammars* to deal with complex combinations of exclusive-or expressions. Maude-NPA uses inductive techniques to recognize terms that are unlearnable by the intruder, and generates grammars that describe these terms. It works well with most theories but occasionally has problems with XOR and Abelian group theories, especially when they occur many times in a protocol, as they do in IBM CCA. We are currently reassessing our grammar generation techniques in the light of our experience with IBM CCA.

The fact that Maude-NPA is taking a long time to complete, even when it does not produce that many states, means that it is generating many states which are subsequently rejected as unreachable or redundant using the state space reduction techniques. Paradoxically, this behavior could be improved by improving the state space reduction techniques, since if unreachable states are removed earlier, less states are generated later on. Performance may also be improved by sound and complete transformations to simpler protocols. For example, Küster's and Truderung's transformations to XOR-linear protocols generally resulted in protocols that were easier for Maude-NPA to analyze, and although they were done manually, they use a general strategy that could possibly be automated. We could perhaps employ similar techniques to produce protocols that are "small" with respect to an XOR-complexity metric, rather than XOR-linear.

Finally, we note the contribution made to this work by never patterns. As far as we know, this is the first work that considers their effect on soundness and completeness. Completeness-preserving never patterns have the potential to be a valuable tool for use as an additional state space reduction technique. Indeed it should be fairly straightforward to prove that a pattern unreachable and add it automatically to a specification as a completeness-preserving never pattern; this was indeed a feature of the NPA tool [24] that preceded Maude-NPA. A more ambitious plan would be to search automatically for completeness-preserving never patterns, e.g. by finding patterns that keep on occurring in unreachable states, testing them for unreachability, and adding them as never patterns if they past the test.

---

[4] It is not clear from [9], whether performance was the chief factor in not choosing a higher bound.

7   Conclusions

We have specified, analyzed and compared different versions of the IBM CCA API protocols. This is to the best of our knowledge the first application of a general-purpose unbounded session cryptographic protocol analysis tool that directly models the properties of XOR to XOR-based cryptographic APIs. We have identified the bottlenecks and performance issues and have outlined plans for handling them. Finally, we have introduced the notion of *completeness-* and *attack-preserving* never patterns as a new means of controlling the size of the search space, and have outlined plans for automating their use.

References

1. Abadi, M., Blanchet, B., Fournet, C.: Just fast keying in the pi calculus. ACM Trans. Inf. Syst. Secur. 10(3) (2007)
2. Blanchet, B.: An Efficient Cryptographic Protocol Verifier Based on Prolog Rules. In: 14th IEEE Computer Security Foundations Workshop (CSFW 2014), Cape Breton, Nova Scotia, Canada, June 2001, pp. 82–96. IEEE Computer Society (2014)
3. Bond, M.: Attacks on cryptoprocessor transaction sets. In: Koç, Ç.K., Naccache, D., Paar, C. (eds.) CHES 2001. LNCS, vol. 2162, pp. 220–234. Springer, Heidelberg (2001)
4. Butler, F., Cervesato, I., Jaggard, A.D., Scedrov, A.: A formal analysis of some properties of kerberos 5 using msr. In: CSFW, pp. 175–1790. IEEE Computer Society (2002)
5. Cachin, C., Chandran, N.: A secure cryptographic token interface. In: Proceedings of the 22nd IEEE Computer Security Foundations Symposium, CSF 2009, Port Jefferson, New York, USA, July 8-10, pp. 141–153 (2009)
6. Chevalier, Y., Küsters, R., Rusinowitch, M., Turuani, M.: An NP decision procedure for protocol insecurity with XOR. In: 18th Annual IEEE Symposium on Logic in Computer Science, LICS 2003 (2003)
7. Comon-Lundh, H., Shmatikov, V.: Intruder deductions, constraint solving and insecurity decision in presence of exclusive-or. In: 18th Annual IEEE Symposium on Logic in Computer Science (LICS 2003), pp. 271–280 (2003)
8. Comon-Lundh, H., Cortier, V.: New decidability results for fragments of first-order logic and application to cryptographic protocols. In: Nieuwenhuis, R. (ed.) RTA 2003. LNCS, vol. 2706, pp. 148–164. Springer, Heidelberg (2003)
9. Cortier, V., Keighren, G., Steel, G.: Automatic analysis of the aecurity of XOR-based key management schemes. In: Grumberg, O., Huth, M. (eds.) TACAS 2007. LNCS, vol. 4424, pp. 538–552. Springer, Heidelberg (2007)
10. Cortier, V., Steel, G.: A generic security API for symmetric key management on cryptographic devices. In: Backes, M., Ning, P. (eds.) ESORICS 2009. LNCS, vol. 5789, pp. 605–620. Springer, Heidelberg (2009)
11. Erbatur, S., et al.: Effective Symbolic Protocol Analysis via Equational Irreducibility Conditions. In: Foresti, S., Yung, M., Martinelli, F. (eds.) ESORICS 2012. LNCS, vol. 7459, pp. 73–90. Springer, Heidelberg (2012)

12. Escobar, S., Meadows, C., Meseguer, J.: Maude-NPA: Cryptographic Protocol Analysis Modulo Equational Properties. In: Aldini, A., Barthe, G., Gorrieri, R. (eds.) FOSAD 2007/2008/2009. LNCS, vol. 5705, pp. 1–50. Springer, Heidelberg (2007)
13. Escobar, S., Meadows, C., Meseguer, J., Santiago, S.: Sequential Protocol Composition in Maude-NPA. In: Gritzalis, D., Preneel, B., Theoharidou, M. (eds.) ESORICS 2010. LNCS, vol. 6345, pp. 303–318. Springer, Heidelberg (2010)
14. Thayer Fabrega, F.J., Herzog, J., Guttman, J.: Strand Spaces: What Makes a Security Protocol Correct? Journal of Computer Security 7, 191–230 (1999)
15. González-Burgueño, A.: Protocol Analysis Modulo Exclusive-Or Theories: A Case study in Maude-NPA. Master's thesis, Universitat Politècnica de València (March 2014), https://angonbur.webs.upv.es/Previous_work/Master_Thesis.pdf
16. IBM. Comment on Mike's Bond paper A Chosen Key Difference Attack on Control Vectors (2001),
    http://www.cl.cam.ac.uk/~mkb23/research/CVDif-Response.pdf
17. IBM. CCA basic services reference and guide: CCA basic services reference and guide for the IBM 4758 PCI and IBM 4764 (2001),
    http://www-03.ibm.com/security/cryptocards/pdfs/bs327.pdf.2008
18. Keighren, G.: Model Checking IBM's Common Cryptographic Architecture API. Technical Report 862, University of Edinburgh (October 2006)
19. Kemmerer, R.A.: Using formal verification techniques to analyze encryption protocols. In: IEEE Symposium on Security and Privacy, pp. 134–139. IEEE Computer Society (1987)
20. Küsters, R., Truderung, T.: Reducing protocol analysis with xor to the xor-free case in the horn theory based approach. J. Autom. Reasoning 46(3-4), 325–352 (2011)
21. Linn, J.: Generic security service application program interface version 2, update 1. IETF RFC 2743 (2000), https://datatracker.ietf.org/doc/rfc2743
22. Longley, D., Rigby, S.: An automatic search for security flaws in key management schemes. Computers & Security 11(1), 75–89 (1992)
23. Meadows, C.: Applying formal methods to the analysis of a key management protocol. Journal of Computer Security 1(1) (1992)
24. Meadows, C.: The NRL protocol analyzer: An overview. Journal of Logic Programming 26(2), 113–131 (1996)
25. Meadows, C., Cervesato, I., Syverson, P.: Specification and Analysis of the Group Domain of Interpretation Protocol using NPATRL and the NRL Protocol Analyzer. Journal of Computer Security 12(6), 893–932 (2004)
26. Meadows, C.: Analysis of the internet key exchange protocol using the nrl protocol analyzer. In: IEEE Symposium on Security and Privacy, pp. 216–231. IEEE Computer Society (1999)
27. Meier, S., Schmidt, B., Cremers, C., Basin, D.: The TAMARIN prover for the symbolic snalysis of security protocols. In: Sharygina, N., Veith, H. (eds.) CAV 2013. LNCS, vol. 8044, pp. 696–701. Springer, Heidelberg (2013)
28. Mukhamedov, A., Gordon, A.D., Ryan, M.: Towards a verified reference implementation of a trusted platform module. In: Christianson, B., Malcolm, J.A., Matyáš, V., Roe, M. (eds.) Security Protocols 2009. LNCS, vol. 7028, pp. 69–81. Springer, Heidelberg (2013)
29. National Institute of Standards and Technology. FIPS PUB 46-3: Data Encryption Standard (DES), supersedes FIPS 46-2 (October 1999)

30. Nieuwenhuis, R. (ed.): CADE 2005. LNCS (LNAI), vol. 3632. Springer, Heidelberg (2005)
31. Steel, G.: Deduction with xor constraints in security api modelling. In: Nieuwenhuis (ed.) [30], pp. 322–336
32. Verma, K.N., Seidl, H., Schwentick, T.: On the complexity of equational horn clauses. In: Nieuwenhuis (ed.) [30], pp. 337–352

Robustness Modelling and Verification of a Mix Net Protocol

Efstathios Stathakidis[1], Steve Schneider[1], and James Heather[2]

[1] Computing Department, University of Surrey, Guildford, UK
{e.stathakidis,s.schneider}@surrey.ac.uk
[2] Chiastic Security Ltd, Guildford, UK
james@chiastic-security.co.uk

Abstract. Re-encryption Mix Nets are used to provide anonymity by passing encrypted messages through a collection of servers which each permute and re-encrypt messages. They are used in secure electronic voting protocols because they provide a combination of anonymity and verifiability. The use of several peers also provides for robustness, since a Mix Net can run even in the presence of a minority of dishonest or incorrectly behaving peers. However, in practice the protocols for peers to decide when to exclude a peer are complex distributed algorithms, and it is non-trivial to gain confidence that the Mix Net will be robust and live in the presence of faulty or malicious peers. In this paper we model and analyse the algorithm used by *Ximix*, a particular Mix Net implementation, using the CSP process algebra and the FDR model checker. We model and analyse the protocol in the presence of a realistic intruder based on Roscoe and Goldsmith's perfect *Spy* [1]. We show that in the current implementation the protocol does not satisfy the robustness requirement. Finally, we propose a method of making it robust, and verify in FDR that the proposed solution is sound and provides this robustness. Along the way, we highlight the omissions and deviations from the original RPC proposal; Mix Net protocols are extremely fragile, and small and seemingly benign changes may result in security flaws. Our experimental results show that, with our modification, Ximix guarantees termination and produces a correct output in the presence of an intruder who can corrupt a minority of mix servers.

Keywords: Mix Nets, formal methods, model-checking, CSP, FDR.

1 Introduction

Since ancient times, elections have been the most important aspect in ensuring democracy. A voting system should provide voters with the assurance that her vote has been cast as intended and included in the tally without being modified, whilst guaranteeing the secrecy of the vote. Recent proposals for secure electronic voting aim to provide *end-to-end verifiability* using cryptographic techniques, and can use *anonymising Mix Nets* to provide secrecy of the ballot (by anonymising which voter has cast any particular vote) while also providing the

L. Chen and C. Mitchell (Eds.): SSR 2014, LNCS 8893, pp. 131–150, 2014.

assurance that the votes have been decrypted correctly. The upcoming elections in the State of Victoria, Australia, will be the world's first large-scale political elections where a verifiable electronic voting system will be used. A key component in achieving this is the *Mix Net* and, of course, this should be robust and produce its required output. However, this key liveness property is generally not analysed in the literature, and it is one that the Victorian system's Mix Net (Ximix) is required to provide.

A *Mix Net* is a cryptographic protocol, which *unlinks* the correspondence between its input vector of encrypted values and the permuted vector of decrypted values given as output, thus providing anonymity to the communicating entities. The first Mix Net was introduced by Chaum [2] for constructing anonymous mail systems. In its general construction, a Mix Net consists of a sequence of servers $M_1 \ldots M_n$, also called mix servers, that collectively execute a protocol. Based on the way the mix servers operate on the input ciphertexts, Mix Nets are classified as *decryption* and *re-encryption* Mix Nets. However, most of those proposed in the literature fall into the second category. We briefly explain how a re-encryption Mix Net works; for more details about decryption and re-encryption Mix Nets, we refer the reader to [2, 3].

The first re-encryption Mix Net was introduced by Park *et. al* [4]. In this type of Mix Net, a joint public-key is generated by combining the public-keys of the mix servers. The inputs are encrypted under the joint public-key and then submitted to the Mix Net. Each mix server, in turn, re-encrypts its inputs, shuffles them using its own secret permutation and fresh randomnesses and then posts them onto a publicly accessible Web Bulletin Board ($\mathcal{WBB}$). Once all the mix servers have finished their mixing, the decryption phase starts, where the final list of ciphertexts is decrypted in a distributed manner to achieve robustness and then posted on the $\mathcal{WBB}$ for public verification. To ensure correctness of the execution, each mix server produces a zero-knowledge proof, which is posted on the $\mathcal{WBB}$ alongside the mixed and decrypted messages.

Owing to their importance in providing anonymity to the communicating parties, Mix Nets play a significant role in building systems where security requirements, such as privacy, should hold. Their main application is in electronic-voting [5–8], but they have been also used in other real-life applications, such as: electronic cash payments, RFID tags and anonymous Web browsing. In electronic voting schemes, Mix Nets are used to ensure that no one can track and reveal a voter's vote, thereby guaranteeing the privacy of the vote and the anonymity of the voter. However, this is not always enough; a well constructed Mix Net should fulfill a number of security and safety requirements, such as robustness, correctness and public verifiability. A Mix Net is called *robust* if it terminates and produces a proof of the correctness of the operation in the presence of (a limited number of) faulty or malicious mix servers. *Correctness* guarantees that the output is, indeed, a valid permutation of the input ciphertexts. Additionally, it is crucial for misbehaviour to be detectable by anyone who is interested in checking the correctness of the execution, a property called *public verifiability*.

In this work we are interested in the liveness properties that a Mix Net should meet; a Mix Net which does not produce any output is of no interest.

For the first time in the literature, we construct here a formal model and present an automated verification of the Ximix[1] Mix Net, which will be used in the real large-scale elections in Victoria State, Australia, in November 2014. The Victorian Electoral Commission's (VEC) vVote voting system, which uses Ximix, is based on the Prêt à Voter [6] voting scheme and how it works is presented in [7, 8]. It is a requirement that Ximix be robust and be able to produce a correct result provided that a threshold number of mix servers are available and follow the protocol without deviating from it. Auditing in Ximix is performed according to the Randomised Partial Checking (RPC) auditing technique [9] and its source code is available at http://www.cryptoworkshop.com/ximix/doku.php.

There are numerous different Mix Net proposals existing in the literature. Each of those designs follows a different method in constructing such schemes, and it is not always clear which of the desired security requirements are met, or how best to resolve some of the implementation questions such as what to do if a mix server fails. For example, a Mix Net based on randomized partial checking for verification does not normally indicate where the random challenges come from, yet this is a subtle issue in practice and an inappropriate approach can undermine the security of the Mix Net. Standardisation of Mix Nets would address issues such as this, providing some clear direction for developers and thus confidence for users.

Once such questions are addressed, the advantages and benefits of Mix Nets being standardised will be apparent. Having a standard for Mix Nets would provide a reference point for future implementations and the properties that they provide. For example, Verificatum [10], the most advanced Mix Net implementation up to date, has been used in small-scale elections in Norway and Israel. Hence, in the case where Verificatum would become a standard, techniques like RPC verification would also become standardised and not leave up to the constructor to decide. The benefits of Mix Nets having a standard is also apparent from the subject matter of this paper: the approach taken to the development of Ximix, the Mix Net under analysis, was to combine results from a number of different research papers, and to provide additional implementation detail, resulting in a system which required further formal analysis. Its correctness relies entirely on the programmers and the way they interpret and implement the proposed techniques. In the presence of a Mix net standard, we might hope that these opacities could have been avoided.

2   Preliminaries

Here, for convenience, names, sets, data and functions used throughout the model are detailed. The set of all mix servers is denoted by $\mathcal{P}$ and is defined to be $\mathcal{H} \cup \mathcal{D}$,

---

[1] The modelling has been performed on the snapshot of the source code taken on 1st April 2014. The authors of this paper have checked the published code and can confirm that any changes since then have no effect in this work.

where $\mathcal{H}$ (resp. $\mathcal{D}$) denotes the set of all honest (resp. dishonest) mix servers. By $\mathbb{P}\mathcal{A}$ we denote the powerset function as applied to a set $\mathcal{A}$. By $\mathbf{c}$, we denote the unmixed vector of inputs and by $M_j(m)$, we denote the vector m mixed twice using mix server j's secret permutation values, where m is either the unmixed vector $\mathbf{c}$ or some received signed mixed vector thereof. The set of all commitments to the secret permutation values is denoted by $\mathcal{C}$. When sending to a mix server, these commitments are individually hashed, thereby creating then a hash value, here $h(commit)$. Messages have the form $\mathsf{S}_j(sk_j, M_j(m))$, where sk_j is the jth mix server's secret key. The set of all messages that can be feasibly sent and received in a protocol run is denoted by $\mathcal{M}$. For example, consider the channel name $comm$ of type $\mathcal{P} \times \mathbb{P}\mathcal{P} \times \mathcal{M} \times \mathcal{C}$. In this scenario, $comm.A.\{B, C\}.\mathsf{S}_A(sk_A, M_A(m)).\mathsf{S}_A(sk_A, h(commit))$ may be an event indicating that a vector of messages, m, has been sent by the mix server A to $\{B, C\}$. Additionally, the length of a message m is denoted by $\#m$ and is calculated by counting the layers of signatures. The outer signatory of a message is verified using the corresponding public key and we use a function $outer(m)$ to return the signatory of a given message m. Similarly, the function $seq(m)$ returns the outer mix sequence of m. The function $val(id)$ returns the index of a given mix server in the shuffle plan. The function $prev(id)$ returns the identity of the preceding mix server. The set of all possible partially decrypted mixed messages, $\mathcal{L}$, is taken to be $\{\mathsf{P}_j(pt_j, seq(m), zkp) \mid m \in \mathcal{M}\}$, where pt_j is the jth mix server's share of the distributed secret key and zkp is the associated zero-knowledge proof, thus proving the correctness of the partial decryption. Finally, the set of all possible fully decrypted messages, $\mathcal{O}$, is considered to be $\{dec(m) \mid m \in \mathcal{L}\}$, where $dec(m)$ is a function that decrypts a permuted vector of ciphertexts into a permuted vector of plaintexts.

3   Ximix Mix Net Outline

Ximix is an Elliptic Curve El Gamal [11] based re-encryption Mix Net written in Java, where the main idea behind its design follows the RPC auditing technique. It consists of a number of execution phases: (i) *initialisation*; (ii) *mixing*; (iii) *checking* and; (iv) *decryption*. We will analyse each of these separately. There are two more important components of Ximix:

1. the Command Service;
2. a Transient Board for each mix server.

The *Command Service* ($\mathcal{CS}$) is the central trusted component of the system, which is responsible for talking to the mix servers and instructing them to mix, transmit their output list of ciphertexts to another mix server and to create *Transient Boards* ($\mathcal{TB}$) to host their produced data. The $\mathcal{CS}$, which is under the control of the VEC, has a great deal of power, controls the data flow, controls the whole process and specifies the execution *plan* (also called *shuffle plan*) used to emulate the RPC pairing of mix servers. In a setting with four mix servers A, B, C and D, the shuffle plan looks like $\langle AA, BB, CC, DD \rangle$. Robustness in the Ximix implementation relies heavily on the CS, as it is a single point of trust and a single point of failure. Any misbehaviour could potentially lead to the Mix Net's crash, thus violating the robustness requirement. Additionally, in Ximix, the mix servers exist to provide services, including board hosting. That is, as the shuffle plan progresses, a new TB is created on the appropriate mix server (as specified in the shuffle plan) to host the intermediate and output data (shuffled messages and commitments to the secret permutation). The *primary* mix server is responsible for sending the unmixed data to the first mix server in the shuffle plan and it is also responsible for maintaining a *Visible Board* (VB), which differs from its internal TB in that all the mix servers have read access to this. At the end of the process, the contents of the last mix server's TB, as well as all the partially decrypted messages, are posted to the VB. How these components interact with each other, in the case where all faithfully follow the protocol, is illustrated (for the case of two mix servers) in Figure 1. One can note that actions internal to the mix servers and the CS, such as receiving a message, mixing and then posting it onto the corresponding TB have been abstracted away and shown as self messages. Additionally, for clarity, all the instructions the CS can send and the mix servers are willing to accept, have been presented as *instruct* events.

4   Ximix Message Communication Diagram

For clarity, the diagram of Figure 1 illustrates how the data flows in a faithful run of Ximix with two mix servers.

Initialisation Phase. When the execution starts, the CS selects the shuffle plan and chooses the primary mix server. The original unmixed data are handed to the Mix Net by the VEC authorities and the CS instructs the primary mix server to transmit them to the head of the shuffle plan. The mixing phase now begins.

Mixing Phase. The mixing phase starts with the first mix server in the shuffle plan re-encrypting and shuffling its input data using fresh randomness and secret permutation values. As the shuffle progresses, each mix server is asked by the CS to create a TB to store its own shuffled messages and commitments to secret permutations. Once the data have been stored on the TB, the CS instructs the mix server that is currently mixing to transmit them to the next mix server in the plan. When the mixing phase is complete, only the shuffled data of the last transient board in the plan are copied onto the VB. Currently, a significant omission is that neither the data transmitted between the mix servers nor the final sequence of mixes is posted anywhere for public verification. Consequently, Ximix does not currently provide universal verifiability.

Checking Phase. The Mix Net uses static verification, i.e. each mix server maintains separate sets of commitments to the secret permutation values. At the end of the mixing phase, these commitments are downloaded by the CS and checked

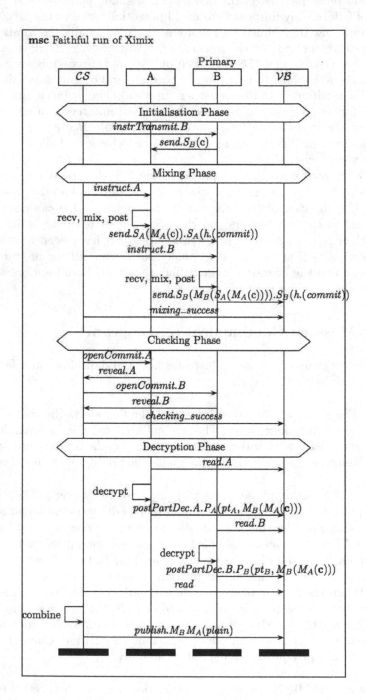

Fig. 1. Ximix protocol with two servers

for consistency. No attempt to verify anything is made during the actual mixing phase (verification *not in-phase* with mixing). This stands in contrast to the approach taken by Verificatum, where checking is *in-phase* with mixing and each mix server checks the received data before proceeding to the mixing. In Ximix, the checking is done in the order of the mixing, but after all commitments have been downloaded. The CS issues periodic RPC challenges as it goes, and up to half the input-output relationships are revealed for each mix server. This is done using an interactive zero-knowledge protocol [12] between the CS, but the produced proof convinces only the verifier (CS), which interacts with the prover (mix servers). As a consequence, the CS cannot prove to a third party the correctness of the execution, even if all the mix servers have faithfully followed the protocol. If cheating is found, the whole process stops and an operator's intervention is required; the corrupt mix server is replaced and the whole process restarts. Otherwise, the execution proceeds with the decryption phase.

Decryption Phase. Assuming checking success, the data from the VB are downloaded by the CS and decrypted, by a quorum (threshold number) of mix servers, using El Gamal threshold decryption, whereby each is asked to provide a partial decryption of the final list of encrypted data. In Ximix, mix servers act as decryption servers, in that they partially decrypt what has been output by the last mix server, but they send their partial decryptions only to the CS. Once the CS has received a threshold number of partial decryptions for each message, it assembles them into the plaintext messages and publishes the fully decrypted message(s). As part of the decryption process, periodic challenges are issued against the partial decryptions, using an interactive zero-knowledge proof protocol (IZKP) between the CS and the mix servers.

5 Modelling and Formal Analysis in CSP

In the preceding section, Ximix was described based on how it operates and processes the input messages. In this section, we present the processes modelling the individual components of Ximix, and how these are composed into models to be checked for robustness. Before proceeding to the modelling, it is pertinent to discuss the modelling decisions and assumptions we will make.

Modelling Decisions and Assumptions. We follow the typical approach to modelling of security properties, treating cryptographic primitives, such as digital signatures, encryption and decryption, as *symbolic operations* with the appropriate algebraic properties. Each component is modelled as an individual process; the communications between them are *synchronous* and over *authenticated* channels. We assume that the *checking* phase will always succeed. No participant wishes to be expelled from the protocol, and for this reason they provide the checker with valid commitments to secret permutations when asked to reveal a subset of them.

Faithful Model of Ximix. Ximix will be used under the strong assumption that all its components faithfully follow the protocol without deviating from it. Additionally, VEC can tolerate failure of one mix server, which provides us with our *threshold* value. Before the execution starts, the shuffle plan is fed to the mix servers so they know their position and the neighbouring mix servers. The complete script for this model can be found at http://www.tvsproject.org/csp/Ximix_faithful.csp.

Honest mix servers. An honest mix server waits for an instruction, sent by the CS, to be received and then starts to operate. It is willing to receive any structurally correct message signed by the previous mix server in the plan and then posts the received message on its TB. Obviously, the head of the plan is willing to receive the unmixed data signed by the primary mix server. This is represented in HON_SRVR by the external choice over the set of the messages signed by the sender and the set of the hashed commitments. Then, CS instructs the first mix server to mix the received data and the execution proceeds to the mixing phase.

$$HON\_SRVR(id) =$$
$$\text{if } id == primaryServer \text{ then}$$
$$\quad \underset{\substack{m\in\{m'\in\mathcal{M}\mid\#m'=0\},\\ commit\in C}}{\Box} \left(\begin{array}{l} send.S_{id}(sk_{id}, m).S_{id}(sk_{id}, h(commit)) \rightarrow \\ MIX(id, 1, m) \end{array} \right)$$
$$\text{else}$$
$$\quad CS\_instructsToCopy.id \rightarrow$$
$$\quad \underset{\substack{m\in\{m'\in\mathcal{M}\mid outer(m')=prev(id)\},\\ commit\in C}}{\Box} \left(\begin{array}{l} recv.m.commit \rightarrow \\ postInData!id!m!commit \rightarrow \\ CS\_instructsToShuffle.id \rightarrow \\ MIX(id, val(id), m) \end{array} \right)$$

The mix server mixes the received data twice (in order to emulate the RPC pairing) and the CS then instructs it to create a TB to store the output data for future verification. Once this has been done, the CS requests the mix server to transmit only the output message and the commitments to the secret permutations to the next mix server in the shuffle plan. This is modelled in the MIX process below. Here, the *toBeMixed* value indicates the message received from the previous mix server, which will be mixed twice by the current one. The *rnd* symbolises the round number of the mixing; a mix server proceeds to the mixing phase only when this value equals its own number in the sequence.

Command Service. The CS sends instructions to the mix servers and announces the phase's success or failure. The instructions are sent sequentially and at the end of the execution of each phase and a *done* event indicates its successful completion.

$$CS(\langle\rangle) = mixing\_done \rightarrow STOP$$
$$CS(\langle id \rangle \frown ids) =$$
$$\left(\begin{array}{l} CS\_instructsToCopy.id \rightarrow \\ CS\_instructsToShuffle.id \rightarrow \\ CS\_instructsToCreateTB.id \rightarrow \\ CS(ids) \end{array} \right)$$

$$MIX(id, rnd, toBeMixed) =$$
$$\left(\begin{array}{l} CS\_instructsToCreateTB.id \rightarrow \\ postInterData!id!interData!interCommit \rightarrow \\ postOutData!id!outData!finalCommit \rightarrow \\ send.S_{id}(sk_{id}, outData).S_{id}(sk_{id}, h(finalCommit)) \rightarrow \\ SKIP \end{array} \right)$$

Transient and Visible Board. As we have described earlier, each mix server maintains a $\mathcal{TB}$, which always allows *post* and *read* requests from the owner mix server. The other mix servers are blind on what has been posted to the other $\mathcal{TB}$s and when the execution starts, they are both empty.

$$TB(id, inData, interData, outData, interCommit, outCommit) =$$
$$\square_{\substack{inD,interD,outD \in \mathcal{M} \\ inC,outC \in \mathcal{C}}} \left(\begin{array}{l} postInData!id.inD.inC \rightarrow \\ postInterData!id.interD.interC \rightarrow \\ postOutData!id.outD.outC \rightarrow \\ TB(id, inData \cup inD, interD, outD, interC, outCommit \cup inC) \end{array} \right)$$

The $\mathcal{VB}$ is visible to all mix servers as well as to the $\mathcal{CS}$. The mix servers *read* what has been posted there and *post* their partial decryptions. The $\mathcal{CS}$ reads all the partially decrypted messages, combines them and *outputs* (posts) the fully decrypted messages on the $\mathcal{VB}$. (For brevity, the partial decryption and the combination processes are run in the $\mathcal{CS}$'s side and are not presented here; we refer the reader to the full script.)

$$VB(primaryServer, outData, outCommit, partDec) =$$
$$\square_{\substack{id \in \mathcal{P}, outD \in \mathcal{M}, \\ outC \in \mathcal{C}, newPartDec \in \mathcal{L}}} \left(\begin{array}{l} postDataToVB.id!outD!outC \rightarrow VB(\dots, outD, outC, \dots) \\ \square \; readFromVBData.id!outData \rightarrow VB(\dots) \\ \square \; postPartDec.id!newPartDec \rightarrow VB(\dots, partDec \cup newPartDec) \\ \square \; CSreadFromVBPartDec!partDec \rightarrow VB(\dots) \end{array} \right)$$

Putting the Network Together. Based on the modelling assumptions presented in Section 5, and connecting the channels of the various processes so they can synchronise, we can produce our final $SYSTEM_{Ximix}$ process, which is defined in terms of the parallel composition of all the following processes, synchronised on their common events. The $MIXING$ process is defined as the parallel composition of the HON_SRVR and the TB processes.

$$SYSTEM_{Ximix} = MIXING \parallel CHECKING \parallel DECRYPT \parallel COMBINE \parallel VB(\dots) \parallel CS(ShufflePlan)$$

We should now verify whether our system satisfies the robustness requirement or not and to check this, we need to determine whether a fully decrypted message is always output. In [13], the output of the system was modelled as a synchronous *agree* event among the majority of the mix servers, whilst here, the $\mathcal{CS}$ is responsible for combining all the partial decryptions and publishing the fully decrypted message(s). No consensus among the mix servers is provided for. For this purpose, we create the specification process $RBST$, which always performs an *output* event, and use the Failures/Divergences refinement model ($\mathcal{FD}$), to check in FDR that the following assertion holds:

$$RBST = output \rightarrow RBST, \quad RBST \sqsubseteq_{FD} (SYSTEM_{Ximix} \setminus \Sigma \backslash \{| \; publish \; |\}) [\![^{output}/_{publish.m} \; |_{m \in \mathcal{O}}]\!]$$

To perform a rigorous analysis of the system, we must include an intruder model. To this end, in the next section, we introduce our threat model and place it in parallel with the $SYSTEM_{Ximix}$ process defined here.

6   Adapting the Intruder

In the previous section, we modelled and verified Ximix under the assumption that all the components are honest. However, this is a strong assumption; a system consisting of honest participants is of little interest. In this section, we use Roscoe and Goldsmith's perfect *Spy* [1, 14] as the basis of our threat model and investigate whether Ximix still meets the robustness requirement. The description of the *Spy* presented in the first half of this section it will be used as well in the second, subject to some minor modifications, in order to accommodate the behaviour of the proposed scheme. However, our Spy is not as powerful as in the original Roscoe and Goldsmith version, which is in complete control of the whole network. Obviously, that would be pointless in the current case, as it would clearly violate robustness.

Here, the *Spy* plays the role of a mix server that can receive ingoing messages over a *learn* channel, *infer* events based on received messages and its initial knowledge and then *say* messages that it has inferred. This intruder model provides active attacks against the system, by blocking outgoing messages and sending those that deviate from the protocol. The Spy is constructed using the same approach as taken by Roscoe and Goldsmith, with respect to the messages of the Ximix system.

The initial knowledge of the intruder consists of all the mix servers' identities and public-keys, the initial unmixed vector of values and, of course, its own secret key as well as the assigned share of the secret key.

Apart from the channels *learn*, *say* and *infer*, that allow messages to be received, sent and internally inferred, the *Spy* has in its alphabet all the events that an honest mix server can perform, so it can communicate with the other components. Furthermore, the *Spy* can ignore the instructions sent by the CS by *absorbing* them and carrying on its operation.

In this model of Ximix, the intruder *learns* messages sent to him only from the previous mix server and *says* to the next one in the shuffle plan. In the case where he is the last mix server in the plan, an additional *say* event is renamed to *postDataToVB* and can post any message that has been learnt. In the same fashion, a *say* event is renamed to *postPartDec*, which means that he can post partial decryptions of messages that he has received and, of course, these are of length larger than or equal to the threshold. In the next section, we shall show that the intruder *hears* all the messages sent to any of the honest mix servers and he can choose to send different messages to different mix servers, so each message can be potentially be sent to an individual mix server, that is, to the singleton $\{x\}$, where x is the identity of some honest mix server. The new $SYSTEM_{Spy}$ process is now defined in terms of the parallel composition of the $SYSTEM_{Ximix}$ and the *renSpy* processes, as:

$$SYSTEM_{Spy} = SYSTEM_{Ximix}\ _{\alpha SYSTEM_{Ximix}}\|_{\alpha renSpy}\ renSpy$$

where, $\alpha SYSTEM_{Ximix}$ and $\alpha renSpy$, are the alphabets containing all the events these processes can engage in. Using the assertion presented in Section 5, we check that $SYSTEM_{Spy}$ does not satisfy its liveness property and in Section 7 we present some traces illustrating this behaviour.

The *Spy* can also mount a Denial of Service (*DoS*) attack, by perpetually posting messages to the $\mathcal{VB}$, thus leading the system to an unstable condition (divergence, in CSP terms) and one defence against this is to add an extra constraint that allows only one post (per event) for each mix server. This is modelled as:

$$DIVERGE = VB(primaryServer, empty, empty, empty) \,\big|\big|\big|\,(post.x?\_ \rightarrow STOP \mid x \in \mathcal{D})$$

where *post* is an abbreviation for the *postDataToVB* and *postPartDec* events the intruder can perform and "_" allows any message of the appropriate type.

A misbehaving $\mathcal{CS}$ can also break robustness. For example, it can refuse to send instructions to a specific mix server, making it wait indefinitely, and causing deadlock. We describe this attack in Section 7. Additionally, upon receiving a threshold number of partially decrypted data, it can either refuse to combine them or output an unrelated message.

So far, we have seen that the proposed Ximix is not resilient in the current implementation, for it is vulnerable to attacks carried out by an intruder and a dishonest $\mathcal{CS}$. FDR confirms that the *RBST* assertion specified in Section 5 does not hold and the complete script for this model can be found at http://www.tvsproject.org/csp/Ximix_Spy.csp. In the following section we show how to make the system robust.

Robust Ximix. Here, we describe the changes required to guarantee successful termination in the presence of the intruder introduced above. In our modified Ximix, upon termination, a valid and fully decrypted message is published and any external party interested in verifying its correctness can do so.

One of the purposes of a Mix Net is to *distribute the trust* among the mix servers, so that the whole system does not rely on the integrity of a single component. However, Ximix relies critically on the availability and honesty of the $\mathcal{CS}$. Hence, the first step in making Ximix robust is to remove the $\mathcal{CS}$ and instead, allow the mix servers to broadcast their messages. In this context, an honest mix server sends the same message to all the mix servers, while a dishonest one may send different messages to different mix servers, or refuse to send to some of them. We also allow any external party interested in checking and combining the partial decryptions to do so, in order to check the fully decrypted messages. A few changes to the previous model are needed to accommodate these decisions, which we describe below and the complete script can be found at http://www.tvsproject.org/csp/Ximix_robust.csp.

Honest mix servers. An honest mix server is willing to receive messages from another mix server; he can, however, *timeout* before or after receiving it. Adding this behaviour in the system, we allow the execution to continue even when a dishonest mix server refuses to send messages to some of the honest ones (simply times out without performing any action). In this case, the honest mix servers will absorb the timeout and continue to operate. At the end of the mixing phase, the honest mix servers will have posted on their $\mathcal{TB}$s at least one message of threshold length, mixed by the majority of the mix servers. The requisite changes in the HON_SRVR and $renHON\_SRVR$ processes are shown below. Additionally, an extra $timeout$ is added after the $send$ event in the MIX process presented in Section 5.

$$HON\_SRVR(id) =$$
$$\text{if } id == primaryServer \text{ then}$$
$$\square \quad \begin{pmatrix} send.S_{id}(sk_{id}, m).S_{id}(sk_{id}, h(commit)) \to \\ timeout \to \\ MIX(id, 1, m) \end{pmatrix}$$
$$\quad {}_{m \in \{m' \in \mathcal{M} | \#m' = 0\},}_{commit \in \mathcal{C}}$$
$$\text{else}$$
$$timeout \to MIX(id, val(id), m)$$
$$\square \quad \square \quad \begin{pmatrix} recv.m.commit \to \\ postInData!id!m!commit \to \\ timeout \to \\ MIX(id, val(id), m) \end{pmatrix}$$
$$\quad {}_{m \in \{m' \in \mathcal{M} | outer(m') = prev(id)\},}_{commit \in \mathcal{C}}$$

The (non-replicated) external choice in the HON_SRVR process models the ability of an honest mix server to timeout before receiving a message. This will only occur where a dishonest mix server refuses to send anything. Minor changes are required when renaming the HON_SRVR process:

$$renHON\_SRVR(id) = HON\_SRVR(id) [\![{}^{comm.id.\mathcal{P}\backslash\{id\}.m.c} /_{send.m.c} | m \in \mathcal{M}, c \in \mathcal{C}]\!]$$
$$[\![{}^{comm.\mathcal{P}\backslash\{id\}.z.m.c} /_{recv.m.c} | m \in \mathcal{M}, c \in \mathcal{C}, z \in \mathcal{P}]\!]$$

Here, each $send$ is renamed to a $comm.id.\mathcal{P}\backslash\{id\}$, broadcasting a message to all mix servers other than id. Each $recv$ is renamed to an incoming $comm.\mathcal{P}\backslash\{id\}.z$, where z can be either a singleton containing only id or a set of some mix servers.

Dishonest mix servers. A dishonest mix server learns all messages sent to any of the honest mix servers, so *learn* events are renamed to incoming $comm.x.\mathcal{P}\backslash\{x\}$, where x is an honest mix server. The renaming of *say* events is slightly more complicated in that we allow the *Spy* to send messages only to individual servers (singleton sets). This simplifies the modelling, but without loss of generality: if he wants to send different messages to different mix servers then he can now do so, and if he wishes to send the same message to all mix servers, he can send it to each of them separately.

$$renSpy = Spy [\![{}^{comm.x.\mathcal{P}\backslash\{x\}.m.c} /_{learn.m.c} | m \in \mathcal{M}, x \in \mathcal{H}, c \in \mathcal{C}]\!]$$
$$[\![{}^{comm.y.\{x\}.m.c} /_{say.m.c} | m \in \mathcal{M}, x \in \mathcal{H}, y \in \mathcal{D}, c \in \mathcal{C}]\!]$$
$$[\![{}^{postDataToVB.y.m.c} /_{say.m.c} | m \in \mathcal{M}, y \in \mathcal{D}, c \in \mathcal{C}]\!]$$
$$[\![{}^{postPartDec.y.m} /_{say.m.c} | m \in \mathcal{M}, y \in \mathcal{D}, c \in \mathcal{C}, \#m >= threhold]\!]$$

Putting the Network Together. Our final $SYSTEM_{XimixNoCS}$ process is now defined in terms of the parallel composition of the new $renSpy$ and the processes defined in the earlier sections, as:

$$SYSTEM_{XimixNoCS} = MIXING \parallel CHECKING \parallel DECRYPT \parallel COMBINE \parallel DIVERGE \parallel renSpy$$

Eventually, using the $RBST$ assertion defined in Section 5, we check that the robustness of the system is maintained and the traces described in Section 7 support this argument.

7 Results and Analysis

In this section, we verify the protocols against the liveness property and we present interesting traces illustrating their behaviour. Each of the traces discussed in this section were obtained via simulating the models in ProBE, a CSP animator (built into FDR3), which allows the user to explore the behaviour of a process. Due to the number of events occurring in the traces, we keep only those highlighting the importance of the trace. For clarity, throughout this section, we give traces using three mix servers; the shuffle plan will always be $\langle AA, BB, CC \rangle$ and B the primary mix server. There are numerous possible corruption scenarios that can be modelled and analysed and we have chosen the most representative ones. We start with the $SYSTEM_{Ximix}$ process, which is a faithful model of Ximix where all the components are honest.

$\langle CS\_instructsToCopy.A,$
$comm.B.\{A\}.S_B(\mathbf{c}).S_B(h(commit)),$
$postInData.A.S_B(\mathbf{c}).S_B(h(commit)),$
$CS\_instructsToShuffle.A,$
$CS\_instructsToCreateTB.A,$
$CS\_instructsToCopy.B,$
$comm.A.\{B\}.S_A(M_A(\mathbf{c})).S_A(h(commit)),$
$...$
$comm.B.\{C\}.S_B(M_B(S_A(M_A(\mathbf{c})))).S_B(h(commit)),$
$...$
$postDataToVB.C.S_C(M_C(S_B(M_B(S_A(M_A(\mathbf{c})))))).S_C(h(commit)),$
$mixing\_done, checking\_done,$
$readFromVBData.A.S_C(M_C(S_B(M_B(S_A(M_A(\mathbf{c})))))),$
$postPartDec.A.P_A(pt_A, M_C(M_B(M_A(\mathbf{c})))),$
$...$
$decryption\_done,$
$CSreadFromVBPartDec.\{P_A(pt_A, M_C(M_B(M_A(\mathbf{c})))), ...\},$
$publish.M_C(M_B(M_A(\mathbf{p}))) \rangle$

In general, after the $decryption\_done$ event, the CS combines all the partial decryptions of length greater than or equal to the threshold with preference given to longer messages. Hence, $M_C(M_B(M_A(\mathbf{p})))$ will be considered the published output as it is of length greater than the threshold. Of course, more interesting traces result from checking the protocol under the existence of our Spy.

In $SYSTEM_{Spy}$, consider the case where the intruder is the primary mix server, B; he starts by sending the unmixed data to A. A is honest, mixes twice the original data, produces its own mixed messages and communicates them to B. Now, B decides to act dishonestly and instead of sending to C that received by A, he sends the unmixed data. At this point, C, being unable to distinguish which of the preceding mix servers misbehaved, accepts what B sent and operates on them. As C is the last in the shuffle plan, he posts a mixed message of length 1 (mixed only by C) to the VB. Each server is now instructed by the CS to read the posted message and partially decrypt it. However, none of them will post any partial decryption of the posted message because it is of length being less than the threshold. Subsequently, the CS cannot read or combine any partial decryptions and the whole process stops without publishing anything. Clearly, this violates the robustness requirement. This is demonstrated in the left trace below.

$\langle \ldots$
$comm.B.\{C\}.S_B(\mathbf{c}).S_B(h(commit)),$
$\ldots$
$postDataToVB.C.S_C(\mathsf{M}_C(\mathbf{c})).S_C(h(commit)),$
$readFromVBData.A.S_C(\mathsf{M}_C(\mathbf{c})),$
$postPartDec.A.nothing,$
$\ldots$
$CSreadFromVBPartDec.empty,$
$STOP\rangle$

$\langle \ldots$
$comm.A.\{B\}.S_A(\mathsf{M}_A(\mathbf{c})).S_A(h(commit)),$
$\ldots$
$CS\_instructsToCopy.B,$
$\ldots$
$CS\_instructsToCopy.C,$
$\times\rangle$

A similar behaviour arises when C is the spy. Upon receiving a message from B, he ignores it, mixes the initial data and posts them on the $\mathcal{VB}$. No one can now partially decrypt the posted message and the CS cannot publish anything. In another scenario, (trace above on the right), the intruder, B, sends the unmixed data to A and absorbs the instructions sent by the CS, but refuses to transmit the received messages to C, thus resulting in a deadlock. This is because C is willing to receive a message from B, but it never arrives.

However, as we have seen in Section 6, the CS can also break robustness. The empty trace $\langle \times \rangle$ illustrates the scenario in which the CS does not send a *copy* instruction to A. Although the mix server is always willing to receive and execute it, such an instruction never arrives and the $SYSTEM_{Spy}$ deadlocks without performing even one event.

In our revised system, in the trace below, the intruder, acting as dishonest mix server A, times out without sending any message. However, this does not prevent the other mix servers from continuing to operate.

$\langle comm.B.\{A,C\}.S_B(\mathbf{c}).S_B(h(commit)), timeout,$
$timeout,$
$comm.B.\{A,C\}.S_B(\mathsf{M}_B(\mathbf{c})).S_B(h(commit)), timeout,$
$\ldots$
$comm.C.\{A,B\}.S_C(\mathsf{M}_C(S_B(\mathsf{M}_B(\mathbf{c})))).S_C(h(commit)), timeout$
$\ldots$
$timeout,$
$postPartDec.B.\mathsf{P}_B(pt_B, \mathsf{M}_C(\mathsf{M}_B(\mathbf{c}))),$
$postPartDec.C.\mathsf{P}_C(pt_C, \mathsf{M}_C(\mathsf{M}_B(\mathbf{c}))),$
$readFromVBPartDec.\{\mathsf{P}_C(pt_B, \mathsf{M}_B(\mathsf{M}_B(\mathbf{c}))), \mathsf{P}_C(pt_C, \mathsf{M}_C(\mathsf{M}_B(\mathbf{c})))\},$
$publish.\mathsf{M}_C(\mathsf{M}_B(\mathbf{p}))\rangle$

Having received nothing from A, mix server B mixes the original data (which are visible to everyone on the $\mathcal{VB}$) and broadcasts them to $\{A,C\}$. Honest C operates similar to B and posts its own mixed data onto the $\mathcal{VB}$. All the mix servers proceed now to the partial decryption of the messages that they have received during the mixing phase and post them onto the $\mathcal{VB}$. The intruder can choose either to post a valid partial decryption of a message of correct length or not to post anything. Neither of these can violate the robustness of the system: enough partial decryptions of the same message (posted by the honest mix servers) appear on the $\mathcal{VB}$. Any external party can now combine them and publish the fully decrypted message.

A more interesting behaviour occurs when the intruder sends different messages to different mix servers, or does not send to some of them, in order to cause a dispute among them. More specifically, the intruder acting as dishonest mix server B, refuses to communicate the initial data to A and times out. Honest A reads the initial data from the $\mathcal{VB}$, mixes and broadcasts them to $\{B, C\}$. Now, dishonest B chooses to send different mixed messages to A and C: he sends a mix of the received messages to A and a mix of the initial data to C. At this point, C cannot work out which of A or B has misbehaved. When receiving $S_B(M_B(\mathbf{c}))$ from B, he might think that A did not mix the initial output and simply forwarded them to B (or timed out). On the other hand, he might think that B ignored message from A and that B mixed the initial data and transmitted to him. This is shown in the following trace.

$\langle timeout,$
...
$comm.A.\{B, C\}.S_A(M_A(\mathbf{c})).S_A(h(commit)), timeout,$
...
$comm.B.\{A\}.S_B(M_B(S_A(M_A(\mathbf{c})))).S_B(h(commit)),$
$comm.B.\{C\}.S_B(M_B(\mathbf{c})).S_B(h(commit)), timeout,$
...
$comm.C.\{A, B\}.S_C(M_C(S_B(M_B(\mathbf{c})))).S_C(h(commit)), timeout,$
...
$postPartDec.A.P_A(pt_A, M_C(M_B(\mathbf{c}))),$
$postPartDec.B.P_B(pt_B, M_B(M_A(\mathbf{c}))),$
$postPartDec.C.P_C(pt_C, M_C(M_B(\mathbf{c}))),$
$readFromVBPartDec.\{P_A(pt_A, M_C(M_B(\mathbf{c}))), P_B(pt_B, M_B(M_A(\mathbf{c}))), P_C(pt_C, M_C(M_B(\mathbf{c})))\},$
$publish.M_C(M_B(\mathbf{p}))\rangle$

In the partial decryption phase, dishonest B is able to post two different partially decrypted messages, both of threshold length: $P_B(pt_B, M_B(M_A(\mathbf{c})))$ and $P_B(pt_B, M_C(M_A(\mathbf{c})))$. A has seen two different messages, both of threshold length; he partially decrypts the latest to arrive. C has enough information to partially decrypt only one message and finally, the output is $M_C(M_B(\mathbf{p}))$.

All these traces describe some of the possible behaviours of the system, and might not have been appreciated without this analysis. We saw that it is easy for the intruder to break the robustness of the original protocol by choosing not to perform some specific actions. To recognise our contributions in making the system robust, the above traces show that whatever actions an intruder is willing to perform, the robustness of the protocol remains intact.

Moreover, in order to make sure that our changes for making Ximix robust are sound, we further analysed the system in the presence of a stronger intruder, who controls more than the threshold number of mix servers (that is, two). As expected, and verified in FDR, the modified Ximix is not robust in this setting, and FDR provides a trace where two dishonest servers (out of three) can prevent the Mix Net from completing.

The automated verification of the models, with three mix servers, completes in a matter of minutes in a modern laptop. However, when adding an extra mix server, the state space escalates quickly. Table 1 in Appendix B shows this problem.

8   Previous Work

In this context, the first formal model of a re-encryption Mix Net was provided by Stathakidis *et. al* [13]. In their work, they verify the robustness and privacy properties of a typical $\mathcal{WBB}$-based re-encryption Mix Net in the presence of a realistic intruder, using the FDR model-checker and show the modifications that are needed in order to make such a Mix Net robust. The Mix Net verified in their work was inspired by Verificatum [10], and it has few similarities with Ximix. For example, the partial decryption phase is absent from [13] and the mix servers check the received messages before proceeding to their mixing. As we saw in Section 5, in Ximix, the mix servers are willing to accept any *structurally correct* message and proceed to their mixing without checking their validity. Obviously, this may lead to an incorrect output. Furthermore, a different notion of a $\mathcal{WBB}$ is presented in [13], where it acts as a broadcast channel and keeps a consistent record of all messages being sent between the mix servers. In Ximix, only the messages produced by the last mix server are posted on the $\mathcal{WBB}$ and each of them maintains a local board, called a *Transient Board* ($\mathcal{TB}$). Küsters *et. al* [15], provide a formal security analysis of Chaumian RPC Mix Nets. They propose a new security definition, called *accountability*, which allows one to measure the level of privacy and verifiability of such a Mix Net precisely. Their analysis is interesting, but it is not automated.

9   Conclusion

We have described and conducted a formal modelling and verification of the Ximix Mix Net which will be used in real large-scale elections in Victoria, Australia, in November 2014. It was our aim to be explicit about all the subtleties of the protocol and to apply sufficient rigour in ensuring its robustness. We demonstrated that Ximix is not robust in the presence of an intruder, based on Roscoe and Goldsmith's perfect Spy, and described the modifications that are needed in order to satisfy this liveness requirement. In our revised Ximix, the election does not rely on the integrity of a single component, but instead distributes the trust among them. Our analysis demonstrates that Ximix guarantees completion and produces a valid output in the presence of a dishonest component. Additionally, we explained the impact on the lack of standardisation in Mix Nets and in what extent they can be standardised. Although this standardisation is difficult to be achieved in practice, it would be useful and solve issues occurring when Mix Nets are used in real world applications, such as in constructing trustworthy electronic voting systems.

Acknowledgements. This work was supported by the EPSRC project Trustworthy Voting Systems, project EP/G025797/1. We would like to thank Chris Culnane, University of Surrey, Chris Mitchell, Royal Holloway University of London, and the anonymous reviewers for their pertinent comments.

A   CSP

Communicating Sequential Processes (CSP) is a process algebra designed for describing processes that interact with each other. It was introduced by Hoare [16] in 1978 and since then it has been extensively used for applying the theory of concurrency in practice. The core of the CSP algebra is a *process*, which is described by the way it communicates with its environment. Processes proceed from one state to another by engaging in *events*. An event describes a particular action that can be performed or refused by a process and the set of all possible events is denoted by Σ. In CSP, all the communication events are instantaneous and they happen only when both the processes and the environment agree on their occurrence (*handshaken* communication). At the construction of a process in CSP, the alphabet of a process P, denoted αP, is the set of all visible communication events that this process may perform. For a detailed explanation of CSP and its associated FDR model-checker, we refer the reader to [17–19].

$STOP$ is the simplest CSP process, which does nothing and $SKIP$ is the process indicating successful termination. The process $a \rightarrow P$ is initially willing to communicate a and then behaves like P. $P \,\square\, Q$ can act either as P or Q, the choice of which is in the hands of the environment. Replicated external choice replicates the choice over the set $\mathcal{A}$, and is denoted by $\square_{x \in \mathcal{A}} P(x)$. By $P \parallel_{\mathcal{A}} Q$ we denote generalised parallel, which synchronises P and Q on events lying in the set $\mathcal{A}$. Alphabetised parallel is denoted by $P \,_{\alpha P}\|_{\alpha Q}\, Q$ and synchronises P and Q on events lying in the intersection of αP and αQ. We write $\|_{i \in I} [\alpha P] \, P(i)$ for the replicated alphabetised parallel composition of processes $P(i)$ indexed over I, where each $P(i)$ is allowed to perform events from αP and the processes are synchronised on the common events. In hiding, $P \setminus \mathcal{A}$, the internal events from $\mathcal{A}$ are hidden from the environment. In renaming, $[\![^a/_b]\!]$, the events b occurring in the process are replaced by the events a. $P \,|||\, Q$ is the interleaving process, where the processes P and Q run independently of each other without synchronising on any event.

B   State-Space Explosion Problem

Individual checks were performed for each possible instantiation of the models, with a minority of mix servers, testing the inevitability of an output being produced. For the case where three mix servers where used, six models were checked (three for $SYSTEM_{\text{Spy}}$ and three for $SYSTEM_{\text{XimixNoCS}}$). Similarly, in the case where four mix servers were participating in the process, eight checks, in total, were performed. All checks verified the inevitability of a *publish* event, regardless of the behaviour of the dishonest mix servers. All checks were performed using the refinement checker FDR3 beta 7 on a desktop with an Intel i7 Quad-Core CPU @ 3.6GHz with 8GiB of memory running 64-bit Ubuntu 12.04.

Looking at $SYSTEM_{\text{XimixNoCS}}$, one can see how the state-space grows when the intruder is the last in the shuffle plan. In this case, he *learns*, *infers* and is able to *deduce* and finally *say* many more messages due to the fact that he receives many messages broadcast from all the previous honest mix servers.

Table 1. The FDR verification results for our models of Ximix. As the state-space increases quickly with the number of mix servers, it was not possible for FDR to handle such huge states. These are denoted as "NA" in the table.

System $\mathbb{PH}, \mathbb{PD}$	$SYSTEM_{\text{Ximix}}$		$SYSTEM_{\text{Spy}}$		$SYSTEM_{\text{XimixNoCS}}$	
	RBST	States	RBST	States	RBST	States
{A,B,C}	✓	45	-	-	-	-
{A,B,C,D}	✓	56	-	-	-	-
{B,C}, {A}	-	-	×	5680	✓	289
{A,C}, {B}	-	-	×	13544	✓	2679
{A,B}, {C}	-	-	×	11882	✓	2060
{A}, {B,C}	-	-	×	57869	×	1695
{B,C,D}, {A}	-	-	NA	NA	NA	NA
{A,C,D}, {B}	-	-	NA	NA	NA	NA
{A,B,D}, {C}	-	-	NA	NA	NA	NA
{A,B,C}, {D}	-	-	NA	NA	NA	NA
{A,B}, {C, D}	-	-	NA	NA	NA	NA

However, FDR cannot handle the state-space produced when four mix servers are taking part in the process. An honest mix server is required to post onto (resp. read from) its internal $\mathcal{TB}$ all the received messages, as well as to post (resp. read) the intermediate and ouputted messages. In a similar fashion, the mix servers post their partially decrypted messages onto the $\mathcal{VB}$. Internally, the *Spy*, does not need to maintain a $\mathcal{TB}$, as he is able to say any received (or mixed by him) message. All these communications between the mix servers and the boards are computationally expensive and increase the overall state-space.

In the way the models are constructed, *COMBINE* is the most demanding process. That is, the combiner (the $\mathcal{CS}$ in $SYSTEM_{\text{Spy}}$ and any interested party in $SYSTEM_{\text{XimixNoCS}}$), is responsible for reading all the partial decryptions from the $\mathcal{VB}$, checking the associated generated zero-knowledge proofs, combining them into a fully decrypted plaintext message and publishing it. Hence, it is willing to read *all* the possible sets of partially decrypted messages and the correct way for implementing it in CSP involves the use of powersets. In the case with four mix servers, each of them is able to partially decrypt any received message of length greater than or equal to the threshold (here, three). Using the assumption that messages are in strictly increasing order, i.e. a mix server can mix only a message signed by a preceding mix server, the cardinality of the set containing all the possible partially decrypted messages is 20. Under this circumstance, the combiner is willing to read any possible combination of these 20 messages, that is, $\mathbb{P}(20) = 2^{20} \simeq 1M$, but FDR struggles when performing such calculations.

References

1. Roscoe, A.W., Goldsmith, M.: The perfect spy for model-checking crypto-protocols. In: Proceedings of DIMACS Workshop on the Design and Formal Verification of Crypto-Protocols. Rutgers University (September 1997)
2. Chaum, D.: Untraceable electronic mail, return addresses, and digital pseudonyms. Commun. ACM 24(2), 84–88 (1981)
3. Golle, P., Jakobsson, M., Juels, A., Syverson, P.F.: Universal re-encryption for mixnets. In: Okamoto, T. (ed.) CT-RSA 2004. LNCS, vol. 2964, pp. 163–178. Springer, Heidelberg (2004)
4. Park, C.-s., Itoh, K., Kurosawa, K.: Efficient Anonymous channel and all/nothing election Scheme. In: Helleseth, T. (ed.) EUROCRYPT 1993. LNCS, vol. 765, pp. 248–259. Springer, Heidelberg (1994)
5. Adida, B.: Helios: Web-based open-audit voting. In: Proceedings of the 17th USENIX Security Symposium (Security 2008) (2008)
6. Ryan, P.Y.A., Bismark, D., Heather, J., Schneider, S., Xia, Z.: Prêt à voter: a voter-verifiable voting system. IEEE Transactions on Information Forensics and Security 4(4), 662–673 (2009)
7. Burton, C., Culnane, C., Heather, J., Peacock, T., Ryan, P.Y.A., Schneider, S., Srinivasan, S., Teague, V., Wen, R., Xia, Z.: A supervised verifiable voting protocol for the victorian electoral commission. In: Kripp, M.J., Volkamer, M., Grimm, R. (eds.) Electronic Voting. LNI, vol. 205, pp. 81–94. GI (2012)
8. Burton, C., Culnane, C., Heather, J., Peacock, T., Ryan, P.Y.A., Schneider, S., Srinivasan, S., Teague, V., Wen, R., Xia, Z.: Using prêt à voter in victorian state elections. In: Proceedings of the 2012 International Conference on Electronic Voting Technology/Workshop on Trustworthy Elections, EVT/WOTE 2012, p. 1. USENIX Association, Berkeley (2012)
9. Jakobsson, M., Juels, A., Rivest, R.L.: Making mix nets robust for electronic voting by randomized partial checking. In: Boneh, D. (ed.) USENIX Security Symposium, USENIX, pp. 339–353 (2002)
10. Wikström, D.: Verificatum. Website (2014), http://www.verificatum.org/verificatum/index.htm
11. Gamal, T.E.: A public key cryptosystem and a signature scheme based on discrete logarithms. IEEE Transactions on Information Theory 31(4), 469–472 (1985)
12. Chaum, D., Pedersen, T.P.: Wallet databases with observers. In: Brickell, E.F. (ed.) CRYPTO 1992. LNCS, vol. 740, pp. 89–105. Springer, Heidelberg (1993)
13. Stathakidis, E., Williams, D.M., Heather, J.: Verifying a mix net in csp. In: Proceedings of the 13th International Workshop on Automated Verification of Critical Systems (AVoCS 2013). Electronic Communications of the EASST, vol. 66. European Association of Software Science and Technology (2013)
14. Roscoe, A.W.: The theory and practice of concurrency. Prentice Hall (1998)
15. Küsters, R., Truderung, T., Vogt, A.: Formal analysis of chaumian mix nets with randomized partial checking. IACR Cryptology ePrint Archive 2014, 341 (2014)
16. Hoare, C.A.R.: Communicating sequential processes. Commun. ACM 21(8), 666–677 (1978)

17. Schneider, S.: Concurrent and Real Time Systems: The CSP Approach, 1st edn. John Wiley & Sons, Inc., New York (1999)
18. Roscoe, A.: Understanding Concurrent Systems, 1st edn. Springer-Verlag New York, Inc., New York (2010)
19. Gardiner, P., Goldsmith, M., Hulance, J., Jackson, D., Roscoe, B., Scattergood, B., Armstrong, P.: Fdr2 user manual (2010),
http://www.fsel.com/fdr2_manual.html

Stego Quality Enhancement by Message Size Reduction and Fibonacci Bit-Plane Mapping

Alan A. Abdulla, Harin Sellahewa, and Sabah A. Jassim

Applied Computing Department, University of Buckingham, Buckingham, UK
alananwer@yahoo.com,
{harin.sellahewa,sabah.jassim}@buckingham.ac.uk

Abstract. An efficient 2-step steganography technique is proposed to enhance stego image quality and secret message un-detectability. The first step is a pre-processing algorithm that reduces the size of secret images without losing information. This results in improved stego image quality compared to other existing image steganography methods. The proposed secret image size reduction (SISR) algorithm is an efficient spatial domain technique. The second step is an embedding mechanism that relies on Fibonacci representation of pixel intensities to minimize the effect of embedding on the stego image quality. The improvement is attained by using bit-plane(s) mapping instead of bit-plane(s) replacement for embedding. The proposed embedding mechanism outperforms the binary based LSB randomly embedding in two ways: reduced effect on stego quality and increased robustness against statistical steganalysers. Experimental results demonstrate the benefits of the proposed scheme in terms of: 1) SISR ratio (indirectly results in increased capacity); 2) quality of the stego; and 3) robustness against steganalysers such as RS, and WS. Furthermore, experimental results show that the proposed SISR algorithm can be extended to be applicable on DICOM standard medical images. Future security standardization research is proposed that would focus on evaluating the security, performance, and effectiveness of steganography algorithms.

Keywords: Steganography, LSB, Fibonacci, SISR, bit-plane mapping.

1 Introduction

Steganography is concerned with concealing a secret message by embedding it in another innocuous message during seemingly mundane communication sessions in such a way that only the sender and intended recipient are aware of the secret's existence. Steganography is becoming a common tool in protecting sensitive communications used by: intelligence and law enforcing agencies to prevent crime and terrorism; in health care systems to maintain the privacy of critical information such as medical records; and in financial organizations such as banks to prevent customers' account information from being accessed by illegally.

Security and privacy are important issues when medical images and their patient information are stored and transmitted across public networks. The well-known standard for storage and exchange of medical images is DICOM (Digital Imaging and

L. Chen and C. Mitchell (Eds.): SSR 2014, LNCS 8893, pp. 151–166, 2014.
© Springer International Publishing Switzerland 2014

Communication in Medicine). A historical background and a brief overview of DICOM are reported in [1,2].

In steganography, the secret information (e.g. text, image, video, audio) one wishes to send is called the message. This message is embedded in a cover, which is typically an image, a video or an audio file. After the message is embedded, the cover becomes a stego. Here, we are focused on the scenario whereby both the message and the cover are 8-bit grey scale images. The reason for choosing images as a cover is images usually have a high degree of redundancy, which makes them suitable to embed information without degrading their visual quality. Moreover, images are widely exchanged over the internet than other digital media and they attract little suspicion.

General factors that need to be addressed by embedding techniques are: the quality of the stego, message detectability, payload capacity, and the robustness of the stego against distortion attacks. Least significant bit (LSB) replacement [3,4] and its variants are common embedding techniques in the spatial domain. Embedding in LSB of a pixel (i.e. only the LSB plane of the cover image is overwritten with the secret bit stream), makes it difficult to notice a change in the value of these pixel by naked eye. A review of few embedding techniques will be presented in Section 2.

In recent years, different Fibonacci systems of representing integers have been used as an alternative to the binary representation to improve quality and capacity by embedding in more than one bit-plane. In 8-bit grey scale images, a pixel value in the range 0-255 can be represented by 12 bits using Fibonacci representation [5]. However, unlike the binary based embedding methods, capacity of Fibonacci based embedding depends on the selected cover image -- not every pixel of the cover is a "good candidate" for the embedding. The non-uniqueness of Fibonacci representation of integers means that the embedding procedure cannot comply the Zeckendorf's theorem, i.e. the embedding procedure should not create a two consecutive 1s. To avoid this situation, Fibonacci based schemes may have to skip the current pixel and consider the next candidate pixel for selection [5]. In our paper, the redundancy problem of Fibonacci representation will be avoided the restrictions of Zeckendorf's theorem, and thereby resulting that each pixel of the cover can be used for embedding.

Whilst steganographers aim to design techniques to meet the goals of steganograpy, steganalysers attempt to defeat the goal of steganography by detecting the presence of a hidden message. There are a number of statistical and structural attacks to determine the presence/absence of a hidden message and estimate the size of the embedded secret message. Two well-known steganalysis techniques used to detect and estimate the secret message are Regular Singular groups (RS) [6], and Weighted Stego (WS) [7].

Some steganographic schemes attempt to be robust against steganalysis while others attempt to have better stego quality, but such schemes usually have capacity limitations [8]. Decreasing the capacity increases the un-detectability [9]. There is a trade-off between hiding capacity and quality of stego image. It is a challenge for a steganographer to achieve a good balance among the different steganography requirements. It is worth noting that is no agreed benchmarking standard to evaluate steganography algorithms [10]. An international standard for evaluating the security, performance, and effectiveness of steganography algorithms would be of great benefit to security researchers, and to producers and consumers of steganography products.

This paper is primarily concerned with improving quality and robustness aspects of steganography techniques whilst maintaining a full payload capacity. The contributions of the paper are: 1) we present an algorithm, SISR, to reduce the secret image size as a pre-processing step prior to secret image embedding; and 2) we present a bit-plane mapping technique for Fibonacci based message embedding. As a result of these two proposals, we are able to improve stego image quality and un-detectability without affecting the payload capacity. Although our focus for SISR algorithm is to reduce embedding impact on the stego image, it can also be applied on the well-known DICOM standard images for storage and exchange of medical images. The proposed SISR algorithm is useful for both secure and privacy-preserving archiving/storing and transmission of medical images. Furthermore, the output bits from SISR algorithm that represent the DICOM image can be embedded with less impact on the stego quality.

The rest of the paper is organized as follows: a review of literature on steganography is presented in Section 2. Section 3 presents the proposed SISR algorithm to pre-process secret images before embedding and Section 4 presents the proposed embedding technique. The experimental results are shown in Section 5. Finally, our conclusions and direction future work are given in Section 6.

2   Literature Review

The LSB [3,4], LSB Matching (LSBM) [11], and LSBM Revisited (LSBMR) [12] are basic embedding techniques. Other embedding techniques based on region selection such as randomly embedding across the cover [13], and embedding in edge areas [14,15,16] have been proposed. The sequential LSB replacement increases even pixel values either by one or they are left unchanged, while odd pixel values are decreased by one or are left unchanged. This could create an imbalance in the embedding distortion, referred to in the literature as the *asymmetry* problem, in the stego image that can be exploited to detect the presence of a hidden message. More seriously, the secret message embedded by the sequential LSB replacement is easily detected. This problem can be partially mitigated by the use of pseudorandom number generator (PRNG) to randomly distribute the message across the cover instead of embedding the message in sequential order [13]. This is called LSB randomly embedding. However, the LSB randomly still suffers from the asymmetry problem.

The LSBM scheme is designed to solve the asymmetry problem by checking if the next secret bit does or does not match the LSB of the corresponding cover pixel. If not, then LSBM randomly increases or decrease the pixel value by 1 [11]. LSBM does not suffer from the asymmetry problem, and has the advantage of high payload capacity, as well as good visual imperceptibility property [17]. In these LSB methods, the probability of pixel value change is 0.5 (i.e. they have almost the same stego quality). Mielikainen [12], used a function to develop the LSBM technique called LSB Matching Revisited (LSBMR) to improve the imperceptibility of the stego whereby two secret bits are embedded in a pair of pixels of the cover. Incorporating this function reduces the probability of changing pixel values from 0.5 to 0.375. However, these improvements come at the expense of limited payload capacity because LSBMR algorithm cannot be performed on saturated pixels.

Another approach to secret hiding is to embed a message in selective regions of an image using some criteria. Embedding the secret message in edge area of the cover helps improve imperceptibility and robustness against detection [14]. Sobel, Prewitt, and Canny [15] are the most popular edge detection techniques that have been used in steganography. However, edge based embedding have problems with determining the same edge area by the receiver because the act of embedding in an edge area could change the original edge pixels into non-edge pixels. Therefore, some parts of a secret message may be lost, i.e. the survivability of secret message cannot be guaranteed at the receiving end. Hempstalk, [14], suggested to overcome this problem by first replacing all LSBs of the cover with 0. Then the edge detection technique is applied on the modified cover to select the edge pixels for embedding. The receiver applies the same pre-processing then use the edge detection technique on the modified stego to select edge pixels. In [16], Weiqi et al. attempted to tackle the same problem by limiting the edge pixels involved in embedding using a threshold that depends on size of the secret message, and proposed an embedding algorithm based on edge areas and LSBMR embedding method. Their algorithm is, therefore adaptive, and only sharper edge pixels are used for embedding. All selective regions (variants of edge-based) algorithms have the advantage of improved quality of stego and detectability, but at the expense of limiting payload capacity because the secret bits can be embedded in only edge areas.

Fibonacci bit-plane representation of the cover pixel has been investigated for steganography due to their advantages over binary representation when embedding in higher bit-planes than the LSB. This alternative representation of integers provides more bit-planes than the binary [18]. When embedding in higher bit-planes, the effect on the quality of the stego is less with the Fibonacci than with the binary [5].

Diego et al. [5], produced a Fibonacci embedding technique by strict observation of the Zeckendorf theorem. In their scheme, they first select a pixel then decompose the pixel value into Fibonacci and also select the plane that use for embedding. Then they check if the selected pixel is a good candidate or not, and if it is not, skip it and select the next candidate pixel. If it is a good candidate, then the secret bit is replaced with the agreed bit-plane. They claimed that the same embedding scheme can be also applied to different planes resulting in more robust data hiding and possibly higher visual distortion. As they mentioned, the main aim of their scheme is to investigate the possibility of inserting a secret bit without altering the perceptual quality of stego. Also they claimed that if the secret bits are embedded in the LSB using binary or Fibonacci based embedding, the PSNR of Fibonacci becomes similar or higher comparing to the binary embedding based. The limitation of their algorithm is, not every pixel of the cover is going to be used for embedding.

A generalized Fibonacci decomposition is proposed in [18] as an improvement of [5]. The most common generalization of Fibonacci is the p-number Fibonacci sequence. In order to provide more places for embedding, the authors of [18] investigated the p-number Fibonacci bit-planes to determine which planes are suitable for embedding. In their scheme, they first decompose the selected pixel into bit-planes using p-number Fibonacci. Then the selected plane is chosen for embedding and also the Zeckendorf theorem is applied on the selected pixel. Finally, they did a comparison between the proposed scheme and binary embedding in term of quality and capacity. As a result, even they claimed that the proposed scheme is better than binary in term of capacity and quality, but still their limitation is capacity because some pixel values are not candidates for embedding.

The use of more advanced versions of Fibonacci, and other representations have been investigated [19,20,21] to increase security. Fibonacci-like steganography by bit-plane(s) mapping instead of bit-plane(s) replacement has been proposed to increase embedding capacity by embedding two bits of the message in three bits of a cover pixel. However, the increase of security is achieved at the expense of a marginal loss in stego quality. In this paper, we extend the work in [21] to improve cover quality by embedding one secret bit in a cover pixel whilst maintaining the same robustness against statistical steganalysers. In this paper, the purpose of using Fibonacci representation instead of the binary representation is to gain higher stego quality, i.e. by embedding the obtained bits from SISR algorithm into Fibonacci representation of a cover pixel provides higher PSNR than embedding into binary representation (see Tables 9,11, and 12). This happens because of two reasons, first the obtained stream of bits from the proposed SISR algorithm that represent the secret image contains more 0s than 1s (see Tables 7 and 8) for natural images, the reason is discussed in Section 5.3. The second reason is in the Fibonacci representation complying with Zeckendorf theorem, more cover pixels' LSB have a value of 0 than 1, while in binary representation the number of cover pixels' LSB that have 0 value are almost equal to those have 1 (see Table 10). These two reasons increase the probabilities of similarity/matching between the secret bit and the cover pixel's LSB of the Fibonacci representation. Moreover, unlike the exiting Fibonacci-based bit-plane replacement techniques, the proposed mapping algorithm can be applied on all pixel intensities to embed secret bits.

3   Secret Image Size Reduction (SISR) Algorithm

Here we propose the SISR algorithm to reduce the size of secret images prior to embedding it in a cover image in order to reduce the impact of embedding on the stego quality (i.e. PSNR).

3.1   SISR Algorithm: Encoding

The SISR encoding algorithm works as follows:

- Split the secret image into non-overlapping blocks of size 4x4.
- Let B_{ij} be a block of 16 pixels, and i, j $\in$ {1,.., 4}. For each such block do the following steps:

1. Let $m = \min_{ij} (B_{ij})$, be the minimum pixel value, and let i*, j* be the indices of the element in B_{ij} achieving m with smallest j, then smallest i.
2. Let $D_{ij} = B_{ij} - m$, be the difference between each pixel and m.
3. Let $D_{max} = \max_{ij} (D_{ij})$, be the maximum difference value.
4. Let T be a set of possible thresholds to determine the length of bits that represent D_{ij}.

$$T = \{2^n - 1 \mid 0 \leq n \leq 7\}$$

5. Let $T^* = \min (z)$, where $z \in T$ and $z \geq D_{max}$, in other words, T^* is the closest value in T to D_{max} that is greater than or equal to D_{max}.

When $T^* = 0$, it means that the maximum value of differences D_{ij} between each pixel and the minimum pixel (m) is 0, i.e., all pixels in that block have the same value. When $T^* = 1$, it means that the maximum value of differences D_{ij} between each pixel and the minimum pixel (m) is 1, and so on for other values of T^*.

6. If $T^* = 255$, record 1 and followed by the original pixels' value in 8 bits, otherwise record 0. If $T^* = 0$, the recorded 0 is followed by the minimum pixel value in 8 bits, then the value of T^* in 3 bits (111), and no information is needed to represent the position of the minimum pixel value because all of the pixels have the same value, therefore in this case only 12 bits are needed to represent the block. Otherwise, when $T^* = 1, 3, 7, 15, 31, 63$ or 127, the recorded 0 is followed by:

— 8 bits: indicates the minimum pixel value.
— 3 bits: indicates the value of T^*, should be 110, 101, 011, 100, 010, 001, or 000 representing 1, 3, 7, 15, 31, 63, or 127 respectively.
— 4 bits: indicates as the position of the minimum pixel value (in each block there are 16 pixels) therefore the positions are from 0 to 15.
— From $T^* = 2^L - 1$, L is the length of bits that are needed for representing D_{ij}, where $1 \leq L \leq 7$

Note that the 8 bits that indicate the minimum pixel value and the 3 bits that indicate the value of T^* are needed when the image pixel value is between 0 and 255. It is possible to extend this reduction algorithm to be applicable on different pixel value ranges, for example if the image pixel value is between 0 to 511, we can extend the algorithm to be applicable by representing the minimum pixel value in 9 bits instead of 8 bits and the value of T^* in 4 bits instead 3 bits.

3.2   SISR Algorithm: Recovery

The following steps describe the recovery of the original images from the obtained stream of bits b_i described in Section 3.1:

1. Take the first bit; if it is 1 take each of the following 8 bits and convert them into decimal (repeat it 16 times). If this is the case, terminate here, otherwise follow the next steps.

2. Next 8 bits indicate m.
3. Next 3 bits indicate T^*.
4. Next 4 bits indicate index of m.
5. From value of T^*, the length of the bits are decided so as to represent D_{ij}. For example if $T^* = 15$, 4 bits are needed to represent D_{ij}.
6. $B_{ij} = D_{ij} + m$.

3.3   Example Application of SIRS Algorithm

The example 4x4 block of pixel intensities in Table 1 will be used to illustrate the SISR encoding and recovery steps.

Table 1. Block of 16 pixels

30	25	26	35
35	22	29	28
31	24	22	29
30	34	32	30

SISR Encoding Steps

1. The minimum pixel value is 22 (00010110 in 8 bits), and its index is 5 (0101 in 4 bits).
2. The differences between pixels and the minimum pixel are presented in Table 2.
3. The maximum value of the subtraction, see 3rd and 7th column of the Table 2, is 13.
4. The nearest value in set T that should be equal or greater than the maximum value of the subtraction, which is 13, is 15 (i.e. $T^* = 15$).
5. Now this stream of bits represents the block of Table 1:

 0: indicates that the algorithm has been done on the block, i.e. T* is not equal to 255.
 00010110: represents the minimum pixel value.
 100: represents the value of T*.
 0101: represents the index of the minimum pixel value.
 1000, 0011, 0100, 1101, 1101, 0111, 0110, 1001, 0010, 0000, 0111, 1000, 1100, 1010, 1000: representing values of difference (see 3rd and 7th column of Table 2) 8, 3, 4, 13, 13, 7, 6, 9, 2, 0, 7, 8, 12, 10, 8 respectively.
 Finally, this stream (0 00010110 100 0101 1000 0011 0100 1101 1101 0111 0110 1001 0010 0000 0111 1000 1100 1010 1000) represents the 4x4 block (see Table 1), i.e. 76 bits are representing original 128 bits.

Table 2. Differences between pixels and minimum pixel value

B_{ij}	m	D_{ij}	b	B_{ij}	m	D_{ij}	b
30	22	8	1000	24	22	2	0010
25	22	3	0011	22	22	0	0000
26	22	4	0100	29	22	7	0111
35	22	13	1101	30	22	8	1000
35	22	13	1101	34	22	12	1100
29	22	7	0111	32	22	10	1010
28	22	6	0110	30	22	8	1000
31	22	9	1001				

In Table 2, the 1st and 5th column, B_{ij}, is the values of the pixels in the Table 1 (excluding first minimum pixel value 22). The 2nd and 6th column is the minimum pixel value m, and the 3rd and 7th column, D_{ij}, is the subtraction of the minimum pixel value from each pixel value. The 4th and 8th column, b, represents the result of subtraction in binary form.

SISR Recovery Steps
From the obtained stream above, the original 4x4 block of pixels can be recovered as follows:

1. Take the first bit; which is 0, then go to the next step.
2. Convert the next 8 bits into decimal, which is 22, that represents the minimum pixel value m.
3. The next 3 bits, 100, represent the value of T*, i.e. T*=15.
4. The next 4 bits, 0101, represent the index of m.
5. As T* = 15, take each next 4 bits 15 times and convert them into decimal to represent D_{ij} as illustrated in Table 3.
6. Add m to D_{ij}, then the original pixels B_{ij} are obtained (see 4th and 8th column of Table 3).
7. Sequentially insert each value in the 4th and 8th column of Table 3 to its position in the block; the block is recovered exactly as it is (see Table 1).

Table 3. Producing original pixels from the recovered D_{ij}

b	D_{ij}	m	B_{ij}	b	D_{ij}	m	B_{ij}
1000	8	22	30	0010	2	22	24
0011	3	22	25	0000	0	22	22
0100	4	22	26	0111	7	22	29
1101	13	22	35	1000	8	22	30
1101	13	22	35	1100	12	22	34
0111	7	22	29	1010	10	22	32
0110	6	22	28	1000	8	22	30
1001	9	22	31				

Table 4. Number of obtained bits from proposed SISR algorithm for block size 4x4

T*	Number of bits		
	Obtained	Original	Reduction
0	12	128	116
1	31	128	97
3	46	128	82
7	61	128	67
15	76	128	52
31	91	128	37
63	106	128	22
127	121	128	7
255	129	128	-1

Table 4 illustrates the number of obtained bits and the number of reduced bits depending on the value of T* for the 4x4 block. From this table, one can notice that only in the case T* = 255, the algorithm increases the number of bits by 1; otherwise, the algorithm reduces the number of bits to represent the block of 16 pixels.

Although the main focus in this paper of proposing SISR algorithm is to reduce the secret image size prior to embedding, it can also be used for the image storage and transmission purposes.

4   Proposed Embedding Algorithm

A PRNG is used to randomly select a cover pixel. Converting the cover pixel value into Fibonacci results in more than one representation for each pixel value. Complying with Zeckendorf theorem means that the first three LSBs of a cover pixel in Fibonacci representation belong to the set {000, 001, 010, 100, 101} as illustrated in Table 5. Depending on the set, our mapping embeds a secret bit into a cover pixel by mapping the secret bit onto the first 3 LSB bits according to Table 5.

Table 5. Mapping Algorithm

3 Cover bits	Secret bits	
	0	1
000	000	001
001	000	001
010	010	001
100	100	101
101	100	101

From the structure of Table 5, the secret bits can be recovered by extracting the 1st LSB of the selected stego pixel using the same PRNG used at the embedding stage.

5   Experimental Results

Two sets of experiments were performed to evaluate the proposed algorithms: one to evaluate the proposed SISR algorithm and one to evaluate the performance of the proposed mapping based embedding technique.

5.1   Experimental Setup

The following databases were used in our experiments:

1. The Miscellaneous volume of Signal and Image Processing Institute (SIPI) database of University of Southern California [22]. This database consists of 44 images of which 16 are color and 28 are monochrome images. We resized these 44 images to 512 x 512 and convert them into gray level images with 8 bits per pixel. These images will be considered as cover images. Also, the original 44 images were resized to 256 x 128 and converted to gray scale with 8 bits per pixel to be used as secret images. The reason for resizing these images to 256 x 128 is to make the number of bits that represent a secret image is equal to the number of cover pixels (262144 pixels).

2. Two sample databases of DICOM images: BU001015-V01 database contains 193 gray level images of size 256 x 256 with 16 bits per pixel, and MR database contains 140 gray level images of sizes 256 x 256 or 512 x 512 with 16 bits per pixel [23]. These two databases are used to test the proposed SISR algorithm.

5.2   SISR Algorithm Evaluation

The proposed SISR algorithm was applied on SIPI database images of size 256 x 128, BU001015 database images of size 256 x 256, and MR database images of sizes 256 x 256 and 512 x 512. The original number of bits in each tested image is 128 *256 *8 = 262144 bits, 256 x 256 x 16 = 1048576 bits, and 1048576 bits or 512 x 512 x 16 = 4194304 bits for SIPI, BU001015, and MR database respectively. We used the reduction ratio factor RR in Eq. 1 to evaluate the reduction efficiency of our approach.

$$RR = \frac{Total\ size\ in\ bits\ of\ the\ obtained\ bitstream}{Total\ size\ in\ bits\ of\ the\ input\ image} \tag{1}$$

Table 6 shows RR after applying the SISR algorithm on the three databases for 3 different block sizes (i.e., 4x4, 8x8, and 16x16). From Table 6, we can see that the best RR is achieved with 4x4 blocks. The RR decreases as the block size increases. Furthermore, Table 7 and 8 demonstrates that the SISR algorithm produces a higher ratio of 0s bits than 1s.

Table 6. Average RRs for SISR algorithm for different block sizes

| Databases | Block Size | | | | | |
| | 4x4 | | 8x8 | | 16x16 | |
	Average	Std.	Average	Std.	Average	Std.
SIPI	0.697	0.141	0.754	0.138	0.841	0.111
DICOM(BU001015)	0.442	0.017	0.446	0.020	0.469	0.020
DICOM(MR)	0.426	0.031	0.429	0.034	0.455	0.035

Table 7. Ratio of Zero and One value bits obtained after applying the proposed SISR algorithm on 4x4block size

| Databases | Ratio of 0s | | Ratio of 1s | |
	Average	Std.	Average	Std.
SIPI	0.55	0.037	0.45	0.037
DICOM(BU001015)	0.58	0.004	0.42	0.004
DICOM(MR)	0.58	0.010	0.42	0.010

In order to embed DICOM images, we resized the first 10 images of databases BU001015 and MR to 128 x 128 to embed them in the first 10 cover images of SIPI database. The images were resized to 128 x 128 to make the number of bits that represent secret medical DICOM images equal to the number of cover pixels (262144 pixels). The result of the SISR algorithm is illustrated in Table 8.

Table 8. RRs and the ratio of Zero and One value bits Obtained after applying the proposed SISR algorithm on block size 4x4 for resized DICOM database images

| Databases | RR | | Ratio of 0s | | Ratio of 1s | |
	Average	Std.	Average	Std.	Average	Std.
BU001015	0.403	0.038	0.60	0.005	0.40	0.005
MR	0.438	0.028	0.58	0.005	0.42	0.005

5.3   Embedding Algorithm Evaluation

Two experiments are conducted to evaluate the performance of the proposed embedding technique. The first is to test stego quality and the other is to measure the detectability of secrets. The results are compared with the binary based LSB randomly (LSB binary) and Fibonacci based LSB randomly (LSB Fibonacci) embedding technique. Images of SIPI database (44 images) of size 512 x 512 are used as cover images. For each cover image, 44 secret images of SIPI database of size 256 x 128 are embedded producing (44 x 44 = 1936) stego images for each tested embedding technique including the proposed embedding technique.

Stego Quality. Average PSNR values for the 1936 stego images are shown in Table 9 for LSB binary, LSB Fibonacci and the proposed embedding technique.

Table 9. Stego quality (PSNR) after embedding the original image (SIPI) vs. the reduced image

	Binary LSB		Fibonacci LSB		Fibonacci mapping	
	Original	Reduced	Original	Reduced	Original	Reduced
PSNR	51.15	52.80	52.24	52.92	51.14	52.90

The second column of Table 9 presents the results when the original image bits are embedded in the cover using binary based LSB randomly embedding technique. The third column presents the results when reduced image bits are embedded using binary based LSB randomly embedding technique. The same for the fourth and fifth columns using Fibonacci based embedding technique. Finally, the sixth and seventh columns present the results when original image bits and reduced image bits are embedded using the proposed mapping based embedding technique. Results show that reducing the secret image using the proposed SISR algorithm leads to better stego quality compared to embedding the original image. The reason why the Fibonacci based LSB has higher PSNR value than others is because in the case of Fibonacci based LSB embedding technique, not every pixel is used for embedding, i.e. some pixels are excluded from embedding whereas in the binary based LSB and the proposed Fibonacci based mapping embedding technique, every pixel can be, and is, used for embedding. The average of maximum number of pixels, for the SIPI database, used for embedding for each tested technique is illustrated in figure 1. Moreover, the stego quality is better with the proposed mapping based embedding compared to binary based LSB. This is achieved because: 1) the proposed SISR algorithm always provide a bit stream that contains more 0s than 1s; and 2) in the Fibonacci representation compliant with the Zeckendorf theorem, the probability of pixels that have a value of 0 as the LSB is always greater than those that have 1. Whilst in the binary representation, the probability of pixels' LSB that have a value of 0 is equal to those that have 1. These two factors increase the probability of the secret bit and the LSB of the Fibonacci representation being the same. The reasons of why the proposed reduction algorithm provides a stream of bits that always contains more 0s than 1s are:

1. First recorded bit (see Section 3.1) is 1 when $T^*=255$ otherwise it is 0. As in most blocks of the secret image T* is not equal to 255, the stream has more number of 0s than 1s.
2. There is one possible 3-bit string (i.e., 000, 001, 010, 100, 101, 011, 110) to represent each T* value. The four most frequently occurring T* values are represented by the 3 bits with two or more 0s. Then the three remaining 3-bit strings represents the three least frequent T* value.

Table 10. Ratio of Zeros and Ones of cover pixels' LSB value for SIPI database

	LSB of Binary		LSB of Fibonacci	
	Number of 0s	Number of 1s	Number of 0s	Number of 1s
Average	0.47	0.53	0.60	0.40
Std.	0.095	0.095	0.126	0.126

Table 11. Stego quality (PSNR) after embedding the original DICOM (BU00105) image vs. the reduced image

	Binary LSB		Fibonacci LSB		Proposed	
	Original	Reduced	Original	Reduced	Original	Reduced
PSNR	51.14	55.10	52.30	55.09	51.81	55.30

Table 12. Stego quality (PSNR) after embedding the original DICOM (MR) image vs. the reduced image

	Binary LSB		Fibonacci LSB		Proposed	
	Original	Reduced	Original	Reduced	Original	Reduced
PSNR	51.14	54.73	52.30	54.72	51.70	54.90

In Table 11 and 12, the first 10 images of databases BU001015 and MR are resized to 128 x 128 and embedded in each of the first 10 cover images of size 512 x 512 of SIPI database. Results in each cell are the average of 100 stego images. For the same reasons that we discussed before, we can see that the proposed embedding technique has higher quality than binary based LSB randomly embedding and Fibonacci LSB randomly embedding, and for the same reasons discussed before, the Fibonacci LSB has higher quality than binary based LSB randomly embedding and proposed embedding technique when the original secret image is embedded.

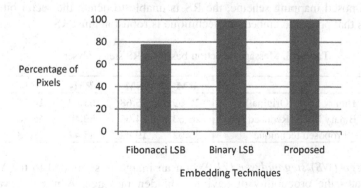

Fig. 1. Embedding Capacity

From figure 1, we can see that the Binary LSB and proposed embedding techniques have 100% of embedding capacity, i.e. every cover pixel can be used to embed the secret; while in Fibonacci LSB, only 78% of the cover pixels on average can be used for embedding the secret.

Detectability. Two steganalysers have been used to evaluate the detectability of the proposed embedding technique. Results are compared with binary based LSB randomly embedding technique.

In Table 13 and 14, binary LSB (Original) means the secret image is embedded using binary based LSB randomly; Binary LSB (Reduced) means the obtained bits from Table 6 (2nd column) is embedded using binary based LSB randomly; and proposed technique means the obtained bits from Table 6 (2nd column) is embedded using the proposed embedding technique.

Regular and Singular (RS) steganalyser [6]. Fridrich et al. found that the RS ratio of a typical image should satisfy the rule: (RM $\cong$ RM- and SM $\cong$ SM-) through large number of experiments. When LSB of the cover is changed, the difference between RM and RM- and the difference between SM and SM- increases. Then, the rule is violated; therefore, one could conclude that the tested image carries a secret message. Note that when payload capacity p = 0 %, i.e. cover without an embedded message, the value of RM is close to RM-, and the value of SM is close to SM-. Depending on the description above, we can discuss the results in Table 13, which presents the average values of RM, RM-, SM, and SM- based on embedding 44 secret images of the SIPI database in each of 44 different covers of SIPI database resulting 1936 stego images. In the case of using binary LSB, embedding higher rate of secret bits lead to an increase in the differences between RM and RM-, SM and SM-, indicating the presence of a secret message. Whereas in the case of the proposed embedding technique, there are no such differences. This indicates that these images are non-stegos. Therefore, the proposed technique is robust against the RS steganalyser. When reduced-size messages are embedded using binary LSB, the RS is still able to detect the secret bits but with lower message size estimation. When embedding the reduced bits using the proposed mapping scheme, the RS is unable to detect the secret bits. This demonstrates that proposed embedding technique is robust against RS.

Table 13. Message detection based on RS steganalyser

	RM	SM	RM-	SM-
Binary LSB (Original)	25.69	25.68	55.42	14.21
Binary LSB (Reduced)	28.63	24.06	53.40	15.56
Proposed technique	39.02	19.82	43.42	19.48

Weighted Stego (WS) steganalyser [7]. When an image is submitted to the WS, its output indicates the probability of having a hidden message. A negative value is treated as 0 and any number >1 is an indication of full 100% secret load. The average result of 1936 stegos is displayed for each embedding technique. Table 14 shows the estimation results of the secret message length of two steganographic techniques, including our proposed by this steganalyser. From Table 14, it is noticeable that the proposed scheme mostly scores a value close to zero indicating the absence of an embedded secret. On the other hand, when embedded using the binary LSB, the steganalyser detects the secret message with high probability. This demonstrates that proposed embedding technique is robust against WS.

Table 14. WS Steganalyser

	Message length
LSB random (original)	0.970
LSB random (Reduced)	0.734
Proposed technique	0.151

6   Conclusion

A two-step efficient steganography scheme has been proposed to enhance stego quality and secret un-detectability. The first step of reducing the number bits required to represent a secret image (i.e., SISR) prior to embedding led to improved stego quality. The second step is the Fibonacci-based mapping technique to embed the message bits. This technique further improved the stego quality by embedding the obtained bits from the SISR algorithm in the cover by bit-plane(s) mapping instead of bit-plane(s) replacement. The proposed embedding mechanism outperforms the binary based LSB randomly embedding in two aspects: stego quality and robustness against steganalysers. Furthermore, this scheme overcomes the limitation imposed by the Zeckendorf theorem which improves capacity as well as the stego quality. The improvement in stego quality is the result of the combined effect of the SISR which results in secret message bit stream that has more 0s than 1s and the nature of the Fibonacci representation of cover pixel values which results in more 0s at the LSB than 1s. The experimental results demonstrated that the proposed embedding scheme is also secure against RS, and WS steganalyser attacks.

Our future work directions are: 1) investigate all existing pixel value decomposition techniques such as Prime, Natural, Lucas, Catalan in terms of providing higher ratio of cover pixels' LSB having the zero value and also propose a new pixel values decomposition technique that provides higher ratio of cover pixels' LSB having zero value; and 2) find a mechanism to improve the SISR algorithm in order to provide higher ratio of obtained bits that their values are zeros than 1s. Thus, these two factors are increase the probability of similarity between the secret bits and cover pixels' LSB results in better stego quality and less message detectability.

References

1. Mildenberger, P., Eichelberg, M., Martin, E.: Introduction to the DICOM standard. European Radiology 12, 920–927 (2002)

2. Gibaud, B.: The Dicom Standard: A Brief Overview. In: Molecular Imaging: Computer Reconstruction and Practice, pp. 229–238. Springer (2008)
3. Chan, C., Cheng, L.M.: Hiding data in images by simple LSB substitution. Pattern Recognition 37, 469–474 (2004)
4. Thien, C., Lin, J.: A simple and high-hiding capacity method for hiding digit-by-digit data in images based on modulus function. Pattern Recognition 362, 875 (2003)
5. Picione, D.D.L., Battisti, F., Carli, M., Astola, J., Egiazarian, K.: A Fibonacci LSB data hiding technique. In: Proc. 14th EUSIPCO (2006)
6. Fridrich, J., Goljan, M., Du, R.: Reliable detection of LSB steganography in color and grayscale images. In: Proc. ACM Workshop on Multimedia and Security, pp. 27–30 (2001)
7. Fridrich, J., Goljan, M.: On estimation of secret message length in LSB steganography in spatial domain. Electronic Imaging, International Society for Optics and Photonics, 23–34 (2004)
8. Najeena, K.S., Imran, B.M.: An efficient steganographic technique based on chaotic maps and adpative PPM embedding. In: Signal Processing Image Processing & Pattern Recognition (ICSIPR), pp. 293–297. IEEE (2013)
9. Kurtuldu, O., Arica, N.: A new steganography method using image layers. In: ISCIS, pp. 1–4. IEEE (2008)
10. Kraetzer, C.: Visualisation of Benchmarking Results in Digital Watermarking and Steganography. In: ECRYPT, pp. 30–46. Citeseer (2007), IST-2002-507932
11. Sharp, T.: An implementation of key-based digital signal steganography. In: Moskowitz, I.S. (ed.) IH 2001. LNCS, vol. 2137, pp. 13–26. Springer, Heidelberg (2001)
12. Mielikainen, J.: LSB matching revisited. LSP IEEE 13(5), 285–287 (2006)
13. Provos, N., Honeyman, P.: Hide and seek: An introduction to steganography. IEEE Security & Privacy 1, 32–44 (2003)
14. Hempstalk, K.: Hiding behind corners: Using edges in images for better steganography. In: Proceedings of the Computing Women's Congress, Hamilton, New Zealand, pp. 11–19 (2006)
15. Chen, W.J., Chang, C.C., Le, T.: High payload steganography mechanism using hybrid edge detector 37(4), 3292–3301 (2010)
16. Weiqi, L., Fangjun, H., Huang, J.: Edge Adaptive Image Steganography Based on LSB Matching Revisited. TIFS 5(2), 201–214 (2010)
17. Yi, L., Xiaolong, L., Bin, Y.: Locating steganographic payload for LSB matching embedding. In: ICME, pp. 1–6 (2011)
18. Battisti, F., Carli, M., Neri, A., Egiaziarian, K.: A Generalized Fibonacci LSB Data Hiding Technique. In: CODEC, pp. 671–683 (2006)
19. Patsakis, C., Fountas, E.: Extending Fibonacci LSB data hiding technique to more integer bases. In: ICACTE, vol. 4, pp. 18–27 (2010)
20. Mammi, E., Battisti, F., Carli, M., Neri, A., Egiazarian, K.: A novel spatial data hiding scheme based on generalized Fibonacci sequences. Proc. SPIE 6982, 1–7 (2008)
21. Abdulla, A.A., Jassim, S., Sellahewa, H.: Efficient high-capacity steganography techniques. Proc. SPIE 8755, 875508-1–875508-11 (2013)
22. http://sipi.usc.edu/database/database.php?volume=misc (last access date was Spetember 15, 2014)
23. http://www.microdicom.com/downloads.html (last access date was September 15, 2014)

Secure Modular Password Authentication for the Web Using Channel Bindings

Mark Manulis[1], Douglas Stebila[2], and Nick Denham[2]

[1] Surrey Centre for Cyber Security, University of Surrey, UK
mark@manulis.eu
[2] Queensland University of Technology, Brisbane, Australia
{stebila,n.denham}@qut.edu.au

Abstract. Secure protocols for password-based user authentication are well-studied in the cryptographic literature but have failed to see widespread adoption on the Internet; most proposals to date require extensive modifications to the Transport Layer Security (TLS) protocol, making deployment challenging. Recently, a few modular designs have been proposed in which a cryptographically secure password-based mutual authentication protocol is run inside a confidential (but not necessarily authenticated) channel such as TLS; the password protocol is bound to the established channel to prevent active attacks. Such protocols are useful in practice for a variety of reasons: security no longer relies on users' ability to validate server certificates and can potentially be implemented with no modifications to the secure channel protocol library.

We provide a systematic study of such authentication protocols. Building on recent advances in modelling TLS, we give a formal definition of the intended security goal, which we call *password-authenticated and confidential channel establishment* (PACCE). We show generically that combining a secure channel protocol, such as TLS, with a password authentication protocol, where the two protocols are bound together using either the transcript of the secure channel's handshake or the server's certificate, results in a secure PACCE protocol. Our prototype based on TLS is available as a cross-platform client-side Firefox browser extension and a server-side web application which can easily be installed on deployed web browsers and servers.

Keywords: password authentication, Transport Layer Security, channel binding.

1 Introduction

Authentication using passwords is perhaps the most prominent and human-friendly user authentication mechanism widely deployed on the Web. In this ubiquitous approach, which we refer to as *HTML-forms-over-TLS*, the user's password is sent encrypted over an established server-authenticated Transport Layer Security (TLS, previously known as Secure Sockets Layer (SSL)) channel in response to a received HTML form. This approach is subject to many threats:

L. Chen and C. Mitchell (Eds.): SSR 2014, LNCS 8893, pp. 167–189, 2014.

the main problems with this technique are that security fully relies on a functional X.509 public key infrastructure (PKI) and on users correctly validating the server's X.509 certificate. In practice, these assumptions are unreliable due to a variety of reasons: the many reported problems with the trustworthiness of certification authorities (CAs), inadequate deployment of certificate revocation checking, ongoing threats from phishing attacks, and the poor ability of the users to understand and validate certificates [1,2]. Hypertext Transport Protocol (HTTP) basic and digest access authentication [3] has been standardized, and digest authentication offers limited protection for passwords, but usage is rare. Public-key authentication of users, e.g. using X.509 certificates, is also rare.

1.1   Password-Authenticated Key Exchange (PAKE)

Password-authenticated key exchange (PAKE) protocols, which were introduced by Bellovin and Merritt [4], and the security of which was formalized in several settings [5,6,7], could mitigate many of the risks of the HTML-forms-over-TLS approach as they do not rely on any PKI and offer stronger protection for client passwords against server impersonation attacks, such as phishing. PAKE protocols allow two parties determine whether they both know a particular string while cryptographically hiding any information about the string. They are resistant to offline-dictionary attacks: an adversary who observes or participates in the protocol cannot test many passwords against the transcript. Successful execution of a PAKE protocol also provides parties with secure session keys which can be used for encryption.

Despite the many benefits of PAKE, and the presence of a variety of existing protocols in the academic literature and in standards [8,9,10], PAKE-based approaches for client authentication have not been adopted in practice. There is no PAKE standard that has been agreed upon and implemented in existing web browser and server technologies. This is due to several practical obstacles, including: patents covering PAKE in general (some of which have recently expired in the US), patents on proposed standards such as the Secure Remote Password (SRP) protocol [11], lack of agreement on the appropriate layer within the networking stack for the integration of PAKE [12], complexity of backwards-compatible deployment with TLS, and user-interface challenges.

There have been a few proposals to integrate PAKE into TLS by adding password-based ciphersuites as an alternative to public-key authenticated ciphersuites. SRP has been standardized as a TLS ciphersuite [13] and has several reference implementations but none in major web browsers or servers. Abdalla et al. [14] proposed the provably secure Simple Open Key Exchange (SOKE) ciphersuite, which uses a variant of the PAKE protocol from [15] that is part of the IEEE-P1363.2 standard [10]. The J-PAKE protocol [16] is used in a few custom applications. Common to all PAKE ciphersuite approaches is that the execution of PAKE becomes part of the TLS handshake protocol: the key output by PAKE is treated as the TLS pre-master secret, which is then used to derive further encryption keys according to the TLS specification. An advantage of this approach is that secure password authentication could subsequently be used in any application that makes use of TLS, and that standard TLS mechanisms for key derivation and secure record-layer communication can continue to be used. However, a major disadvantage is that any new ciphersuites in TLS require substantial vendor-side modifications of the web browser and server software. This is problematic for modern web server application architectures within large organizations, where a TLS accelerator immediately handles the TLS handshake and encryption, then hands the plaintext off to the first of many application servers; requiring the TLS accelerator to have access to the list of valid usernames and passwords may mean a substantial re-architecting. Moreover, using solely PAKE in TLS means abandoning the web public key infrastructure.

1.2   Running PAKE at the Application Layer

A better approach for realizing secure password-based authentication on the web may be to rely on existing TLS implementations to provide confidential communication between clients and servers, and integrate application-level PAKE for password-based authentication, without requiring any modifications to the TLS specification or implementation; in particular, without proposing any new TLS ciphersuites or changing any of the steps of TLS handshake protocol.

However, if the TLS channel is only assumed to provide confidentiality, not authentication, then one must use an alternative mechanism to rule out man-in-the-middle attacks on the TLS channel. Since it is the password-based protocol that provides mutual authentication, there should be a binding between the TLS channel and the password-based protocol. There are several potential values which might be used for binding: the transcript of the TLS handshake protocol, the TLS master secret key (or a value derived from it, such as the TLS Finished message), or even the server's certificate. A recent standard [17] describes three TLS channel bindings, two of which are relevant to us: tls-unique in which the binding string is the Finished message, and tls-server-end-point in which the binding string is the hash of the server's certificate. Notably, TLS channel bindings do not change the TLS protocol itself: all TLS protocol messages, ciphersuites, data transmitted, and all other values are entirely unchanged. Rather, TLS channel bindings expose an additional value to the application that can be obtained locally, thereby requiring minimal changes to TLS implementations.

The high-level approach of running PAKE at the application level is given in Figure 1. Using PAKE at the application-level supplements, rather than replaces, the use of public key certificates.

Several recent works have proposed protocols of this form. Oiwa et al. [18,19,20,21] published an Internet-Draft that employs an ISO-standardized PAKE protocol (KAM3 [8, §6.3],[22]) and binds it to the TLS channel using either the server's certificate or the TLS master secret key, but no formal justification is given for security of the combined construction.

Dacosta et al. [23] proposed the DVCert protocol which aims to achieve direct validation of the TLS server certificates by using a modification of the protocol from [24] for secure server-to-client password-based authentication. Dacosta et al. used an automated cryptographic protocol verifier, ProVerif, to demonstrate that their protocol does not leak password information without addressing any further security properties of secure channels. In particular, the analysis carried out in [23] is insufficient for showing the actual benefit of using PAKE protocols to strengthen the security of the TLS channel: Dacosta et al. simply show that the protocol does not leak password information, which is not surprising since they build on a PAKE protocol which already did not leak information. Rather, the security goals of the overall channel must be fully modelled to provide a complete analysis. This is where our model for PACCE fills the gap: it explains the expected security goals from the combination of PAKE and TLS. While we have no reason to expect that the protocol from [23] is insecure, our model would enable such an analysis of that protocol.

Outside of the world of TLS, the Off-the-Record Messaging protocol [25,26] uses a PAKE-like password authentication functionality that is based on secure two-party computation techniques to authenticate the long-term public keys used in establishing confidential (yet deniable) channel, but with no justification for the security of the combined construction.

1.3   Contributions

We analyze the modular approach to secure password authentication on the web, in which a secure channel protocol such as TLS is combined—in a black box way—with a password-authentication protocol. The black-box approach enables smooth integration of PAKE functionality with secure channels such as TLS without requiring any modification to the original channel protocol specification, nor requiring abandoning public key certificates for server authentication.

At a high level, in our approach, a normal secure channel is established with no assumptions on correct validation of certificates; then a PAKE protocol is run *within* the secure channel to demonstrate (a) mutual knowledge of the password, and (b) absence of a man-in-the-middle attack on the channel; once the PAKE succeeds, the parties continue application communication. Notably, the session key established by the PAKE is not used for symmetric encryption; while it *could* replace the secure channel session key, in practice, such as in TLS, there is no standardized mechanism to do so. Moreover, using the existing channel session key is fine provided we bind the execution of the PAKE and the original channel establishment in a provably secure way. We show how to realize this binding using available TLS standards.

Our formal approach is as follows.

First, we define formally in Section 2 the security property we aim for in our protocol: since the most suitable definition for the security of the combined TLS handshake and record layer protocols is the *authenticated and confidential channel establishment (ACCE)* model of Jager et al. [27], we give a corresponding *password-based ACCE (PACCE)* notion. We apply our PACCE model to analyzing strong passwords in TLS; as ACCE has been used to analyze many real world protocols (SSH [28], EMV [29], and QUIC), PACCE should be suitable for strong password variants of those.

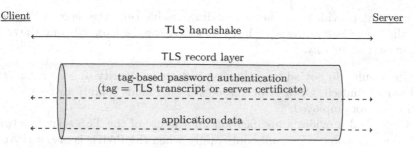

Fig. 1. High-level approach for combining secure password authentication with a TLS channel to establish a single password-authenticated secure channel

Next, we define the various primitives we employ to achieve this goal; these include the original ACCE notion, as well as an unauthenticated *confidential channel establishment (CCE)* protocol. Rather than using a PAKE protocol, we actually employ a newly defined *tag-based password authentication (tPAuth)* protocol, which provides mutual authentication based on knowledge of a shared password, with acceptance only if both parties input the same, possibly public, *tag* to the protocol.

Then, when we run the tPAuth protocol inside the established confidential channel, we bind the two protocols together by setting the tag to be either the transcript of the channel establishment or the long-term public key used by the server in the channel establishment. We prove in Sections 4 and 5 that both of these tags are sufficient to achieve the end result of a password-based authenticated and confidential channel establishment (PACCE) protocol. Our two security theorems provide a qualitative distinction between those two tags as binding mechanisms because they use slightly different assumptions on the confidential channel establishment protocol: when the tag used is the transcript, it suffices to use a CCE protocol, but when the tag used is the server's public key, we require the underlying protocol to be an ACCE protocol. (These results give the first security justification of two standardized TLS channel binding mechanisms, `tls-unique` and `tls-server-end-point` [17].)

Applicability of results to other approaches. Our results, by employing the CCE/ACCE frameworks, are generic and could be applicable to constructions employing a wide variety of protocols, not just TLS. These results justify the *general approach* of the recent proposals of Oiwa et al.'s protocol [18,19,20,21] and Dacosta et al.'s DVCert protocol [23]. We caution that our theorems do not immediately imply security of those particular protocols for several reasons.

Our theorems depend on a formal construction called a tag-based password authentication (tPAuth) protocol. Only the tSOKE protocol we describe in Appendix B is known to be a secure tPAuth. For most PAKE protocols, such as the PAK protocol employed by Dacosta et al., it seems not hard to modify the PAKE security proof to demonstrate tPAuth, but this requires additional work for each particular protocol.

Oiwa et al. provide three channel binding mechanisms: the server's TLS certificate, the server host string (e.g., `http://www.example.com:80`), and the server's TLS master secret key.

- Our results do not address the cryptographic security of server host string binding; indeed, Oiwa et al. intend for this mode mainly to be used when TLS is not employed.
- Section 5.3 provides a justification for the use of the TLS server certificate, though as noted our results only apply when the PAKE is also a tPAuth.
- Section 4.4 provides a justification for the use of the TLS `Finished` message, whereas Oiwa et al. allow use of the TLS master secret key. Since the TLS handshake will not complete unless both parties compute the same `Finished` messages, our results also justify the use of the TLS master secret key for channel binding, though security of the whole construction again is not implied by our results due to the caveat above that our results only apply with the PAKE is also shown to be a tPAuth.

Dacosta et al.'s DVCert protocol only provides server-to-client password-based authentication, whereas our PACCE notion provides mutual authentication.

Implementation. In Section 6, we demonstrate the practical merit of our approach with a reference implementation for the Mozilla Firefox web browser and the Apache web server that takes advantage of the modularity of the construction.

- On the client side, our implementation is achieved entirely as a Firefox *extension*: it is a cross-platform Javascript-based bundle that can be installed by the user at run-time, without any modifications to the source code of the Firefox browser or its TLS library, Network Security Services (NSS).
- On the server side, our implementation is achieved entirely as a cross-platform PHP application: it can be added at run-time without any modifications to the source code of the Apache web server or its TLS library, OpenSSL.

Binding the password authentication protocol to the TLS channel is achieved using the server's TLS certificate as the tag; both Firefox and Apache have APIs exposing the server's certificate to the extension and PHP application, respectively; the server certificate is one of two channel binding mechanisms standardized for TLS [17]. The source code size of our implementation is quite small. Our pure Javascript implementation is completely cross-platform and provides tolerable performance, with total round-trip time under half a second on a laptop, while our native C implementation using OpenSSL libraries provides high client and server performance with a total protocol execution time, including network latency in a corporate network, around 109ms. Our implementation is available for immediate download.

Though the ultimate goal of this line of work would be for such a protocol to be built into the browser, our Javascript extension may be amenable to gradual deployment to seed adoption while still achieving good performance, especially when making use of native libraries.

2   Password-Authenticated Confidential Channels

The security goal for our main construction is that it be a secure *password-authenticated and confidential channel establishment (PACCE)* protocol, which is a new password-based variant of the ACCE model of Jager et al. [27]. ACCE seems to be the most suitable for describing the security requirements of real-world secure channel protocols such as TLS [27,30,31,32] and SSH [28], and so it is natural to adapt it to the password setting.

A PACCE protocol is a two-party protocol that proceeds in two stages: in the *handshake* stage both participants perform an initial cryptographic handshake to establish session keys which are then used in the *record layer* stage to authenticate and encrypt the transmitted session data.[1] At some time during execution, the parties may accept the session as being legitimately authenticated, or reject. The main difference in PACCE compared to the original ACCE model is the use of passwords instead of long-term public keys for authentication.

At a high level, a PACCE protocol is secure if the adversary cannot break authentication, meaning it cannot cause a party to accept without having interacted with its intended partner, and cannot break the confidential channel, meaning it cannot read or inject ciphertexts.

We consider the standard client-server communication model where a party is either a *client* C or a *server* S. For each client-server pair (C, S) there exists a corresponding password $pw_{C,S}$ drawn from a dictionary $\mathcal{D}$.

An instance of party $U \in \{C, S\}$ in a session s is denoted as Π_U^s. Each instance Π_U^s records several variables:

- $\Pi_U^s.\text{pid}$: the *partner identity* with which Π_U^s believes to be interacting in the protocol session.
- $\Pi_U^s.\rho \in \{\text{init}, \text{resp}\}$: the *role* of this instance in the session, either initiator or responder. $\text{init}(\Pi_U^s)$ and $\text{resp}(\Pi_U^s)$ denote Π_U^s's view of who the initiator and responder are in the session, namely $(U, \Pi_U^s.\text{pid})$ when $\Pi_U^s.\rho = \text{init}$, and $(\Pi_U^s.\text{pid}, U)$ when $\Pi_U^s.\rho = \text{resp}$.
- $\Pi_U^s.T$: a *transcript* composed of all messages sent and received by the instance in temporal order.
- $\Pi_U^s.\alpha \in \{\text{active}, \text{accept}, \text{reject}\}$: the *status* of this instance.
- $\Pi_U^s.k$: the *session key* computed by this stage; initially set to empty $\emptyset$; when non-empty, it consists of two symmetric keys $\Pi_U^s.k^{\text{enc}}$ and and $\Pi_U^s.k^{\text{dec}}$ for encryption and decryption with some stateful length-hiding authenticated encryption scheme [27] used to provide confidentiality in the record layer stage. If $\Pi_U^s.\alpha = \text{accept}$, then $\Pi_U^s.k \neq \emptyset$
- $\Pi_U^s.b \in \{0, 1\}$: a randomly sampled bit used in the Encrypt oracle.

---

[1] In the original ACCE model, these stages were called the *pre-accept* and *post-accept* stages respectively. In PACCE, the parties may start sending encrypted data before accepting, so we have renamed the stages to handshake and record layer, which is suggestive of TLS, but of course can be used to model any appropriate protocol.

Two instances Π_U^s and $\Pi_{U'}^{s'}$ are said to be *partnered* if and only if $\Pi_U^s.\text{pid} = U'$, $\Pi_{U'}^{s'}.\text{pid} = U$, $\Pi_U^s.\rho \neq \Pi_{U'}^{s'}.\rho$, and their transcripts form *matching conversations* [27], denoted $\Pi_U^s.T \approx \Pi_{U'}^{s'}.T$.

The adversary $\mathcal{A}$ controls all communications and can interact with parties using certain oracle queries. Normal operation of the protocol is modelled by the following queries:

- $\mathsf{Send}^{\texttt{pre}}(\Pi_U^s, m)$: This query is answered as long as $\Pi_U^s.k = \emptyset$. In response the incoming message m is processed by Π_U^s and any outgoing message which is generated as a result of this processing is given to $\mathcal{A}$. Special messages $m = (\texttt{init}, U')$ and $m = (\texttt{resp}, U')$ are used to initialize the instance as initiator or responder, respectively, and to specify the identity of the intended partner U'. Note that processing of m may eventually lead to the end of the handshake stage, in which case Π_U^s either computes $\Pi_U^s.k$ and switches to the record layer stage or terminates with a failure.
- $\mathsf{Encrypt}(\Pi_U^s, m_0, m_1, \texttt{len}, \texttt{head})$,
- $\mathsf{Decrypt}(\Pi_U^s, C, \texttt{head})$: The Encrypt and Decrypt queries proceed as defined in the original ACCE definition [27] and are omitted due to space restrictions. Note that, compared with ACCE, we allow protocol messages to be sent on the encrypted channel: If $\Pi_U^s.\alpha = \texttt{active}$, then the returned plaintext message m is processed as a protocol message; the resulting outgoing message m' is encrypted using $\mathsf{Encrypt}(\Pi_U^s, m', m', \texttt{len}(m'), \texttt{head})$ and the resulting ciphertext C is returned to $\mathcal{A}$. Otherwise, when $\Pi_U^s.\alpha = \texttt{accept}$, the output of Decrypt is returned to $\mathcal{A}$.

Furthermore, the adversary may obtain some secret information:

- $\mathsf{RevealSK}(\Pi_U^s)$: Return $\Pi_U^s.k$.
- $\mathsf{Corrupt}(C, S)$: Return $\mathsf{pw}_{C,S}$.

Note that, compared with AKE models like the eCK model [33] that use public key authentication, password-based protocols cannot tolerate ephemeral key leakage while maintaining resistence to offline dictionary attacks, hence we do not include an ephemeral key leakage query.

Definition 1 (PACCE security). *An adversary $\mathcal{A}$ is said to (t, ϵ)-break a PACCE protocol if $\mathcal{A}$ runs in time t and at least one of the following two conditions hold:*

1. *$\mathcal{A}$ breaks authentication: When $\mathcal{A}$ terminates, then with probability at least $\epsilon + O(n/|\mathcal{D}|)$ where n is the number of initialized PACCE instances there exists an instance Π_U^s such that*
   (a) $\Pi_U^s.\alpha = \texttt{accept}$, and
   (b) $\mathcal{A}$ did not issue $\mathsf{Corrupt}(\texttt{init}(\Pi_U^s), \texttt{resp}(\Pi_U^s))$ before Π_U^s accepted, and
   (c) $\mathcal{A}$ did not issue $\mathsf{RevealSK}(\Pi_U^s)$ or $\mathsf{RevealSK}(\Pi_{U'}^{s'})$ for any $\Pi_{U'}^{s'}$ that is partnered to Π_U^s, and
   (d) there is no unique instance $\Pi_{U'}^{s'}$ that is partnered to Π_U^s.

2. $\mathcal{A}$ breaks authenticated encryption: *When $\mathcal{A}$ terminates and outputs a triple (U, s, b') such that conditions (a)–(c) from above hold, then we have that* $\left|\Pr\left[b' = \Pi_U^s.b\right] - \frac{1}{2}\right| \geq \epsilon + O(n/|\mathcal{D}|)$.

A PACCE protocol is (t, ϵ)-secure if there is no $\mathcal{A}$ that (t, ϵ)-breaks it; it is secure if it is (t, ϵ)-secure for all polynomial t and negligible ϵ in security parameter κ.

Observe that Definition 1 accounts for online dictionary attacks against PACCE protocols by using a lower bound $\epsilon + O(n/|\mathcal{D}|)$ for the adversarial success probability, which models $\mathcal{A}$'s ability to test at most one password (or a constant number) from the dictionary $\mathcal{D}$ in a single session.

3   Generic Construction

Our generic construction for secure PACCE between a client C and a server S from password-based authentication and a secure channel is as follows. First, the channel establishment protocol (CCE or ACCE) is run until it accepts. Then, using the secure channel, the two parties run a tag-based password authentication (tPAuth) protocol where the tag is a binding value from the secure channel; when the tPAuth protocol accepts, then the parties accept in the overall PACCE protocol, and then continue to use the channel for communication. Our two constructions in Sections 4 and 5 differ only in the way the tPAuth is bound to the established channel: the tag is either the transcript of the channel establishment protocol or the long-term public key of the server.

We now concretely describe the combination of a tPAuth protocol with TLS, as detailed in Figure 2. C and S first build up a standard TLS channel: that is, they execute a normal TLS handshake, then exchange ChangeCipherSpec messages to start authenticated encryption within the TLS record layer, and then exchange their Finished messages for explicit key confirmation. Once Finished messages are successfully exchanged, the parties continue using the authenticated encryption mechanism of the TLS record layer to communicate messages of the tag-based password-authentication protocol tPAuth. In our first construction this binding is achieved by using the Finished messages as the tag; note that Finished messages depend on the (hash of the) entire TLS handshake transcript. In our second construction the tag is the server's certificate (which includes the server's public key) that was communicated by S in its Certificate message of the TLS handshake. Upon successful completion of the password authentication phase both parties continue using session keys and authenticated encryption mechanism of the established TLS channel for secure communication.

4   Construction #1: Binding Using CCE Transcript

Our first generic PACCE protocol $\Gamma_T := \Gamma_T(\pi, \xi)$ is constructed as in Section 3 from a confidential channel establishment (CCE) protocol π and a tag-based password authentication (tPAuth) protocol ξ where the tag τ used is the transcript T of the CCE handshake stage. We will see that, because we are using

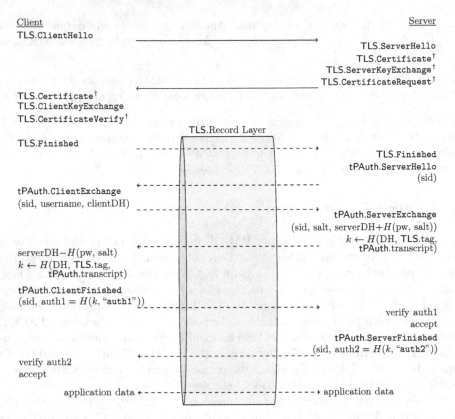

Fig. 2. Protocol message diagram for TLS with tunnelled tag-based password authentication, with example messages for the tSOKE protocol. TLS.tag is either TLS.Finished or TLS.Certificate. [†] denotes optional messages.

the full transcript from the channel establishment to bind the two protocols together, we need not rely on any authenticity properties of the channel, and thus can use a CCE protocol, not an ACCE protocol.

4.1   Building Block: CCE

As a building block in our analysis we use the notion of *confidential channel establishment (CCE)* that differs from (P)ACCE in that it is supposed to guarantee only confidentiality (and integrity) of the established channel, but not authentication of partners; hence, security is only assured for sessions in which the adversary $\mathcal{A}$ remains passive during the handshake stage. We thus model CCE by slightly modifying the PACCE model from Section 2. The differences are explained in Appendix A. Every secure (P)ACCE protocol is also CCE-secure: if we ignore the authentication aspects, then we still get confidential channel establishment in sessions where the adversary is passive during the handshake.

TLS does, when public keys are managed and used properly, provide strong authentication based on public keys, and can be proven ACCE-secure [27]. But, as we observed in the introduction, practice suggests we cannot rely on the web PKI to provide ideal authentic distribution and mapping of public keys to identities. TLS can in practice be seen as a CCE protocol: even though long-term public keys may be used in TLS, we are not confident in their authenticity, so we only take TLS to provide CCE security.

4.2 Building Block: Tag-Based Password Authentication

Tag-based authentication [34] accounts for the use of auxiliary, possibly public, strings (tags) in authentication protocols — each party uses a tag, in addition to the authentication factor, and the protocol guarantees that if parties accept then their tags match. This concept was introduced in [34] for public key-based authentication protocols and then generalized in [35] for other types of authentication factors, including passwords and biometrics. In our analysis we will use a *tag-based password authentication* protocol, denoted tPAuth.

The model of tPAuth can be described using the setting of PACCE protocols from Section 2. A tPAuth session is executed between a client instance Π_C^s and a server instance $\Pi_S^{s'}$ on input the corresponding password $\mathsf{pw}_{C,S}$ from the dictionary $\mathcal{D}$ and some tag $\tau \in \{0,1\}^*$. A tPAuth session is successful if both instances use the same password $\mathsf{pw}_{C,S}$ and tag τ as their input. The security of tPAuth protocols, extends the traditional password authentication requirement that accounts for online dictionary attacks with the requirement of tag equality, and is achieved formally by an extended definition of partnering. The details of tPAuth security appear in Appendix B.

In Appendix B, we present tSOKE, a tag-based variant of the Simple Open Key Exchange (SOKE) variant [14]. An example of the tSOKE message flow can be seen in Figure 2. Since SOKE is a password-authenticated key exchange protocol, it does establish a secure session key; however, we only use its properties of *mutual authentication* and *resistance to dictionary attacks*. The tag is inserted into the key used for explicit key confirmation; it does not need resistance to dictionary attacks since it is known to the adversary, and thus is not used in the Diffie–Hellman portion like the password.

4.3 Security Analysis of Construction #1

Theorem 1 (CCE + tPAuth$_{\tau=T_{CCE}}$ $\implies$ PACCE). *The generic construction of a PACCE protocol $\Gamma_T(\pi, \xi)$ from a CCE protocol π and a tPAuth protocol ξ, with the tag equal to the transcript T_{CCE} from the CCE handshake stage, is PACCE-secure, assuming the underlying protocols are secure.*

The proof consists of a sequence of games. In the first game, the simulator continues to simulate the CCE portion of the protocol but undetectably replaces the tPAuth simulation with that of a real tPAuth challenger. Next, the simulator aborts if any of its instances accept without a partnered instance existing; this will correspond to a violation of authentication in the underlying tPAuth challenger. In the third game, the simulator now undetectably replaces the CCE simulation with that of a real CCE challenger. An adversary who can win against the resulting PACCE simulator can be used to win against the underlying CCE challenger. The proof appears in the full version [36].

4.4   Using `tls-unique` Channel Binding

The `tls-unique` channel binding mechanism [17] can be used to instantiate construction #1. For `tls-unique` the channel binding string is the first `Finished` message, which is the output of a pseudorandom function on the hash of the TLS handshake transcript.

It is straightforward to see that if the `Finished` message is used as the tag for channel binding instead of the full transcript in an analogous generic construction $\Gamma_{fin}(\pi, \xi)$, and the hash function H is collision-resistant and the pseudorandom function PRF is secure, then Γ_{fin} is a secure PACCE. This follows by noting that, except with negligible probability, the parties must use the same transcript in order to arrive at the same tag, and then the proof of Theorem 1 applies.

5   Construction #2: Binding Using Server Public Key

Our second generic PACCE protocol $\Gamma_{pk} := \Gamma_{pk}(\pi, \xi)$ is constructed as in Section 3 from an authenticated and confidential channel establishment (ACCE) protocol π and a tPAuth protocol ξ where the tag τ used is the long-term public key used by the server in the ACCE protocol.

Because we are using only the server's long-term public key, and not the full transcript from the channel establishment, to bind the two protocols together, we now must rely on some authenticity properties of the channel. However, we will not be relying on *users* to correctly validate the server's public key or decide which long-term server public key corresponds with which password: from the external perspective, the protocol is still a PACCE protocol, with authentication only coming from passwords, not from long-term server public keys.

5.1   Building Block: ACCE (With Key Registration)

ACCE [27] is currently the most complete model for the security properties of the core TLS protocol. We use a variant of ACCE as a building block in our generic construction. The first variation is that we allow for either server-only or mutual authentication, as in Giesen et al. [32]; when server-only authentication is used, only client instances are legitimate targets for breaking authentication. The second variation is in how static public keys are distributed. In typical AKE and ACCE models, it is simply assumed that parties have authentic copies of all static public keys, abstracting the problem away. Since we will use ACCE as a building block under the assumption that the "static" public keys are not to be trusted as authentic, we allow the adversary to cause any public key to be accepted as a static public key using a Register query; only sessions where the key is not an adversary-registered key are legitimate targets for breaking. The formal differences between the ACCE model with key registration and the original ACCE model appear in the full version [36].

5.2   Security Analysis of Construction #2

Theorem 2 (ACCE + tPAuth$_{\tau=pk}$ $\implies$ PACCE). *The generic construction of a PACCE protocol $\Gamma_T(\pi, \xi)$ from an ACCE protocol π and a tPAuth protocol ξ, with tag set to the server's public key pk from the ACCE handshake stage, is PACCE-secure, assuming the underlying protocols are secure.*

In the first game, the simulator simulates the ACCE portion of the protocol and undetectably replaces the tPAuth portion using messages that it obtains from a real tPAuth challenger. Next, the simulator aborts if any of its instances accept without an instance whose tPAuth transcripts and the input tags match, which corresponds to an attack against the tPAuth protocol. In the third game, the simulator will use messages obtained from a real ACCE challenger such that it can use any adversary who wins against the resulting PACCE simulator to break the security of the ACCE protocol; when the adversary uses its own long-term public keys (which is allowed since they are not authenticated), we use the key registration functionality of the (modified) ACCE challenger. Due to space constraints, the proof appears in the full version [36].

5.3   Using `tls-server-end-point` Channel Binding

The `tls-server-end-point` channel binding mechanism [17] can be used to instantiate construction #2. For `tls-server-end-point` the channel binding string is the hash of the server's X.509 certificate. Note that the certificate contains the server's public key as a canonically identifiable substring.

It is straightforward to see that if the hash of the certificate is used as the tag for channel binding instead of the raw public key in an analogous generic construction $\Gamma_{cert}(\pi, \xi)$, and the hash function is second-preimage-resistant, then Γ_{cert} is a secure PACCE. This follows by noting that an active adversary must use a certificate that hashes to the same value as the server's certificate.

6   Implementation

As an important motivation for our modular protocol design was the ability to modularly implement the protocol, we produced a prototype to demonstrate this.

For the tag-based password authentication protocol, we propose tSOKE, a tag-based version of the SOKE protocol [14], a highly efficient Diffie–Hellman based PAKE; see Appendix B for details of tSOKE. We used the NIST P-192 elliptic curve group [37].

We implemented the client side of the protocol as an extension for Mozilla Firefox (320 lines of custom Javascript, plus libraries), and the server side of the protocol as a PHP application (210 lines of custom PHP code, plus libraries) on an Apache web server. No modifications to the source code of the underlying web browser (Firefox) or the underlying web server (Apache with OpenSSL) were required—in particular, we did not have to alter the SSL/TLS implementation and we did not have to recompile Firefox or Apache. Since the mechanism that the server code uses to obtain the certificate of the TLS connection is an Apache CGI (Common Gateway Interface) variable, any server-side language would work, not just PHP.

Our implementation is available online under an open-source license at `http://www.douglas.stebila.ca/research/papers/MSD14/` and `http://eprints.qut.edu.au/76270`.

6.1   Firefox Extension

The client-side Firefox extension is written in Javascript and uses an existing Firefox API to obtain the certificate of the TLS connection. The client implementation of our protocol (excluding underlying cryptographic primitives) is just 320 lines of Javascript code. Cryptographic operations can be done in either pure Javascript (relying on about 1400 lines of code from Wu's Javascript elliptic curve cryptography and big integer arithmetic implementation[2] and about 6KB of minified Javascript from the Stanford Javascript Crypto Library[3] for the PBKDF2 algorithm) or can make use of native C OpenSSL libraries using Firefox's js-ctypes API[4].

When the extension detects (using an appropriate triggering mechanism; see the full version [36]) a page that supports the protocol, it displays a notification bar that secure password authentication is supported (Figure 3(a)). The user then clicks on the "Login" button in the notification bar to bring up the password entry dialog box (Figure 3(b)). Note that the notification bar is displayed using Firefox's API for notifications, similar to how alerts are rendered for missing plugins. By using the standard notification mechanism, we provide a trusted UI path to the notification bar, and then, through the login button on that bar, a trusted UI path to the dialog box, somewhat mitigating concerns about the difficulty of providing a trusted UI path in browser-based secure password authentication [38,12]. The status of the mutual authentication is displayed in the notification bar (Figures 3(c), (d)); if successful the browser is redirected to the URL indicated by the server.

At present, Firefox is the only web browser whose extension APIs offer partial implementation of the channel bindings for TLS from RFC5929 [17], providing access to the certificate of the page's TLS connection. The APIs for Google Chrome and Apple Safari extensions do not seem to permit this ability so far, nor does the API for Microsoft Internet Explorer browser toolbars. However, our modular approach is still validated, in that Chrome, Safari, and IE would only need to implement the recommendations from [17] such that our extension could do the rest of the protocol, rather than requiring the full protocol be implemented within the core browser source code as many other approaches require.

---

[2] `http://www-cs-students.stanford.edu/~tjw/jsbn/`

[3] `http://crypto.stanford.edu/sjcl/`

[4] `https://developer.mozilla.org/en-US/docs/Mozilla/js-ctypes`

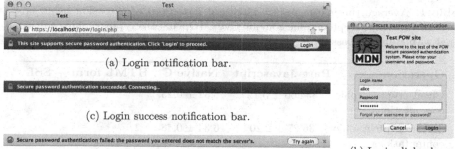

(a) Login notification bar.

(c) Login success notification bar.

(d) Login failure notification bar.

(b) Login dialog box.

Fig. 3. User interface for Mozilla Firefox extension

Branding. Our prototype allows the server to specify some limited "branding" customizations to the login dialog box, including displaying a logo and explanatory text, as can be seen in Figure 3(b). A common objection to server-specified branding is that the protocol becomes insecure due to phishing attacks: while it is true that an attacker could put in a different logo or text in Figure 3(b), the attacker *gains nothing* in doing so: the protocol cryptographically protects the password, even when the user's browser runs the protocol with attacker's server. At best, the attacker can interrupt communication, but will gain no information. Our limited branding does not give the attacker enough power to completely spoof the user interface and trick the user into using the attacker's own dialog box, due to the trusted UI path via the browser notfication bar.

6.2 Performance

In Table 1, we report timings for our implementation. Timings reported are an average of 10 timings, with standard deviation. The total runtime of the protocol includes the network latency for the communication within a corporate network where the client and server machines were located. The average ping time on the network was 48.55 ms (stdev. 37.76).

We report two different sets of timings: "cross-platform" timings, using pure Javascript on the client side and PHP with built-in GMP libraries on the server side; and "native" timings, using calls to OpenSSL for cryptographic operations on both the client side and the server side.

The average total runtime from when the user clicked "Login" after entering their password until the protocol completes was 487.72 ms (stdev. 49.93) using cross-platform code, and 109.04 ms (stdev. 47.96) using native code. In comparison, the average total runtime of password authentication based on an HTML form over TLS in our setting takes 66.16 ms (stdev. 27.80)). The difference of about 40ms with our native code implementation corresponds to one additional round trip and is unlikely to be perceptible by the users. Our native cryptographic code is further comparable to Dacosta et al.'s reported performance of DVCert on laptops [23]. Our protocol implementation includes a variety of

Table 1. Average runtime in ms ($\pm$ standard deviation) of extension using cross-platform Javascript cryptographic code, native C (OpenSSL) cryptographic code, compared with standard passwords submitted using an HTML form over SSL

Operation	Pure Javascript	Native C	HTML form + SSL
Client cryptographic computations	354.06 ± 5.12	5.32 ± 0.23	—
Server cryptographic computations	36.27 ± 2.20	6.34 ± 0.48	—
Total runtime	487.72 ± 49.93	109.04 ± 47.96	66.16 ± 27.80

Software: Mozilla Firefox 21.0, Apache 2.2.22, PHP 5.4.14, GMP 5.1.1, MySQL 5.5.28, OpenSSL 1.0.1e, Mac OS X 10.8.3.
Hardware: 2.6 GHz Intel Core i7 (3720QM), 16 GB of RAM.
Network: Corporate network, ping time 48.55 ± 37.76ms

operations beyond cryptographic computations, so the total runtime is greater than the sum of cryptographic runtime and communication time.

7   Discussion

7.1   TLS Channel Bindings

The tls-unique channel binding works with all TLS ciphersuites, whereas the tls-server-end-point binding only works with TLS ciphersuites that use certificate-based server authentication, though these are most widely used in practice.

tls-server-end-point may be easier to deploy on the server side since the server certificate is often fixed for long periods, and thus more suitable for multi-server architectures where for example an SSL accelerator handles the TLS connection and then passes the plaintext onto one of potentially many layers of application servers. tls-server-end-point is also easily deployable on the client side: for example, the Firefox extension API already makes the server certificate, but not the Finished message, available.

In some sense tls-unique is a stronger channel binding string: with it, we can achieve security of our generic construction using only CCE security of the TLS channel, whereas with tls-server-end-point we rely on the stronger ACCE security notion of TLS. In the end, both allow us to achieve our goal.

TLS keying material exporters [39] are another option for binding to the TLS channel, as they allow an application to obtain keying material derived from the master secret key for a given label. However, TLS channel bindings appear to be the preferred mechanism, and so we focus on them.

7.2 Challenges with PAKE

Although PAKE protocols have been known in the literature since their invention 1992, they have seen almost no deployed adoption for user authentication in real-world protocols and implementations, with the exception of the use of the socialist millionaires' protocol in the Off-the-Record Messaging (OTR) protocol for private instant messaging [26]. Engler et al. [12] recently identified several challenges—divided into two classes, user interface and deployment challenges— to adopting cryptographic protocols for password authentication in the web. It has also been noted that the myriad patents related to PAKE have had a negative impact on adoption [40].

Deployment challenges. This modular architecture may address certain deployment challenges. Engler et al. ask "What is the appropriate layer in the networking stack to integrate PAKE protocols?" They compare two proposed options: TLS-SRP [13] and an earlier draft of the HTTPS-PAKE approach of Oiwa et al. [20]. Adding SRP as a TLS ciphersuite has benefits in that, once implemented, allows multiple applications to use the same TLS implementation. But many drawbacks are identified by Engler et al., including: (i) the need to integrate the application layer with the TLS layer on both the client side (necessitating a complex API between the TLS library and the web browser, for example) and on the server side (which could negatively affect the ability of HTTPS load balancers to terminate TLS connections and then hand them off to web application servers); and (ii) the difficulty of supporting multiple authentication realms within the same domain. HTTPS-PAKE, running as an HTTP authentication mechanism at the application layer, avoids both of these problems. The version of HTTPS-PAKE reviewed by Engler et al. did not have cryptographically strong binding between the two protocols and thus could not prevent man-in-the-middle attacks, but later revisions addressed that issue. Our modular approach avoids the problems that Engler et al. identify for TLS-SRP.

Our approach also better handles the transition from unauthenticated encrypted browsing to authenticated encrypted browsing: a user may browse an HTTPS site for a while before logging in; with TLS-SRP, a new TLS connection is required (and the mechanism for triggering a new TLS connection is unclear); it is much easier to trigger the authentication at the application layer when it is required.

User interface challenges. Engler et al. [12] identify several user interface challenges. We do not aim to fully solve all these challenges in our prototype, as demonstrating a convincing solution to these challenges requires critical examination by usability experts and appropriate user studies. Nonetheless, we have endeavoured to follow some best practices that may at least partially address the identified UI challenges.

It is essential for the security of PAKE protocols that the user always enter their password into a secure dialog box. If the entry mechanism can be spoofed by an attacking website, then the user could be tricked into entering their password directly into a textfield controlled by the attacker. Thus, there must be a *trusted path* to the dialog box in the UI, usually achieved by placing the password entry visibly in the browser chrome (i.e., the parts of the window that make up the browser UI, such as the location bar, rather than the page content). In our prototype, we follow this practice by using Firefox's notification bar. It has been suggested that permitting users to customize notification bars helps to reduce spoofing attacks [38].

The second and third of Engler et al.'s UI challenges are about how to train users to use the system in the first place, and how to communicate failures to users in a way that they do not fall back on insecure methods. Both of these remain a challenge for usability designers, though again, delivering failure notifications via Firefox's trusted path for notifications may provide some benefit. Providing forgotten password resets securely remains an open challenge both in practice and in theory and is outside the scope of our goals.

The final challenge noted by Engler et al. is on how to allow website designers to customize and brand the login dialog without compromising security; it has been suggested that lack of customization and branding was a contributing factor to the lack of adoption of HTTP basic and digest authentication. Our prototype allows the server to provide a few customizations to the login dialog box, including a logo, some explanatory text, as shown in Figure 3(b). While an attacker could use stolen images, the benefit to the attacker is minimal since the password entry will be cryptographically protected.

Adoption challenges. A final challenge for any new security technology is facilitating widespread adoption. Such protocols see a "network effect": it is only useful for a client Alice to use the technology if there are many Bobs who support it, and vice versa. In the end, any secure password authentication technology will be most successful once built in to all major web browsers and web application frameworks. In the meantime, the modular approach in this paper is suitable for gradual deployment. For example, an organization can internally standardize on the use of this approach by deploying an extension to all of its users' browsers without needing to wait for the browser vendor to support the protocol. The more adoption via extension, the more evidence for interest in the technology, and the greater incentive for vendors to provide a native implementation.

Acknowledgements. Mark Manulis was supported by the German Research Foundation (DFG), project PRIMAKE (MA 4957). Douglas Stebila acknowledges funding from Australian Research Council (ARC) Discovery Project DP130104304.

References

1. Schechter, S.E., Dhamija, R., Ozment, A., Fischer, I.: The emperor's new security indicators. In: 2007 IEEE Symposium on Security and Privacy, pp. 51–65. IEEE Computer Society Press (2007)
2. Sunshine, J., Egelman, S., Almuhimedi, H., Atri, N., Cranor, L.F.: Crying wolf: An empirical study of SSL warning effectiveness. In: USENIX Security 2009 (2009)
3. Franks, J., Hallam-Baker, P., Hostetler, J., Lawrence, S., Leach, P., Luotonen, A., Stewart, L.: HTTP Authentication: Basic and Digest Access Authentication. RFC 2617 (Draft Standard), Updated by RFC 7235 (1999)
4. Bellovin, S.M., Merritt, M.: Encrypted key exchange: Password-based protocols secure against dictionary attacks. In: 1992 IEEE Symposium on Security and Privacy, pp. 72–84. IEEE Computer Society Press (1992)
5. Bellare, M., Pointcheval, D., Rogaway, P.: Authenticated key exchange secure against dictionary attacks. In: Preneel, B. (ed.) EUROCRYPT 2000. LNCS, vol. 1807, pp. 139–155. Springer, Heidelberg (2000)
6. Canetti, R., Halevi, S., Katz, J., Lindell, Y., MacKenzie, P.: Universally composable password-based key exchange. In: Cramer, R. (ed.) EUROCRYPT 2005. LNCS, vol. 3494, pp. 404–421. Springer, Heidelberg (2005)
7. Abdalla, M., Catalano, D., Chevalier, C., Pointcheval, D.: Efficient two-party password-based key exchange protocols in the UC framework. In: Malkin, T. (ed.) CT-RSA 2008. LNCS, vol. 4964, pp. 335–351. Springer, Heidelberg (2008)
8. International Organization for Standardization (ISO): ISO/IEC 11770-4: Information technology — security techniques — key management — part 4: Mechanisms based on weak secrets (2006)
9. ITU-T X.1035: Password-authenticated key exchange (PAK) protocol (2007)
10. IEEE P1363.2: Standard specifications for password-based public-key cryptographic techniques (2008)
11. Wu, T.D.: The secure remote password protocol. In: NDSS 1998. The Internet Society (1998)
12. Engler, J., Karlof, C., Shi, E., Song, D.: Is it too late for PAKE? In: Web 2.0 Security and Privacy (W2SP) 2009 (2009)
13. Taylor, D., Wu, T., Mavrogiannopoulos, N., Perrin, T.: Using the Secure Remote Password (SRP) Protocol for TLS Authentication. RFC 5054, Informational (2007)
14. Abdalla, M., Bresson, E., Chevassut, O., Möller, B., Pointcheval, D.: Provably secure password-based authentication in TLS. In: Lin, F.C., Lee, D.T., Lin, B.S., Shieh, S., Jajodia, S. (eds.) ASIACCS 2006, pp. 35–45. ACM Press (2006)
15. Abdalla, M., Pointcheval, D.: Simple password-based encrypted key exchange protocols. In: Menezes, A. (ed.) CT-RSA 2005. LNCS, vol. 3376, pp. 191–208. Springer, Heidelberg (2005)
16. Hao, F., Ryan, P.Y.A.: Password authenticated key exchange by juggling. In: Christianson, B., Malcolm, J.A., Matyas, V., Roe, M. (eds.) Security Protocols 2008. LNCS, vol. 6615, pp. 159–171. Springer, Heidelberg (2011)
17. Altman, J., Williams, N., Zhu, L.: Channel Bindings for TLS. RFC 5929 (Proposed Standard) (2010)
18. Oiwa, Y., Takagi, H., Watanabe, H., Suzuki, H.: PAKE-based mutual HTTP authentication for preventing phishing attacks. In: Maarek, Y., Nejdl, W. (eds.) Proc. 18th International World Wide Web Conference (WWW 2009), pp. 1143–1144. ACM (2009)

19. Oiwa, Y., Watanabe, H., Takagi, H.: PAKE-based mutual HTTP authentication for preventing phishing attacks (2009), http://arxiv.org/abs/0911.5230

20. Oiwa, Y., Watanabe, H., Takagi, H., Ioku, Y., Hayashi, T.: Mutual authentication protocol for HTTP (2012), Internet-Draft, http://tools.ietf.org/html/draft-oiwa-http-mutualauth-12

21. AIST Research Center for Information Security: (Mutual authentication protocol for HTTP), https://www.rcis.aist.go.jp/special/MutualAuth

22. Kwon, T.: Authentication and key agreement via memorable passwords. In: NDSS 2001. The Internet Society (2001)

23. Dacosta, I., Ahamad, M., Traynor, P.: Trust no one else: Detecting MITM attacks against SSL/TLS without third-parties. In: Foresti, S., Yung, M., Martinelli, F. (eds.) ESORICS 2012. LNCS, vol. 7459, pp. 199–216. Springer, Heidelberg (2012)

24. Boyko, V., MacKenzie, P.D., Patel, S.: Provably secure password-authenticated key exchange using diffie-hellman. In: Preneel, B. (ed.) EUROCRYPT 2000. LNCS, vol. 1807, pp. 156–171. Springer, Heidelberg (2000)

25. Borisov, N., Goldberg, I., Brewer, E.A.: Off-the-record communication, or, why not to use PGP. In: ACM Workshop on Privacy in Electronic Society (WPES 2004), pp. 77–84. ACM Press (2004)

26. Alexander, C., Goldberg, I.: Improved user authentication in Off-The-Record messaging. In: Yu, T. (ed.) ACM Workshop on Privacy in Electronic Society (WPES 2007), pp. 41–47. ACM Press (2007)

27. Jager, T., Kohlar, F., Schäge, S., Schwenk, J.: On the security of TLS-DHE in the standard model. In: Safavi-Naini, R., Canetti, R. (eds.) CRYPTO 2012. LNCS, vol. 7417, pp. 273–293. Springer, Heidelberg (2012)

28. Bergsma, F., Dowling, B., Kohlar, F., Schwenk, J., Stebila, D.: Multi-ciphersuite security of the Secure Shell (SSH) protocol. In: Yung, M., Li, N. (eds.) ACM CCS 2014. ACM Press (2014)

29. Brzuska, C., Smart, N.P., Warinschi, B., Watson, G.J.: An analysis of the EMV channel establishment protocol. In: Sadeghi, A.R., Gligor, V.D., Yung, M. (eds.) ACM CCS 2013, pp. 373–386. ACM Press (2013)

30. Krawczyk, H., Paterson, K.G., Wee, H.: On the security of the TLS protocol: A systematic analysis. In: Canetti, R., Garay, J.A. (eds.) CRYPTO 2013, Part I. LNCS, vol. 8042, pp. 429–448. Springer, Heidelberg (2013)

31. Kohlar, F., Schäge, S., Schwenk, J.: On the security of TLS-DH and TLS-RSA in the standard model. Cryptology ePrint Archive, Report 2013/367 (2013), http://eprint.iacr.org/2013/367

32. Giesen, F., Kohlar, F., Stebila, D.: On the security of TLS renegotiation. In: Sadeghi, A.R., Gligor, V.D., Yung, M. (eds.) ACM CCS 2013, pp. 387–398. ACM Press (2013)

33. LaMacchia, B.A., Lauter, K., Mityagin, A.: Stronger security of authenticated key exchange. In: Susilo, W., Liu, J.K., Mu, Y. (eds.) ProvSec 2007. LNCS, vol. 4784, pp. 1–16. Springer, Heidelberg (2007)

34. Jager, T., Kohlar, F., Schäge, S., Schwenk, J.: Generic compilers for authenticated key exchange. In: Abe, M. (ed.) ASIACRYPT 2010. LNCS, vol. 6477, pp. 232–249. Springer, Heidelberg (2010)

35. Fleischhacker, N., Manulis, M., Azodi, A.: A Modular Framework for Multi-Factor Authentication and Key Exchange. Cryptology ePrint Archive, Report 2012/181 (2012), http://eprint.iacr.org/2012/181

36. Manulis, M., Stebila, D., Denham, N.: Secure modular password authentication for the web using channel bindings (full version). Cryptology ePrint Archive, Report 2014/731 (2014), http://eprint.iacr.org/2014/731

37. National Institute of Standards and Technology: Recommended elliptic curves for federal government use (1999),
http://csrc.nist.gov/groups/ST/toolkit/documents/dss/NISTReCur.pdf

38. Dhamija, R., Tygar, J.D.: The battle against phishing: Dynamic security skins. In: Cranor, L.F., Zurko, M.E. (eds.) Symposium on Usable Privacy and Security (SOUPS 2005), pp. 77–88. ACM Press (2005)

39. Rescorla, E.: Keying Material Exporters for Transport Layer Security (TLS). RFC 5705 (Proposed Standard) (2010)

40. Abdalla, M., Bresson, E., Chevassut, O., Möller, B., Pointcheval, D.: Strong password-based authentication in TLS using the three-party group Diffie–Hellman protocol. International Journal of Security and Networks 2, 284–296 (2007)

41. Certicom Research: SEC 1: Elliptic curve cryptography, Version 2.0 (2009)

A   Confidential Channel Establishment (CCE)

The model of CCE protocols can be described using the setting of PACCE protocols from Section 2, with a few differences. The first difference is that there are no passwords or identities involved; hence no Corrupt oracle is needed, nor is the U' parameter required in the initialization in the Send^{pre} query. Further, the security condition is adjusted so that only sessions where the adversary was passive in the handshake stage are considered. The oracles RevealSK, Encrypt and Decrypt remain unchanged. The following definition of CCE security is obtained from Definition 1 by considering the above mentioned modifications.

Definition 2 (CCE security). *An adversary $\mathcal{A}$ is said to (t, ϵ)-break a CCE protocol if $\mathcal{A}$ runs in time t and, when $\mathcal{A}$ terminates and outputs a triple (U, s, b') such that*

(a) $\Pi_U^s.\alpha = \text{accept}$, and

(b) there exists an instance $\Pi_{U'}^{s'}$ that is partnered to Π_U^s, and

(c) $\mathcal{A}$ did not issue RevealSK(Π_U^s) or RevealSK($\Pi_{U'}^{s'}$) for any $\Pi_{U'}^{s'}$ that is partnered to Π_U^s,

then $\left| \Pr\left[b' = \Pi_U^s.b \right] - \frac{1}{2} \right| \geq \epsilon$.

B   Tag-Based Password Authentication (tPAuth)

The model of tPAuth can be described using the setting of PACCE protocols from Section 2. A tPAuth session is executed between a client instance Π_C^s and a server instance $\Pi_S^{s'}$ on input the corresponding password $\text{pw}_{C,S}$ from the dictionary $\mathcal{D}$ and some tag $\tau \in \{0,1\}^*$. A tPAuth session is successful if both instances use the same password $\text{pw}_{C,S}$ and tag τ as their input. The requirement on tag equality leads to the extended definition of partnering: two instances Π_C^s and $\Pi_S^{s'}$ are *partnered* if $\Pi_C^s.\text{pid} = S$, $\Pi_S^{s'}.\text{pid} = C$, $\Pi_C^s.T \approx \Pi_{S'}^s.T$ (matching transcripts), and $\Pi_C^s.\tau = \Pi_S^{s'}.\tau$ (equal tags).

A tPAuth adversary $\mathcal{A}$ is active and interacts with instances of $U \in \{C, S\}$ using the following oracles:

- $\mathsf{Send}(\Pi_U^s, m)$: This query is identical to $\mathsf{Send}^{\mathtt{pre}}$ from the PACCE model except for one important difference — when $\mathcal{A}$ initializes some instance Π_U^s using the special messages $m = (\mathtt{init}, U', \tau)$ or $m = (\mathtt{resp}, U', \tau)$ then it additionally provides as input a tag τ which will be used by the instance in the tPAuth session. This essentially gives $\mathcal{A}$ full control over the tags that are used in the protocol.
- $\mathsf{Corrupt}(C, S)$: This query reveals the corresponding password $\mathsf{pw}_{C,S}$.

The security of tPAuth protocols, defined in the following, extends the traditional password authentication requirement that accounts for online dictionary attacks with the requirement of tag equality, which is implied by condition 3 due to the extended definition of partnering.

Definition 3 (tPAuth security). *An adversary $\mathcal{A}$ is said to (t, ϵ)-break a tPAuth protocol if after the termination of $\mathcal{A}$ that runs in time t with probability at least $\epsilon + O(n/|\mathcal{D}|)$ where n is the number of initialized tPAuth instances there exists an instance Π_U^s such that*

1. $\Pi_U^s.\alpha = \mathtt{accept}$, and
2. $\mathcal{A}$ did not issue $\mathsf{Corrupt}(\mathtt{init}(\Pi_U^s), \mathtt{resp}(\Pi_U^s))$ before Π_U^s accepted, and
3. there is no unique instance $\Pi_{U'}^{s'}$ that is partnered to Π_U^s.

A tPAuth protocol is (t, ϵ)-secure if there is no $\mathcal{A}$ that (t, ϵ)-breaks it; it is secure if it is (t, ϵ)-secure for all polynomial t and negligible ϵ in security parameter κ.

Our tagged variant tSOKE of the Simple Open Key Exchange (SOKE) protocol [14] is shown in Figure 4, and is a secure tPAuth protocol (see full version [36]).

System parameters

Elliptic curve group nistp192 with generator G of order n
Second generator G' constructed verifiably at random with $\langle G \rangle = \langle G' \rangle$
$G' = (\texttt{0x8da36f68628a18107650b306f22b41448cb60fe5712dd57a},$
    $\texttt{0x1f64a649852124528a09455de6aad151b4c0a9a8c2e8269c})$

(constructed verifiably at random [41, §3.1.3.2] with seed string "This is the seed string for the web passwords protocol.")

Registration stage (takes place over a secure channel)

Client A	**Server** B
1. Enter username id_A.	
2. Enter password pw_{AB}.	
3. $\mathsf{salt} \leftarrow \{0,1\}^{128}$.	
4. Choose iteration counter $c \in \mathbb{N}^+$.	
5. $h \leftarrow \mathsf{PBKDF2}(\mathsf{SHA\text{-}256}, \mathsf{pw}_{AB}, \mathsf{salt}, c, 256)$	
6. $\xrightarrow{\ \mathrm{id}_A, \mathsf{salt}, c, h\ }$	
7.	Store salt, c, h for id_A.

Login stage

Client A	**Server** B
Input: tag τ	Input: tag τ
1. Enter username id_A.	
2. Enter password pw_{AB}.	
3. $x \leftarrow_R \{2, \ldots, n-1\}$	
4. $X \leftarrow xG$	
5. $\xrightarrow{\ \mathrm{id}_A, X\ }$	
6.	Look up salt, c, h for id_A.
7.	$y \leftarrow_R \{2, \ldots, n-1\}$
8.	$Y \leftarrow yG$
9.	$Y^* \leftarrow Y + (h \bmod n)G'$
10. $\xleftarrow{\ \mathsf{salt}, c, Y^*\ }$	
11. $h \leftarrow \mathsf{PBKDF2}(\mathsf{SHA\text{-}256}, \mathsf{pw}_{AB}, \mathsf{salt}, c, 256)$	
12. $Y \leftarrow Y^* - (h \bmod n)G'$	
13. $Z \leftarrow xY$	$Z' \leftarrow yX$
14. $pms \leftarrow \mathsf{SHA\text{-}256}(\mathrm{id}_A, h, \tau, X, Y^*, Z)$	$K \leftarrow \mathsf{SHA\text{-}256}(\mathrm{id}_A, h, \tau, X, Y^*, Z)$
15. $A_1 \leftarrow \mathsf{SHA\text{-}256}(pms, \text{"auth1"})$	
16. $\xrightarrow{\quad A_1 \quad}$	
17.	Abort if $A_1 \neq \mathsf{SHA\text{-}256}(pms, \text{"auth1"})$
18.	$A_2 \leftarrow \mathsf{SHA\text{-}256}(pms, \text{"auth2"})$
19. $\xleftarrow{\quad A_2 \quad}$	
20. Abort if $A_2 \neq \mathsf{SHA\text{-}256}(pms, \text{"auth2"})$	
21. Accept	Accept

Fig. 4. tSOKE protocol registration and login stages

A Modular Framework for Multi-Factor Authentication and Key Exchange

Nils Fleischhacker[1], Mark Manulis[2], and Amir Azodi[3]

[1] Saarland University, Germany
`fleischhacker@cs.uni-saarland.de`
[2] Surrey Centre for Cyber Security, University of Surrey, UK
`mark@manulis.eu`
[3] Hasso Plattner Institute, Germany
`amir.azodi@hpi.de`

Abstract. Multi-Factor Authentication (MFA), often coupled with Key Exchange (KE), offers very strong protection for secure communication and has been recommended by many major governmental and industrial bodies for use in highly sensitive applications. Over the past few years many companies started to offer various MFA services to their users and this trend is ongoing.

The MFAKE protocol framework presented in this paper offers *à la carte* design of multi-factor authentication and key exchange protocols by mixing multiple *types* and *quantities* of authentication factors in a secure way: MFAKE protocols designed using our framework can combine any subset of multiple low-entropy (one-time) passwords/PINs, high-entropy private/public keys, and biometric factors. This combination is obtained in a modular way from efficient single-factor password-based, public key-based, and biometric-based authentication-only protocols that can be executed in concurrent sessions and bound to a single session of an unauthenticated key exchange protocol to guarantee forward secrecy.

The modular approach used in the framework is particularly attractive for MFAKE solutions that require backward compatibility with existing single-factor authentication solutions or where new factors should be introduced gradually over some period of time. The framework is proven secure using the state-of-the art game-based security definitions where specifics of authentication factors such as dictionary attacks on passwords and imperfectness of the biometric matching processes are taken into account.

Keywords: two-factor, multi-factor authentication, tag-based authentication, key exchange, framework, modular design.

1 Introduction

Authentication Factors. An *authentication factor* is used to produce some evidence that an entity at the end of the communication channel is the one

L. Chen and C. Mitchell (Eds.): SSR 2014, LNCS 8893, pp. 190–214, 2014.

which it claims to be. Modern computer security knows different types of authentication factors, all of which are widely used in practice. Their standard classification considers three main groups (see e.g. [18]), characterized by the nature of provided evidence: knowledge, possession, and physical presence. The evidence of knowledge is typically offered by low-entropy *passwords*. These include memorizable (long-term) passwords or PINs, e.g. for login purposes and banking ATMs, and one-time passwords that are common to many online banking and e-commerce transactions. The evidence of possession corresponds to physical devices such as smart cards, tokens, or TPMs, equipped with long-term (high-entropy) *secret keys* and some cryptographic functionality. These devices have tamper-resistance to protect secret keys from exposure. The evidence of physical presence refers to unique *biometric identifiers* of human beings.

A different approach might be needed for an attacker to compromise a particular factor, depending on its type and use. For instance, passwords are susceptible to social engineering (e.g. phishing) and dictionary attacks. Digital devices can be lost or stolen. Those offering tamper-resistance may nonetheless fall to reverse-engineering [20,21], side-channel attacks [17], and trojans (e.g. recent Sykipot Trojan attacks against smart cards). Biometric data can be obtained from a physical contact with the human or copied if available in a digital form. Since the number of personal biometrics that permit efficient use in security technologies is limited, their wide use across different application domains makes it even harder to keep those factors private.

Multi-Factor Authentication (with Key Exchange). The strength of Multi-Factor Authentication (MFA) is based on the assumption that if an entity has many authentication factors, regardless of their nature, then it is hard for the attacker to compromise them all. That is, by combining different factors within a single authentication process, MFA aims at higher assurance in comparison to single-factor schemes. MFA has found its way into practice[1], most notable are combinations of long-term passwords with secret keys, possibly stored in tokens (e.g. Two-Factor SSH with USB sticks) or any of these with one-time passwords (e.g. OATH HOTP/TOTP, RSA SecurID, Google Authenticator). Many companies, e.g. Google, Facebook, Yahoo are now offering their users optional two-factor authentication mechanisms based on one-time passwords. The increasing use of smart phones to access services and the recent progress by Apple and Samsung to equip smart phones with fingerprint readers is expected to further boost the practical deployment of the MFA technology. Since MFA is

---

[1] MFA definitions and usage in practice are not consistent. For example, according to [1, Sec. 8.3], for two-factor authentication it suffices to deploy RADIUS authentication or use a single tamper-proof hardware token or a VPN access with individual certificate, whereas using two factors of the same type is not regarded as a two-factor solution. [2, Level 3] explicitly requires hardware tokens and some additional factor, e.g. password or biometric. This is in line with the perception of MFA where authentication with a certificate alone is considered single-factor [33] but deployment of two or more passwords multi-factor [35]. For the purpose of generality, we regard any approach with at least two factors irrespective of their type as MFA.

mostly used to authenticate a client/user to a remote server the authentication of the client becomes its main security goal. The server-side authentication in MFA protocols offers further protection and is typically performed without using multiple factors on the server side.

The concept of Multi-Factor Authenticated Key Exchange (MFAKE), formalized in [33], extends MFA with establishment of secure session keys. In addition to authentication goals it aims at key secrecy, usually modeled in terms of (Bellare-Rogaway style) AKE-security [7,9,14]. Earlier MFAKE protocols focused mostly on two factors and were often unsuccessful: for instance, password-token combination from [31] was broken in [37] which itself was broken in [28], the scheme from [34] was cryptanalyzed in [36], and a biometric-token combination from [29] has fallen in [30]. Partially, these attacks were due to the missing modeling and analysis in those works.

A formal approach to MFAKE introduced in [33] was the first to account simultaneously for all three types of authentication factors. Most notable is their modeling of biometric factors. Unlike some previous single-factor biometric schemes, e.g. [16,10], that regarded biometrics as low- or high-entropy secrets, [33] drops biometric secrecy in favor of the *liveness assumption* (see also [13,12]) aiming at physical presence of a user. The protocol from [33] has recently been cryptanalysed in [24], who showed how an adversary that steals user's password and impersonates the server can essentially compromise all other authentication factors of the client. The model in [33] didn't consider server authentication and the only way to prevent the above attack against the protocol is to require mandatory authentication on the server side. The protocol would remain insecure if server authentication is left optional (as intended by the model) due to the way in which client messages bind different authentication factors together, as also exploited in [24].

MFAKE protocols may differ not only in nature of factors but also in their quantity. To this end, [35] introduced Multi-Factor Password AKE (MFPAKE), extending the PAKE setting [6], where arbitrary many low-entropy passwords (long-term and one-time) can be combined to authenticate the client. Their protocol further offers public-key based server-side authentication and supports verifier-based PAKE setting from [22,23].

Generalized and Modular MFAKE Approach? Various problems in the design of secure MFAKE protocols, coupled with the fact that existing protocols differ in nature and quantity of deployed factors and that perception of MFA varies across products, standards, and research literature, motivates the need for a simpler and modular MFAKE approach.

Our goal is to build secure MFAKE protocols out of well-known and understood concepts behind existing single-factor solutions. We argue that in general this approach though not necessarily more efficient helps to avoid caveats, arising in the combination of factors and results in a cleaner, less error-prone protocol design. The generality of the approach can further be used to formally explain the relationships between MFAKE and single-factor authentication schemes, and its modularity is beneficial for an independent accommodation of other factors, e.g. for social authentication [11,15].

A general MFAKE protocol can be built from different types of single-factor AKE protocols that are then combined in a smart way into a secure MFAKE solution. The feasibility of this approach and its formal correctness is implied by our work. A direct combination of different black-box AKE schemes is suboptimal since it would include some redundancy in the computations of forward-secure session keys. Therefore, our approach for a general MFAKE is to use single-factor *authentication-only* protocols (to avoid computation of multiple session keys) and derive one forward-secure session key at the end of the protocol.

1.1   Contributions and Organization

General MFAKE Model. We introduce and model a general framework for (α, β, γ)-MFAKE, including its MFA-only version, building on the three-factor AKE model from [33]. In a standard client-server setting we admit arbitrary *quantities* and *combinations* of low-entropy passwords (long-term and one-time), high-entropy secret keys (possibly with corresponding public keys), and biometric factors (with explicit and implicit matching). We model dictionary attacks on passwords and also account for the imperfect matching process of biometric templates. When modeling biometrics we follow the *liveness assumption* of [33] and do not treat biometric distributions as secret. We discuss why this assumption is realistic from the practical point of view.

Remark 1. In the full version of this paper [19] we further relate our (α, β, γ)-MFAKE framework to several existing authentication models and protocols. By varying the parameters α, β, and γ we can show that many current single-factor and multi-factor settings can be seen as special cases of our general framework: $(1, 0, 0)$-MFAKE implies PAKE models from [6,22], $(0, 1, 0)$-MFAKE implies two-party AKE models from [7,9], $(1, 1, 1)$-MFAKE subsumes the three-factor client authentication approach from [33], while $(\alpha, 0, 0)$-MFAKE is related to the MF-PAKE protocol introduced in [35].

Modular (α, β, γ)-MFAKE Framework. We give a simple generic (α, β, γ)-MFAKE protocol construction, based on sub-protocols that can be instantiated from a wide range of existing, well-understood and efficient authentication-only schemes. More precisely we consider arbitrary many independent runs of efficient authentication-only protocols that rely on passwords, secret keys, and biometrics and link them to a single independent session of an Unauthenticated Key Exchange (UKE) in a way that generically binds authentication and key establishment and results in an AKE-secure MFAKE protocol (with forward secrecy) that offers MFA for the client and strong (optional) authentication of the server.

To this end, we define a generalized notion of *tag-based* MFA, extending the preliminary concepts from [25] that considered the use of tags (auxiliary strings) in public key-based challenge-response scenarios. For all types of single-factor authentication-only protocols we demonstrate existence of efficient tag-based flavors and discuss their generic extensions with tags. We show how to use tags in an (α, β, γ)-MFAKE protocol to bind all independent (black-box) sub-protocols in a secure way. (In this way, for example, we avoid the type of problems identified in [24] for the protocol in [33].)

ORGANIZATION. Generalized (α, β, γ)-MFAKE, its MFA-only version, and security goals are modeled in Section 2. Our modular and generic construction of (α, β, γ)-MFAKE is specified and analyzed in Section 3, along with the underlying sub-protocols and their instantiations.

2   Generalized MFAKE: Definitions and Security

Our definitions of generalized MFAKE extend the model from [33], which in turn builds on the models from [7,6].

2.1   System Model and Correctness

Participants, Sessions, and Authentication Factors. An MFAKE protocol is executed between two participants: a *client* C and a *server* S. Several instances of any participant can exist at a time. This models multiple concurrent protocol sessions. An instance of participant $U \in \{C, S\}$ in session s is denoted as $[U, s]$. The session id s is the transcript of all messages sent and received by the instance, except for the very last protocol message. At the end of the protocol each instance either accepts or rejects.

By $\mathsf{pid}([U, s])$ we denote *partner identity* with which $[U, s]$ is interacting in the protocol session. Two instances $[U, s]$ and $[U', s']$ are said to be *partnered* if and only if $\mathsf{pid}([U, s]) = U'$, $\mathsf{pid}([U', s']) = U$, and their session ids form *matching conversations* [7,9], denoted $s = s'$.

Each client C may have arbitrary types and quantities of authentication factors that it may use in multiple protocol sessions as detailed in the following.

PASSWORDS. A client C may hold an array of α *passwords*, denoted $\boldsymbol{pwd}_C$. Each password $\boldsymbol{pwd}_C[i]$, $i = 1, \dots, \alpha$ is assumed to have low entropy, chosen from a dictionary $\mathcal{D}_{\mathsf{pwd}}$. Passwords can be used across multiple sessions, in which case they are considered to be long-term. We also allow for one-time passwords [3,32] that have been previously registered with the server. Our setting can be extended to deal with verifier-based password authentication, e.g. [23,8,26], where the server stores some non-trivial function of $\boldsymbol{pwd}_C[i]$ for better protection against server compromise attacks.

CLIENT SECRET KEYS. A client C may hold an array of β *secret keys*, denoted $\boldsymbol{sk}_C$. Each secret key $\boldsymbol{sk}_C[i] \in \mathsf{KeySp}$, $i = 1, \dots, \beta$ is assumed to have high entropy. In case of public key-based client authentication there exists an array of corresponding public keys, denoted $\boldsymbol{pk}_C$, which is assumed to be known system-wide. Any $\boldsymbol{sk}_C[i]$ can be stored in a secure hardware token (e.g. in a smartcard or TPM), in which case its usage in the protocol assumes client's access to the corresponding device, i.e., our model doesn't distinguish between hardware tokens and private keys of the client.

BIOMETRICS For each client C there are γ public biometric distributions $\mathtt{Dist}_{C,i}$, $i = 1, \ldots, \gamma$. The process of measuring some biometric (being it face, any particular finger, or iris) is comprehended as drawing a *biometric template* $W_{C,i}$ according to $\mathtt{Dist}_{C,i}$. Upon the enrollment of the client an array $\boldsymbol{W}_C$ containing γ biometric templates $\boldsymbol{W}_C[i]$, $i = 1, \ldots, \gamma$ is created and will be used as a reference for the session-dependent matching process on the server's side. We do not need to require that $\boldsymbol{W}_C$ is stored in clear on the server's side. Our model admits the case, where the server stores some non-trivial transformation of $\boldsymbol{W}_C$, e.g. using secure sketches [16,10].

Functionality of biometric data matching is modeled through an algorithm BioMatch, which takes as input a candidate template W^* and a reference template W, which may also be given *implicitly* in a transformed form, and outputs 1 indicating that W^* matches W and 0 otherwise. For example, BioMatch can require that the Hamming distance between W and W^* remains below some threshold, an approach used, e.g. in [10,33]. We also take into account that biometric measurements are *not* perfect:

- For any client C,

$$\Pr\left[\mathsf{BioMatch}(W^*_{C,i}, \boldsymbol{W}_C[i]) = 1 \mid W^*_{C,i} \leftarrow \mathtt{Dist}_{C,i}, i \in [1,\gamma]\right] \geq 1 - \mathsf{false}^{\mathrm{neg}}_i,$$

where $\mathsf{false}^{\mathrm{neg}}_i$ is the probability with which ith biometric of C is *falsely rejected*.
- For any two clients C', C with $C' \neq C$,

$$\Pr\left[\mathsf{BioMatch}(W^*_{C',i}, \boldsymbol{W}_C[i]) = 0 \mid W^*_{C',i} \leftarrow \mathtt{Dist}_{C',i}, i \in [1,\gamma]\right] \geq 1 - \mathsf{false}^{\mathrm{pos}}_i,$$

where $\mathsf{false}^{\mathrm{pos}}_i$ is the probability with which ith biometric of C' is *falsely accepted*.

While false rejection is important for MFA correctness, false acceptance impacts the lower bounds of the protocol's security.

SERVER SECRET KEY We assume that server S may have a high-entropy secret key sk_S with the corresponding system-wide known public key pk_S.

Generalized MFAKE. We define generalized MFAKE and its correctness property.

Definition 1 ((α, β, γ)-MFAKE). *A multi-factor authenticated key exchange protocol (α, β, γ)-MFAKE(C, S) is a two-party protocol, executed between a client instance $[C, s]$ with α passwords, β secret keys, and γ biometric templates and a server instance $[S, s']$ such that at the end of their interaction each instance either accepts or rejects. The correctness property of the protocol requires that for all $\kappa \in \mathbb{N}$, if at the end of the protocol session $[C, s]$ accepts holding session key k_C and $[S, s]$ accepts holding session key k_S, and $[C, s]$ and $[S, s]$ are partnered, then $\Pr[k_C = k_S] = 1$.*

In *authentication-only* MFA protocols parties either reject or accept their communication partner *without* computing any session keys. The following definition of client's MFA towards the server accounts for imperfect biometric matching process, where servers may falsely reject clients.

Definition 2 $((\alpha, \beta, \gamma)$**-MFA**$)$. *A multi-factor authentication-only protocol* (α, β, γ)*-MFA is a two-party protocol, executed between a client instance* $[C, s]$ *with* α *passwords,* β *secret keys, and* γ *biometric templates and a server instance* $[S, s']$ *such that at the end of their interaction the server instance either accepts* C *as its communication partner or rejects. Let 'acc C' denote the event that* $[S, s]$ *accepts the client. The correctness property of the* (α, β, γ)*-MFA protocol requires that* $\Pr[\text{acc } C] \geq 1 - \sum_{i=1}^{\gamma} \mathsf{false}_i^{\mathrm{neg}}$.

For *server-side authentication* the multi-factor aspect is typically irrelevant, i.e., the client decides whether to accept the server based on pk_S. The correctness property in this case is perfect.

2.2 Security Goals: AKE-Security and Mutual Authentication

MFAKE protocols must guarantee standard goals with respect to session key security and mutual authentication against any probabilistic polynomial-time adversary $\mathcal{A}$. Due to asymmetry with regard to the use of multiple factors on the client side and typically one factor (secret key) on the server's side, mutual authentication is dealt with separately for clients and servers.

Liveness Assumption for Biometrics. We assume that biometric data is public and resort to liveness assumption [33] to ensure physical presence of a client. Liveness of a client C is modeled through a special *biometric computation oracle* $\mathsf{BioComp}([C, s], W_{\mathcal{A},i})$: depending on the state of $[C, s]$ this oracle uses client's secret keys $\boldsymbol{sk}_C$ and passwords $\boldsymbol{pwd}_C$ together with an input biometric template $W_{\mathcal{A},i}$ that must be chosen according to some *adversary-specified* distribution $\mathsf{Dist}_{\mathcal{A},i}$ to perform the required computation step that would otherwise be performed using a template $W^*_{C,i}$ chosen according to the distribution $\mathsf{Dist}_{C,i}$. The crucial condition here is that $\mathsf{Dist}_{\mathcal{A},i}$ must significantly differ from $\mathsf{Dist}_{C,i}$ such that $\Pr[\mathsf{BioMatch}(W_{\mathcal{A},i}, \boldsymbol{W}_C[i]) = 0] \geq 1 - \mathsf{false}_i^{\mathrm{pos}}$ for any $W_{\mathcal{A},i} \leftarrow_R \mathsf{Dist}_{\mathcal{A},i}$. For simplicity, we assume that $\mathcal{A}$ queries $\mathsf{BioComp}$ only with templates $W_{\mathcal{A},i}$ from the distributions $\mathsf{Dist}_{\mathcal{A},i}$, $1 \leq i \leq \gamma$ (alternative modeling of $\mathsf{BioComp}$ would require $\mathcal{A}$ to specify some template generation algorithm with a suitable distribution $\mathsf{Dist}_{\mathcal{A},i}$ which will be invoked within $\mathsf{BioComp}$ on each new query to pick $W_{\mathcal{A},i}$). Liveness assumption requires that any *new* message m, whose computation depends on the ith biometric template of C, must be previously generated by the $\mathsf{BioComp}$ oracle, before an active adversary can make use of it. Using $\mathsf{BioComp}$ oracle $\mathcal{A}$ can test own biometric templates in client's computations. Note that the liveness assumption allows for replay attacks on biometric-dependent messages, i.e. $\mathcal{A}$ can consult the $\mathsf{BioComp}$ oracle to obtain a new message in one session of C and then replay it in another.

Remark 2. Hao and Clarke [24] criticized [33] for the assumption that biometric data is public, arguing that templates that can be obtained by the adversary in practice are often of poor quality so that obtaining high-quality templates should be seen as a corruption of the client. This might be a valid argument in certain use cases, however, for the purpose of generality, it seems more appropriate to assume that biometric data is public and resort to the liveness assumption, when modeling security of biometric-based protocols. Since biometric data is used in many different domains (e.g. e-passports, personal computers, entry access systems, etc.) leakage of high-quality templates is not unlikely. In contrast to private keys, biometric characteristics are produced by nature and are bound to a specific person. From this perspective, their modeling via liveness assumption, aiming at user's *physical presence* seems to be more appropriate. Liveness assumption has also been in the focus of recent standardization initiatives, e.g. ISO/IEC WD 30107 Anti-Spoofing and Liveness Detection Techniques.

Client and Server Corruptions. An active adversary $\mathcal{A}$ may corrupt authentication factors of a client C through its CorruptClient(C, type, i) oracle by indicating the type of the corrupted factor and its position i in the array. Corrupted passwords and secret keys are revealed to $\mathcal{A}$, whereas corrupted biometric factors imply that $\mathcal{A}$ no longer needs to follow restrictions put forth by the liveness assumption on those factors. $\mathcal{A}$ can ask multiple CorruptClient queries for different factors of its choice. This models realistic scenarios, where different factors may require different attacks. Server corruptions are handled through the CorruptServer oracle, which responds with sk_S.

Adversarial Queries. Our security definitions will be given in form of games with a PPT adversary $\mathcal{A}$ that interacts with the instances through a set of oracles, as specified in the following. We assume that U, U' $\in \{C, S\}$.

Invoke(U, U') allows $\mathcal{A}$ to invoke a session at party U with party U'. If U is a client then U' must be a server, and vice versa. In response, a new instance [U, s] with pid([U, s]) = U' is established. [U, s] takes as input the authentication factors of U. If [U, s] is supposed to send a message first then this message is generated and given to $\mathcal{A}$.

Send([U, s], m) allows $\mathcal{A}$ to send messages to the protocol instances (e.g. by forwarding, modifying, or creating new messages). In general, the oracle processes m according to the state of [U, s] and eventually outputs the next message (if any) to $\mathcal{A}$. However, if U = S and m is such that it was not produced by an instance of C = pid([S, s]) but its computation was expected to involve ith biometric of C, then m is processed only if it was output by BioComp([C, ·], $W_{\mathcal{A},i}$) or if $\mathcal{A}$ previously queried CorruptClient(C, 3, i).

BioComp([C, s], $W_{\mathcal{A},i}$) outputs message m (if any) computed based on the internal state of [C, s] using sk_C, pwd_C, and $W_{\mathcal{A},i}$ (from Dist$_{\mathcal{A},i}$ as explained above).

RevealSK([U, s]) gives $\mathcal{A}$ the session key computed by [U, s] (if such key exists).

CorruptClient(C, type, i) allows $\mathcal{A}$ to corrupt authentication factors of C. If type = 1 then $\mathcal{A}$ is given $pwd_C[i]$; if type = 2 then it receives $sk_C[i]$;

if type $= 3$ then $\mathcal{A}$ receives nothing but the liveness assumption for the ith biometric of C is dropped.

CorruptServer(S) gives $\mathcal{A}$ server's S secret key sk_S.

Freshness. The notion of freshness prevents $\mathcal{A}$ from using its oracles to attack the protocol in a trivial way. For instance, key secrecy and authentication goals will require that no protocol participant was fully corrupted during the protocol session: a *client C is fully corrupted* if and only if all existing authentication factors of C have been corrupted via corresponding CorruptClient($C, \cdot, \cdot$) queries; a *server S is fully corrupted* if and only if a CorruptServer(S) query has been asked. Our definition of freshness aims at server instances since $\mathcal{A}$ will be required to break AKE-security for their session keys. This is not a limitation since protocol correctness guarantees that any accepted partnered client instance will compute the same key as the server instance. In protocols without server authentication $\mathcal{A}$ can impersonate the server and compute the same key as the client. An instance $[S, s]$ that has accepted is said to be *fresh* if all of the following holds:

- Upon acceptance of $[S, s]$ neither the server S nor the client $C = \text{pid}([S, s])$ were fully corrupted.
- There has been no RevealSK query to $[S, s]$ or to its partnered client instance (if such instance exists).

The above conditions allow full corruption of parties after the session ends (upon acceptance) and thus capture the property of *forward secrecy* that is equally important for *all* types of authentication factors.

Remark 3. Freshness conditions can be made more complicated to incorporate specialized goals such as security against key compromise impersonation (KCI) and corruptions of ephemeral secrets (cf. [27] and its variants). These goals however are *factor-dependent*. For instance, $(\alpha, 0, 0)$-MFAKE protocols with *shared* passwords typically wouldn't offer KCI-security (which by definition makes sense only in the public key setting). It also seems unlikely that $(\alpha, 0, 0)$-MFAKE can tolerate leakage of ephemeral secrets (the only randomness used in the protocol) without enabling an offline dictionary attack. Our conditions thus offer a common security base for all (α, β, γ)-MFAKE flavors, without narrowing the possibility of extension towards more complex requirements.

Security of Session Keys. Secrecy of session keys is modeled in terms of AKE-security in the Real-or-Random indistinguishability framework [4] where multiple Test queries that can be asked only to fresh instances $[S, s]$. Their answers depend on the value of bit b, which is fixed in the beginning of the game: if $b = 1$ then $\mathcal{A}$ receives the real session key held by $[S, s]$; if $b = 0$ then $\mathcal{A}$ is given a random key chosen uniformly from the set of all possible session keys. At the end of the game $\mathcal{A}$ outputs bit b' aiming to guess b. Let $\text{Succ}_{\text{AKE}}^{\mathcal{A}, (\alpha, \beta, \gamma)\text{-MFAKE}}(\kappa)$ denote the probability of the event $b' = b$ in a game played by $\mathcal{A}$ against the AKE-security of (α, β, γ)-MFAKE. Let q denote the total number of invoked sessions. (α, β, γ)-MFAKE is AKE-secure, if for all PPT adversaries $\mathcal{A}$ the following advantage is negligible in κ:

$$\mathsf{Adv}_{\mathsf{AKE}}^{(\alpha,\beta,\gamma)\text{-MFAKE},\mathcal{A}}(\kappa) = \left|\mathsf{Succ}_{\mathsf{AKE}}^{(\alpha,\beta,\gamma)\text{-MFAKE},\mathcal{A}}(\kappa) - q\left(\frac{\alpha}{|\mathcal{D}_{\mathsf{pwd}}|} + \sum_{i=1}^{\gamma} \mathsf{false}_i^{\mathsf{pos}}\right) - \frac{1}{2}\right|.$$

AKE-security is relevant only for (α, β, γ)-MFAKE protocols from Definition 1. It doesn't apply to (α, β, γ)-MFA protocols from Definition 2 that do not support key establishment.

Authentication Requirements. An (α, β, γ)-MFAKE protocol must further provide authentication, which we treat separately for clients and servers. A protocol which satisfies both offers *mutual authentication*.

CLIENT AUTHENTICATION. Let $\mathcal{A}$ be an adversary against client authentication of (α, β, γ)-MFAKE that interacts with client and server instances using the aforementioned queries (whereby Test queries are irrelevant). $\mathcal{A}$ breaks client authentication if there exists a server instance $[S, s]$ that has accepted a client $C = \mathsf{pid}([S, s])$, for which there exists no client instance that is partnered with $[S, s]$, and neither S nor C were fully corrupted upon the acceptance of $[S, s]$.

Let $\mathsf{Succ}_{\mathsf{CAuth}}^{(\alpha,\beta,\gamma)\text{-MFAKE},\mathcal{A}}(\kappa)$ denote the success probability in breaking client authentication. The protocol is CAuth-secure, if for all PPT adversaries $\mathcal{A}$ the following advantage is negligible (in κ):

$$\mathsf{Adv}_{\mathsf{CAuth}}^{(\alpha,\beta,\gamma)\text{-MFAKE},\mathcal{A}}(\kappa) = \left|\mathsf{Succ}_{\mathsf{CAuth}}^{(\alpha,\beta,\gamma)\text{-MFAKE},\mathcal{A}}(\kappa) - q\left(\frac{\alpha}{|\mathcal{D}_{\mathsf{pwd}}|} + \sum_{i=1}^{\gamma} \mathsf{false}_i^{\mathsf{pos}}\right)\right|.$$

This definition of CAuth-security is directly applicable to (α, β, γ)-MFA protocols from Definition 2. The advantage of $\mathcal{A}$ is denoted then $\mathsf{Adv}_{\mathsf{CAuth}}^{(\alpha,\beta,\gamma)\text{-MFA},\mathcal{A}}(\kappa)$ and its success probability is subject to the same bounds as $\mathsf{Succ}_{\mathsf{CAuth}}^{(\alpha,\beta,\gamma)\text{-MFAKE},\mathcal{A}}(\kappa)$. For (α, β, γ)-MFA protocols CAuth-security is the main property.

Remark 4. The low entropy of passwords and non-perfect biometric matching impose a lower bound $q(\frac{\alpha}{|\mathcal{D}_{\mathsf{pwd}}|} + \sum_{i=1}^{\gamma} \mathsf{false}_i^{\mathsf{pos}})$ on the success probability of a CAuth-adversary. This bound is not imposed on the success probability with regard to server authentication as explained below.

SERVER AUTHENTICATION. An adversary $\mathcal{A}$ against server authentication of (α, β, γ)-MFAKE interacts with client and server instances and breaks server authentication if there exists a client instance $[C, s]$ that has accepted a server $S = \mathsf{pid}([C, s])$, for which there exists no server instance that is partnered with $[C, s]$, and neither C nor S were fully corrupted upon the acceptance of $[C, s]$. (α, β, γ)-MFAKE is SAuth-secure, if for all PPT adversaries $\mathcal{A}$ the probability of breaking server authentication, denoted $\mathsf{Succ}_{\mathsf{SAuth}}^{(\alpha,\beta,\gamma)\text{-MFAKE},\mathcal{A}}(\kappa)$ is negligible in the security parameter κ.

3   Modular Design of MFAKE Protocols

Our general (α, β, γ)-MFAKE protocol is built in a modular way from sub-protocols for different authentication factors, yet with some extensions and optimizations. We start with the main building blocks.

3.1   Tag-Based Authentication

Tag-based Authentication (TbA) [25] accounts for the use of auxiliary, possibly public, strings (tags) in authentication protocols. In TbA each party uses a tag, in addition to the authentication factor, and the protocol guarantees that if parties accept then their tags match. For instance, the server accepts some client in a session if and only if that client was alive during that session and used as input the same tag as the server. For public key-based challenge-response protocols, [25] gave a signature-based compiler with the TbA property. In our work we require a more general TbA notion that in addition to public keys encompasses passwords and biometrics as defined in the following.

Definition 3 (Tag-based MFA). *A tag-based MFA protocol (α, β, γ)-tMFA is an (α, β, γ)-MFA protocol from Definition 2, where in addition the client instance $[C, s]$ takes as input tag t_C, the server instance $[S, s]$ takes as input tag t_S, and if $t_C \neq t_S$ then both parties reject; otherwise, they accept as in the (α, β, γ)-MFA protocol.*

Tag-based CAuth-security: *Let $\mathcal{A}$ be a PPT adversary against client authentication of (α, β, γ)-tMFA that interacts with the instances of C and S using the same oracles as for (α, β, γ)-MFA, except that the Invoke oracle is modified such that it receives tag t as an additional input from $\mathcal{A}$. $\mathcal{A}$ is said to break CAuth-security of (α, β, γ)-tMFA if at the end of its interaction there exists a server instance $[S, s]$ that was invoked with tag t_S and has accepted a client $C = \mathrm{pid}([S, s])$, for which there exists no client instance that was invoked with tag $t_C = t_S$ and is partnered with $[S, s]$, and neither S nor C were fully corrupted upon the acceptance of $[S, s]$. The corresponding advantage of $\mathcal{A}$, denoted $\mathrm{Adv}_{\mathrm{CAuth}}^{(\alpha, \beta, \gamma)\text{-tMFA}, \mathcal{A}}(\kappa)$, is then defined analog to the advantage in (α, β, γ)-MFA.*

$\mathcal{A}$ is allowed to test tags of its own choice, i.e. existence of a partnered client instance that was invoked with a tag $t_C \neq t_S$ leads to a successful attack. Definitions of *tag-based server authentication* in (α, β, γ)-tMFA and success probability $\mathrm{Succ}_{\mathrm{SAuth}}^{(\alpha, \beta, \gamma)\text{-tMFA}, \mathcal{A}}(\kappa)$ are obtained by reversing the roles of C and S, as for (α, β, γ)-MFA in Section 2.2.

3.2   Utilized Sub-protocols and Their Examples

Our framework constructs (α, β, γ)-MFAKE in a modular way from simpler protocols that represent special cases of tag-based MFA. We first describe corresponding (non tag-based) protocols for authentication and provide some examples, including the discussion on how to extend those protocols with tags.

PwA : (Tag-Based) Password-Based Authentication Protocol. The first sub-protocol is for password-based authentication, denoted PwA, in which only one party (in our case the client) authenticates itself to the other party (server). In our generalized MFA model the adversarial advantage against client authentication of PwA becomes $\text{Adv}_{\text{CAuth}}^{\mathcal{A},\text{PwA}}(\kappa) = \text{Adv}_{\text{CAuth}}^{\mathcal{A},(1,0,0)\text{-MFA}}(\kappa)$.

For instance, an AKE-secure PAKE protocol with key confirmation from client to server, which is proven secure in the model from [6] can be used as PwA. On the other hand, those PAKE protocols can be somewhat simplified since we do not require PwA to provide session keys. The following is an example for the PAKE protocol from [5] when only client-side authentication with key confirmation is applied.

PwA EXAMPLE. Let $(\mathbb{G}, g, q)$ be a description of the cyclic group of prime order q with generator g that together with two elements $V, W \in \mathbb{G}$ and a hash function $\mathcal{H} : \{0,1\}^* \mapsto \{0,1\}^\kappa$ build public parameters. Assume that $pwd \in \mathbb{Z}_q$ is shared between C and S. In a PwA session, derived from [5], S sends $Y^* = g^y W^{pwd}$ for some $y \leftarrow_R \mathbb{Z}_q$ to C. C picks $x \leftarrow_R \mathbb{Z}_q$ and responds with $(X^*, h) = (g^x V^{pwd}, \mathcal{H}(C, S, Y^*, X^*, (Y^*/W^{pwd})^x, pwd))$. S checks whether $h = \mathcal{H}(C, S, Y^*, X^*, (X^*/V^{pwd})^y, pwd)$ and accepts the client in this case. It is easy to see that client authentication of this PwA follows from the security of PAKE in [5].

VERIFIER-BASED PwA. The Ω-method introduced in [23,22] transforms any PAKE into a verifier-based (aka. asymmetric or augmented) PAKE where passwords are stored on the server side in a blinded way using a random oracle $\mathcal{H}'$, a symmetric encryption scheme (Gen, Enc, Dec), and an additional pair of signing keys (sk, pk), which are not treated as an authentication factor. For a given password pwd the server stores $(\mathcal{H}'(pwd), \text{Enc}_{pwd}(sk))$. The Ω-method proceeds as follows. First a (symmetric) PAKE session is executed using $\mathcal{H}'(pwd)$ as a password on both sides, resulting in an intermediate PAKE key k. This key is used to derive two independent keys k' and k'' and the client is given $\text{Enc}_{k'}(\text{Enc}_{pwd}(sk))$. C decrypts sk and sends a signature on the entire protocol transcript. If this signature verifies using pk the server accepts the client. The session key of verifier-based PAKE becomes k''. The Ω-method can be applied to obtain verifier-based PwA from plain PAKE protocols, in which case k'' can be omitted.

TAG-BASED (VERIFIER-BASED) PwA. In the symmetric case any PwA protocol can be transformed into a tag-based tPwA as follows. Parties on input their tags t first compute $\mathcal{H}_T(pwd, t)$ using a cryptographic hash function $\mathcal{H}_T$, which then serves as a password for the original PwA. Since in Definition 3 the adversary specifies tags upon invocation of an instance any successful CAuth-adversary against tPwA can either be used to break CAuth-security of PwA or to find a collision for $\mathcal{H}$, i.e. $\text{Adv}_{\text{CAuth}}^{\text{tPwA},\mathcal{A}}(\kappa) \leq \text{Adv}_{\text{CAuth}}^{\text{PwA},\mathcal{A}}(\kappa) + q\epsilon_{\mathcal{H}_T}(\kappa)$ in q protocol sessions. A similar trick can be applied to verified-based PwA constructed using the aforementioned Ω-method — instead of $\mathcal{H}'(pwd)$ in the initial (symmetric) PwA session parties would use $\mathcal{H}_T(\mathcal{H}'(pwd), t)$. Security of such verifier-based tPwA follows from the security of the underlying PwA, the Ω-method, and the collision-resistance of $\mathcal{H}_T$.

PkA: (Tag-Based) Public Key Authentication Protocol. The second sub-protocol is a single-side authentication protocol in the public key setting, denoted PkA, with adversarial advantage against its client authentication defined as $\mathsf{Adv}_{\mathsf{CAuth}}^{\mathcal{A},\mathsf{PkA}}(\kappa) = \mathsf{Adv}_{\mathsf{CAuth}}^{\mathcal{A},(0,1,0)\text{-MFA}}(\kappa)$.

TAG-BASED PkA. Examples of PkA include challenge-response protocols, where S sends a (high-entropy) challenge r to C, and C replies with a function of its secret key, e.g. a signature. A generic extension of such PkA protocols with tags, denoted tPkA, uses a cryptographic hash function $\mathcal{H}_T$ and follows immediately from [25] — the challenge r received by C with tag t_C is transformed into $r'_C = \mathcal{H}_T(r, t_C)$, which is then used to generated response to S where it is verified using $r_S = \mathcal{H}_T(r, t_S)$. As shown in [25] this conversion is applicable to various classes of PkA protocols.

BiA: (Tag-Based) Biometric-Based Authentication Protocol. The third sub-protocol is a biometric-based authentication protocol, denoted BiA, in which C authenticates towards S that holds some (possibly blinded) reference template of C. In line with our model (and [33]) we work with public biometric factors and denote the adversarial advantage against client authentication of BiA as $\mathsf{Adv}_{\mathsf{CAuth}}^{\mathcal{A},\mathsf{BiA}}(\kappa) = \mathsf{Adv}_{\mathsf{CAuth}}^{\mathcal{A},(0,0,1)\text{-MFA}}(\kappa)$.

TAG-BASED BiA EXAMPLE. Let $(\mathbb{G}, g, q)$ be a cyclic group of sufficiently large prime order q. C and S first execute an unauthenticated Diffie-Hellman key exchange in $\mathbb{G}$ by exchanging g^x and g^y. Consider two hash functions $\mathcal{H}_1, \mathcal{H}_2 : \mathbb{G} \mapsto \{0,1\}^\kappa$. Let $W'_{C,i}$ resp. $W_{C,i}$ denote the ith bit of the corresponding template. For each bit i the client computes $h_i = \mathcal{H}_1(g^x, g^y, g^{xy}, W'_{C,i}, i)$ using its version of g^{xy} and sends the resulting set $\{h_i\}_i$ to S. S re-computes corresponding values using its version of g^{xy} and the reference template W_C, and accepts the client if τ or more hash values from $\{h_i\}_i$ match. Note that if liveness assumption is in place then the adversary is prevented from sending any h_i that was not computed beforehand through the BioComp oracle. The tag-based CAuth-security of the protocol follows then directly from the classical CDH assumption in the random oracle model.

UKE: Unauthenticated Key Exchange Observe that tag-based authentication protocols do not offer computation of session keys. In our modular (α, β, γ)-MFAKE protocol we will use an *unauthenticated* key exchange, denoted UKE, as another sub-protocol. We assume that UKE satisfies the following standard definition (see e.g. [25]) tailored to the client-server scenario.

Definition 4 (Unauthenticated KE). *An unauthenticated key exchange protocol, denoted UKE, is a two-party protocol executed between a client instance $[C, s]$ and a server instance $[S, s']$ such that at the end both instances accept holding respective session keys k_C and k_S or reject. Let $s = \mathsf{tr}_C$ and $s' = \mathsf{tr}_S$ be respective communication transcripts of the two instances. An UKE protocol is*

correct if their partnering, i.e. $s = s'$, implies equality of their session keys, i.e.,
$k_C = k_S$.

KE-SECURITY. *Consider the following attack game against some correct* UKE
*protocol: A PPT adversary $\mathcal{A}$ receives as input the security parameter κ and can
query the* Transcript *oracle which is parameterized with a random bit b fixed in the
beginning of the game. On an ith query the* Transcript *oracle executes a protocol
session between two new instances of C and S, and hands its communication
transcript tr_i and a key k_i to $\mathcal{A}$, where k_i is real if $b = 0$ or randomly chosen
(for each new* Transcript *query) if $b = 1$. At some point $\mathcal{A}$ outputs bit b'. An* UKE
protocol is KE-secure if the following advantage is negligible in κ for all $\mathcal{A}$:

$$\text{Adv}_{\text{KE}}^{\text{UKE},\mathcal{A}} = \left| \Pr[b = b'] - \frac{1}{2} \right|.$$

UKE EXAMPLE. The unauthenticated Diffie-Hellman key exchange protocol in
a cyclic group $(\mathbb{G}, g, q)$, where C and S exchange g^x and g^y, respectively, and
derive their session keys via $\mathcal{H}(g^x, g^y, g^{xy})$ offers a straightforward KE-secure
UKE scheme in the random oracle model under the CDH assumption.

3.3   Modular (α, β, γ)-MFAKE Protocol Framework

We now detail the modular design of a generalized (α, β, γ)-MFAKE protocol,
which supports arbitrary combinations of authentication factors, both in type
and quantity. In addition to the sub-protocols from the previous section, its
construction utilizes four hash functions $\mathcal{H}_T, \mathcal{H}_C, \mathcal{H}_S, \mathcal{H}_k : \{0,1\}^* \mapsto \{0,1\}^\kappa$,
modeled as random oracles that are used for the purpose of tag derivation, key
confirmation, and key derivation.

PROTOCOL DESCRIPTION. (α, β, γ)-MFAKE is built from *four* sub-protocols:
UKE, tPwA, tPkA, and tBiA. The design is based on the following idea (see
also Figure 1): first, C and S run one UKE session resulting in unauthenticated session keys k_0 for the client and k_0' for the server, that are then used
by both parties to derive tags (t_C and t_S) through $\mathcal{H}_T$. Then, an appropriate
tag-based sub-protocol is executed independently for each authentication factor
of the client. C and S thus execute α sessions of tPwA, β sessions of tPkA (with
client-side authentication), and γ sessions of tBiA, possibly in parallel. S aborts
the protocol and rejects C if any of those sessions results in the rejection of the
client. The server authentication is optional and is executed through a session of
tPkA (with server-side authentication). After finishing all sub-protocols C and
S hold their so-far transcripts $\{\text{tr}_i\}_{i=0,...,\alpha+\beta+\gamma+1}$ and $\{\text{tr}_i'\}_{i=0,...,\alpha+\beta+\gamma+1}$, respectively, and proceed with the confirmation: C sends a hash value, computed
with $\mathcal{H}_C$, on input its unauthenticated key material from the UKE session and
session identifier s, which comprises its so-far transcripts and the identities of
both parties. S verifies that this hash value is as expected. For the optional
server authentication, S responds with its own hash value, computed using $\mathcal{H}_S$

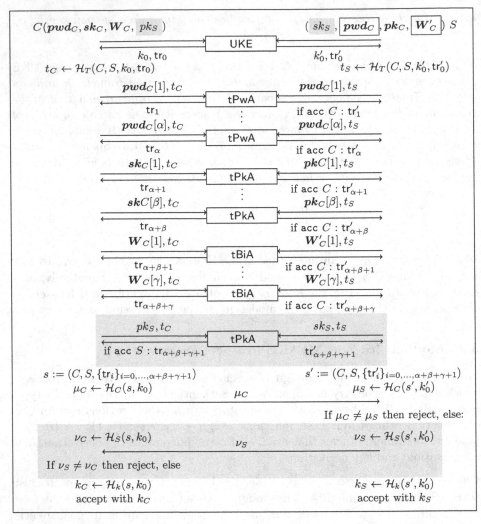

Fig. 1. (α, β, γ)-MFAKE Protocol. The inputs sk_S and pk_S are optional for the case of server authentication and so is the server-authenticated execution of tPkA and the confirmation message ν_S. These optional parts are shown with a light gray background. Boxed input $\boldsymbol{pwd}_C$ on the server side reflects that client's passwords could be stored in some blinded way, in which case tPwA is assumed to follow the steps from [23,8,26]. Boxed input $\boldsymbol{W}'_C$ on the server side means that client's reference templates are not necessarily stored in clear, in which case tBiA must provide implicit matching functionality.

on similar inputs as in the client's case. Upon successful confirmation parties accept with session keys k_C resp. k_S, derived using $\mathcal{H}_k$.

INSTANTIATIONS. Our general (α, β, γ)-MFAKE protocol can be instantiated using concrete sub-protocols from Section 3.2. That is, working in prime-order cyclic groups $(\mathbb{G}, g, q)$, we can use unauthenticated Diffie-Hellman key exchange for UKE, a tag-based password-based authentication protocol PwA obtained from the PAKE protocol in [5] (as detailed in Section 3.2), a suitable tag-based challenge-response protocol for tPkA, e.g. using DSS or Schnorr signatures, and our simple tBiA protocol with explicit matching based on the Hamming distance mentioned in Section 3.2. By using the Ω-method from [23] (as also discussed in Section 3.2) we can obtain a verifier-based version of tPwA and use it in our construction. Finally, as evident from the security analysis in Section 3.4, (α, β, γ)-MFAKE can be instantiated from arbitrary sub-protocols as long as those satisfy the required authentication goals. Moreover, as discussed in Section 3.2, tPwA can be obtained generically from PwA, and for a large class of PkA there exists a generic conversion to tPkA. Hence, all building blocks of (α, β, γ)-MFAKE can be realized using existing efficient (single-factor) authentication solutions.

PERFORMANCE OPTIMIZATIONS. The only dependency amongst the different black-box runs of tag-based authentication sub-protocols is the input tag obtained after the UKE session. Therefore, all subsequent sub-protocol runs can be parallelized, resulting in three generic rounds (UKE, tag-based sub-protocols, and confirmation round). Of course, care should be taken to match client and server messages within each round, in order to account for the potential mismatch in the sending and delivery order of messages in parallel sub-protocol sessions. This can be done by pre-pending labels indicating that a message belongs to the ith session and using these labels to construct matching transcripts on both sides. Further optimizations may include interleaving of messages and using one random challenge of S for all β sessions of tPkA and another one for all γ sessions of tBiA, resulting in a three-pass protocol for MFA-based client authentication and five-pass protocol with further authentication of the server.

3.4   Security Analysis

The initial UKE execution contributes to the forward secrecy of the session keys. In particular, successful key confirmation guarantees that the transcripts tr_0 and tr_0' and the unauthenticated keys k_0 and k_0' match. Independent runs of tag-based authentication-only protocols for each client's factor ensure that C was alive at least during that part of the protocol execution. This is because at least one of those factors must remain uncorrupted prior to the acceptance of the server and all sub-protocol transcripts are linked together in the key confirmation step. Since tr_0 and tr_0' are linked to the transcripts of all authentication-only sub-protocols the key confirmation step further guarantees that C was alive during the UKE session and, hence, the secrecy of unauthenticated keys k_0 and k_0' follows from KE-security of the UKE protocol. The secrecy of k_0 and k_0' carries over to the secrecy of the final session keys k_C and k_S due to the use of independent random oracles. The optional server authentication follows the same reasoning as client authentication using PkA sessions. This intuition is proven in Theorems 1 and 2.

Theorem 1. *Our (α, β, γ)-MFAKE protocol is AKE- and CAuth-secure, in the random oracle model, and*

$$\mathsf{Adv}_{\mathrm{AKE}}^{(\alpha,\beta,\gamma)\text{-MFAKE},\mathcal{A}}(\kappa) \leq \mathsf{Adv}_{\mathrm{KE}}^{\mathrm{UKE},\mathcal{B}}(\kappa) + \alpha \cdot \mathsf{Adv}_{\mathrm{CAuth}}^{\mathrm{tPwA},\mathcal{B}}(\kappa) + \beta \cdot \mathsf{Succ}_{\mathrm{CAuth}}^{\mathrm{tPkA},\mathcal{B}}(\kappa)$$

$$+ \sum_{i=1}^{\gamma} \mathsf{Adv}_{\mathrm{CAuth}}^{\mathrm{tBiA}_i,\mathcal{B}}(\kappa) + (q_{\mathcal{H}_T}^2 + q(q_{\mathcal{H}_C} + q_{\mathcal{H}_K})) \cdot 2^{-\kappa}, \text{and}$$

$$\mathsf{Adv}_{\mathrm{CAuth}}^{(\alpha,\beta,\gamma)\text{-MFA},\mathcal{A}}(\kappa) = \mathsf{Adv}_{\mathrm{AKE}}^{(\alpha,\beta,\gamma)\text{-MFAKE},\mathcal{A}}(\kappa) - q(q_{\mathcal{H}_k} - 1) \cdot 2^{-\kappa}.$$

Proof. We prove this theorem using a series of games that are written for the AKE-security. To the end of the proof we discuss the impact of game hops on the CAuth-security. We denote by $\mathsf{Succ}_{\mathrm{AKE}\text{-}x}^{(\alpha,\beta,\gamma)\text{-MFAKE},\mathcal{A}}(\kappa)$ the success probability of $\mathcal{A}$ in game G_x and define

$$\Delta_x(\kappa) = |\mathsf{Succ}_{\mathrm{AKE}\text{-}x}^{(\alpha,\beta,\gamma)\text{-MFAKE},\mathcal{A}}(\kappa) - \mathsf{Succ}_{\mathrm{AKE}\text{-}(x-1)}^{(\alpha,\beta,\gamma)\text{-MFAKE},\mathcal{A}}(\kappa)|.$$

G_0 This is the original AKE-security game, where the simulator answers the queries of $\mathcal{A}$ on behalf of the instances according to the specification of (α, β, γ)-MFAKE.

G_1 In this game for all simulated server and client instances that have matching UKE transcripts $tr_0 = tr_0'$ the corresponding UKE keys k_0 and k_0' are chosen at random such that $k_0 = k_0'$ holds. Otherwise, k_0 and k_0' are computed as in G_0.

Claim. $\Delta_1(\kappa) \leq \mathsf{Adv}_{\mathrm{KE}}^{\mathrm{UKE},\mathcal{B}}(\kappa)$. *Proof.* For session instances that do not share matching UKE transcripts both games are identical. Any $\mathcal{A}$ that can distinguish between G_1 and G_0 with non-negligible probability can be used to break the KE security from Definition 4. The corresponding KE-adversary $\mathcal{B}$ against UKE would interact with $\mathcal{A}$ and simulate all its (α, β, γ)-MFAKE oracle queries as specified in G_0, except for the messages and keys of the UKE sub-protocol. Assume $\mathcal{A}$ invokes an instance of $U \in \{C, S\}$. If this instance is supposed to send the first message in the UKE session then $\mathcal{B}$ queries its Transcript oracle and uses the first message of the obtained transcript as a response to $\mathcal{A}$. If $\mathcal{A}$ invokes an instance for $U' \neq U$ that is expected to send a message only after having received some incoming message then $\mathcal{B}$ waits for the corresponding Send query of $\mathcal{A}$ and checks whether input message is amongst those output by $\mathcal{B}$ from some transcript that it holds and responds with the next response message from this transcript. If the input message is unexpected then $\mathcal{B}$ runs UKE part on behalf of this instance of U' without consulting its oracle (and will thus be able to compute the UKE key for that session). Once UKE session on behalf of some instance is finished $\mathcal{B}$ has always a key to continue its simulation, either from its own UKE run or from a Transcript query. The way in which $\mathcal{B}$ simulates UKE sessions ensures that the latter type of keys are used in sessions that involve instances with matching UKE transcripts. If Transcript returns real keys then we are in G_0; otherwise in G_1. Hence, $\Delta_1(\kappa) \leq \mathsf{Adv}_{\mathrm{KE}}^{\mathrm{UKE},\mathcal{B}}(\kappa)$.

G_2 In this game the simulator aborts if in the ith tPwA session, for some $i \in \{1, \ldots, \alpha\}$, a server instance $[S, s']$ with tag t_S and (partial) transcript tr'_i accepts client C but there exists no instance of C with matching (partial) transcript tr_i and $t_C = t_S$, and $pwd[i]$ is not corrupted.

Claim. $\Delta_2(\kappa) \leq \alpha \cdot \text{Succ}_{\text{CAuth}}^{\text{tPwA}, \mathcal{B}}(\kappa)$. *Proof.* We prove this with a hybrid argument using sub-games $G_2^{\text{upto}(j)}$, $j = 0, \ldots, \alpha$. Let tPwA$_i$, $i \in \{1, \ldots, \alpha\}$ denote the ith tPwA sub-protocol run. In $G_2^{\text{upto}(j)}$ all tPwA$_i$, $1 \leq i \leq j$ are handled as in G_2 and all tPwA$_i$, $j < i \leq \alpha$ are handled as in G_1. That is, $G_1 = G_2^{\text{upto}(0)}$ and $G_2 = G_2^{\text{upto}(\alpha)}$. As before, we define $\Delta_2^{\text{upto}(j)}(\kappa)$ as the difference in $\mathcal{A}$'s success probability in two consecutive games $G_2^{\text{upto}(j-1)}$ and $G_2^{\text{upto}(j)}$. The difference between the two is that $G_2^{\text{upto}(j)}$ may still abort even if $G_2^{\text{upto}(j-1)}$ does not. Any $\mathcal{A}$ that can distinguish between the games must have successfully caused tPwA$_j$ to abort in $G_2^{\text{upto}(j)}$, in which case an instance $[S, s']$ accepts C in tPwA$_j$ while no partnered client instance with the same tag exists and no CorruptClient$(C, 0, j)$ was asked. Such $\mathcal{A}$ can be used to break CAuth-security of tPwA. The simulator can act as CAuth-adversary $\mathcal{B}$ against tPwA by invoking new instances of the server in the tPwA game using tags of server instances that it simulates in the interaction with $\mathcal{A}$. The simulator relays all tPwA$_j$ related queries of $\mathcal{A}$ as its own queries in the tPwA game and wins if $\mathcal{A}$ causes $G_2^{\text{upto}(j)}$ to abort. Therefore $\Delta_2^{\text{upto}(j)}(\kappa) \leq \text{Succ}_{\text{CAuth}}^{\text{tPwA}, \mathcal{B}}(\kappa)$ and thus $\Delta_2(\kappa) = \sum_{j=1}^{\alpha} \Delta_2^{\text{upto}(j)}(\kappa) \leq \alpha \cdot \text{Succ}_{\text{CAuth}}^{\text{tPwA}, \mathcal{B}}(\kappa)$.

G_3 In this game the simulator aborts if in the ith client side tPkA session, for some $i \in \{1, \ldots, \beta\}$, a server instance $[S, s']$ with tag t_S and (partial) transcript $\text{tr}'_{\alpha+i}$ accepts client C but there exists no instance of C with matching (partial) transcript $\text{tr}_{\alpha+i}$ and $t_C = t_S$, and $sk_C[i]$ is not corrupted.

Claim. $\Delta_3(\kappa) \leq \beta \cdot \text{Succ}_{\text{CAuth}}^{\text{tPkA}, \mathcal{B}}(\kappa)$. *Proof.* We can use essentially the same hybrid argument as in G_2, but for tPkA sessions, and thus build a sequence of β sub-games to show that the difference between any two consecutive sub-games can be upper-bounded by $\text{Succ}_{\text{CAuth}}^{\text{tPkA}, \mathcal{B}}(\kappa)$. This leads to $\Delta_3(\kappa) = \sum_{j=1}^{\beta} \Delta_3^{\text{upto}(j)}(\kappa) \leq \beta \cdot \text{Succ}_{\text{CAuth}}^{\text{tPkA}, \mathcal{B}}(\kappa)$.

G_4 In this game the simulator aborts if in the ith tBiA session, for some $i \in \{1, \ldots, \gamma\}$, a server instance $[S, s']$ with tag t_S and (partial) transcript $\text{tr}'_{\alpha+\beta+i}$ accepts client C but there exists no instance of C with matching (partial) transcript $\text{tr}_{\alpha+\beta+i}$ and $t_C = t_S$, and the ith biometric is not corrupted.

Claim. $\Delta_4(\kappa) \leq \sum_{i=1}^{\gamma} \text{Succ}_{\text{CAuth}}^{\text{tBiA}_i, \mathcal{B}}(\kappa)$. *Proof.* We denote by tBiA$_i$ the tBiA protocol operating on the ith biometric. Again, using the hybrid argument as in G_2, but for tBiA sessions, we can build a sequence of γ sub-games and upper-bound the difference between any two consecutive sub-games $G_4^{\text{upto}(j-1)}$ and $G_4^{\text{upto}(j)}$ with $\text{Succ}_{\text{CAuth}}^{\text{tBiA}_j, \mathcal{B}}(\kappa)$. The simulator can relay all tBiA$_j$ related queries of $\mathcal{A}$ as its own queries in the tBiA game, including those related to the BioComp oracle since all biometric-dependent tBiA messages used in the (α, β, γ)-MFAKE protocol remain identical to those of the tBiA protocol. This leads to $\Delta_4(\kappa) = \sum_{j=1}^{\gamma} \Delta_4^{\text{upto}(j)}(\kappa) \leq \sum_{i=1}^{\gamma} \text{Succ}_{\text{CAuth}}^{\text{tBiA}_i, \mathcal{B}}(\kappa)$.

Remark 5. If the simulation does not abort in this game then it is guaranteed that for each server instance $[S, s']$ that is entering the confirmation round with partial transcripts $\{\text{tr}'_i\}_{1 \leq i \leq \alpha + \beta + \gamma}$ (comprising executions of tPwA, PkA, and tBiA sub-protocols) and tag t_S, and that has not disqualified itself as a candidate for a Test query (i.e. fulfills freshness conditions from Section 2.2), there exists a client instance $[C, s]$ with partial transcripts $\{\text{tr}_i\}_{1 \leq i \leq \alpha + \beta + \gamma}$ such that there exists an index $i, 1 \leq i \leq \alpha + \beta + \gamma$ with $\text{tr}_i = \text{tr}'_i$. Moreover, any such client instance holds tag $t_C = t_S$.

G_5 In this game the simulation aborts if an instance $[S, s']$ enters the confirmation round with partial transcripts tr'_0 and $\{\text{tr}'_i\}_{1 \leq i \leq \alpha + \beta + \gamma}$ and there exists $[C, s]$ with partial transcripts tr_0 and $\{\text{tr}_i\}_{1 \leq i \leq \alpha + \beta + \gamma}$ such that for some index $i : \text{tr}_i = \text{tr}'_i$ but $\text{tr}_0 \neq \text{tr}'_0$.

Claim. $\Delta_5(\kappa) \leq q_{\mathcal{H}_T}^2 2^{-\kappa}$. *Proof.* G_4 already ensures that if $[S, s']$ accepts in all authentication sub-protocols then there exists a client instance with $t_C = t_S$. The only difference between the two games is that G_5 may still abort even if G_4 does not. If $\mathcal{A}$ can distinguish between the games then $\mathcal{A}$ must have successfully caused the simulator to abort in G_5, in which case $[S, s']$ and $[C, s]$ hold tags $t_S = t_C$ but $\text{tr}_0 \neq \text{tr}'_0$. We can thus output a collision for $\mathcal{H}_T$. Since $\mathcal{H}_T$ is a random oracle we get $\Delta_5(\kappa) \leq q_{\mathcal{H}_T}^2 2^{-\kappa}$.

Remark 6. G_5 implies that any instance $[S, s']$ that was not disqualified as a candidate for a Test query (upon entering the confirmation round) has a corresponding client instance $[C, s]$ with the same UKE transcript and at least one matching tag-based sub-protocol transcript.

G_6 This game proceeds as G_5, except that on behalf of an instance $[S, s']$ that is not disqualified as a candidate for a Test query the simulator computes $\mu_S \leftarrow \mathcal{H}'_C(s')$ and $k_S \leftarrow \mathcal{H}'_K(s')$ using two private random oracles $\mathcal{H}'_C$ and $\mathcal{H}'_k$, and sets $\mu_C = \mu_S$ and $k_C = k_S$ for the corresponding $[C, s]$ that has matching UKE transcript and at least one matching tag-based sub-protocol transcript.

Claim. $\Delta_6(\kappa) \leq q(q_{\mathcal{H}_C} + q_{\mathcal{H}_K}) \cdot 2^{-\kappa}$. Considering that in the previous game, confirmation values and session keys of $[S, s']$ were derived through random oracles $\mathcal{H}_C$ and $\mathcal{H}_k$ on input k'_0 (which is random as ensured by G_1) and the transcript s', any $\mathcal{A}$ that can distinguish between the games must ask at some point a query for $\mathcal{H}_C$ or $\mathcal{H}_k$ containing k'_0 and s' for any of the q invoked sessions as input. Therefore, $\Delta_6(\kappa) \leq q(q_{\mathcal{H}_C} + q_{\mathcal{H}_k}) \cdot 2^{-\kappa}$.

G_6 implies that if $[S, s']$ accepts and is not disqualified as a candidate for a Test query then k_S is uniformly distributed in the domain of session keys. Hence, the probability of $\mathcal{A}$ to win in G_6 no longer depends on the key, i.e. $\mathcal{A}$ can win in G_6

only by guessing bit b (with probability $\frac{1}{2}$).

Summarizing the probability differences across all games we obtain

$$\mathsf{Adv}_{\mathsf{AKE}}^{(\alpha,\beta,\gamma)\text{-}\mathsf{MFAKE},\mathcal{A}}(\kappa) = \left| \mathsf{Succ}_{\mathsf{AKE}}^{(\alpha,\beta,\gamma)\text{-}\mathsf{MFAKE},\mathcal{A}}(\kappa) - q\left(\frac{\alpha}{|\mathcal{D}_{\mathsf{pwd}}|} + \sum_{i=1}^{\gamma} \mathsf{false}_i^{\mathsf{pos}} \right) - \frac{1}{2} \right|$$

$$= \left| \sum_{i=1}^{6} \Delta_i(\kappa) + \frac{1}{2} - q\left(\frac{\alpha}{|\mathcal{D}_{\mathsf{pwd}}|} + \sum_{i=1}^{\gamma} \mathsf{false}_i^{\mathsf{pos}} \right) - \frac{1}{2} \right|.$$

Taking into account that

$$\sum_{i=1}^{6} \Delta_i(\kappa) \leq \mathsf{Adv}_{\mathsf{KE}}^{\mathsf{UKE},\mathcal{B}}(\kappa) + \alpha \cdot \mathsf{Succ}_{\mathsf{CAuth}}^{\mathsf{tPwA},\mathcal{B}}(\kappa) + \beta \cdot \mathsf{Succ}_{\mathsf{CAuth}}^{\mathsf{tPkA},\mathcal{B}}(\kappa)$$

$$+ \sum_{i=1}^{\gamma} \mathsf{Succ}_{\mathsf{CAuth}}^{\mathsf{tBiA}_i,\mathcal{B}}(\kappa) + (q_{\mathcal{H}_T}^2 + q(q_{\mathcal{H}_C} + q_{\mathcal{H}_K})) \cdot 2^{-\kappa},$$

and that

$$\mathsf{Adv}_{\mathsf{CAuth}}^{\mathsf{tPwA},\mathcal{B}}(\kappa) = \left| \mathsf{Succ}_{\mathsf{CAuth}}^{\mathsf{tPwA},\mathcal{B}}(\kappa) - \frac{q}{|\mathcal{D}_{\mathsf{pwd}}|} \right|$$

$$\mathsf{Adv}_{\mathsf{CAuth}}^{\mathsf{tBiA}_i,\mathcal{B}}(\kappa) = \left| \mathsf{Succ}_{\mathsf{CAuth}}^{\mathsf{tBiA}_i,\mathcal{B}}(\kappa) - q \cdot \mathsf{false}_i^{\mathsf{pos}} \right|$$

we obtain

$$\mathsf{Adv}_{\mathsf{AKE}}^{(\alpha,\beta,\gamma)\text{-}\mathsf{MFAKE},\mathcal{A}}(\kappa) \leq \mathsf{Adv}_{\mathsf{KE}}^{\mathsf{UKE},\mathcal{B}}(\kappa) + \alpha \cdot \mathsf{Adv}_{\mathsf{CAuth}}^{\mathsf{tPwA},\mathcal{B}}(\kappa) + \beta \cdot \mathsf{Succ}_{\mathsf{CAuth}}^{\mathsf{tPkA},\mathcal{B}}(\kappa)$$

$$+ \sum_{i=1}^{\gamma} \mathsf{Adv}_{\mathsf{CAuth}}^{\mathsf{tBiA}_i,\mathcal{B}}(\kappa) + (q_{\mathcal{H}_T}^2 + q(q_{\mathcal{H}_C} + q_{\mathcal{H}_K})) \cdot 2^{-\kappa},$$

which is negligible by assumptions on UKE, tPwA, tPkA, and tBiA.

Proof for CAuth-security. With regard to client authentication, consider the above game sequence from the perspective of the CAuth-security game and success probability $\mathsf{Succ}_{\mathsf{CAuth}}^{(\alpha,\beta,\gamma)\text{-}\mathsf{MFAKE},\mathcal{A}}(\kappa)$. Freshness conditions regarding server instances encompass the requirements that are relevant for the CAuth-game. Then, Remark 6 implies that in G_5 for each server instance $[S, s']$ for which $\mathcal{A}$ could still win the game there exists a client instance $[C, s]$ with the matching UKE transcript and at least one matching tag-based sub-protocol transcript. In G_6, μ_C and μ_S are computed using private oracle, while for CAuth-security modifications of k_C and k_S are irrelevant. The probability difference to G_5 is thus upper-bounded by $q \cdot q_{\mathcal{H}_C} \cdot 2^{-\kappa}$. Then, $[S, s']$ must have received $\mu_C = \mu_S$ without having a partnered client instance. That is $\mathcal{A}$ must have asked a Send query containing a value that matches a uniformly distributed μ_S. This happens with probability at most $q \cdot 2^{-\kappa}$ for up to q invoked server instances. We thus obtain the following CAuth-success

$$\mathsf{Succ}_{\mathsf{CAuth}}^{(\alpha,\beta,\gamma)\text{-}\mathsf{MFAKE},\mathcal{A}}(\kappa) = \mathsf{Succ}_{\mathsf{AKE}}^{(\alpha,\beta,\gamma)\text{-}\mathsf{MFAKE},\mathcal{A}}(\kappa) - q(q_{\mathcal{H}_k} - 1) \cdot 2^{-\kappa} - \frac{1}{2}.$$

Taking into account that by definition

$$\mathsf{Adv}_{\mathsf{AKE}}^{(\alpha,\beta,\gamma)\text{-MFAKE},\mathcal{A}}(\kappa) = \left| \mathsf{Succ}_{\mathsf{AKE}}^{(\alpha,\beta,\gamma)\text{-MFAKE},\mathcal{A}}(\kappa) - q\left(\frac{\alpha}{|\mathcal{D}_{\mathsf{pwd}}|} + \sum_{i=1}^{\gamma} \mathsf{false}_i^{\mathsf{pos}} \right) - \frac{1}{2} \right|$$

we obtain a negligible CAuth-advantage

$$\begin{aligned}
\mathsf{Adv}_{\mathsf{CAuth}}^{(\alpha,\beta,\gamma)\text{-MFAKE},\mathcal{A}}(\kappa) &= \left| \mathsf{Succ}_{\mathsf{CAuth}}^{(\alpha,\beta,\gamma)\text{-MFAKE},\mathcal{A}}(\kappa) - q\left(\frac{\alpha}{|\mathcal{D}_{\mathsf{pwd}}|} + \sum_{i=1}^{\gamma} \mathsf{false}_i^{\mathsf{pos}} \right) \right| \\
&= \left| \mathsf{Succ}_{\mathsf{AKE}}^{(\alpha,\beta,\gamma)\text{-MFAKE},\mathcal{A}}(\kappa) - q(q_{\mathcal{H}_k} - 1) \cdot 2^{-\kappa} - \frac{1}{2} \right. \\
&\qquad\qquad \left. - q\left(\frac{\alpha}{|\mathcal{D}_{\mathsf{pwd}}|} + \sum_{i=1}^{\gamma} \mathsf{false}_i^{\mathsf{pos}} \right) \right| \\
&= \left| \mathsf{Adv}_{\mathsf{AKE}}^{(\alpha,\beta,\gamma)\text{-MFAKE},\mathcal{A}}(\kappa) - q(q_{\mathcal{H}_k} - 1) \cdot 2^{-\kappa} \right|.
\end{aligned}$$

$\square$

Theorem 2. *Our (α, β, γ)-MFAKE protocol with server authentication is* SAuth-*secure in the random oracle model, and*

$$\mathsf{Succ}_{\mathsf{SAuth}}^{(\alpha,\beta,\gamma)\text{-MFAKE},\mathcal{A}}(\kappa) \leq \mathsf{Adv}_{\mathsf{KE}}^{\mathsf{UKE},\mathcal{B}}(\kappa) + \mathsf{Succ}_{\mathsf{CAuth}}^{\mathsf{tPkA},\mathcal{B}}(\kappa) + (q_{\mathcal{H}_T}^2 + q(q_{\mathcal{H}_S} + 1)) \cdot 2^{-\kappa}.$$

Proof. This proof resembles in part the proof of Theorem 1 and proceeds in a series of similar games. We denote by $\mathsf{Succ}_{\mathsf{SAuth}\text{-}x}^{(\alpha,\beta,\gamma)\text{-MFAKE},\mathcal{A}}(\kappa)$ the success probability of $\mathcal{A}$ in game G_x. For each game G_x, we define $\Delta_x(\kappa)$ as the difference in $\mathcal{A}$'s success probability when playing against the two consecutive games G_{x-1} and G_x, i.e., $\Delta_x(\kappa) = |\mathsf{Succ}_{\mathsf{SAuth}\text{-}x}^{(\alpha,\beta,\gamma)\text{-MFAKE},\mathcal{A}}(\kappa) - \mathsf{Succ}_{\mathsf{SAuth}\text{-}(x-1)}^{(\alpha,\beta,\gamma)\text{-MFAKE},\mathcal{A}}(\kappa)|$.

G_0 This is the original SAuth-security game, where the simulator answers the queries of $\mathcal{A}$ on behalf of the instances according to the specification of (α, β, γ)-MFAKE.

G_1 This game proceeds as G_0, except that for all simulated server and client instances that have matching UKE transcripts $\mathsf{tr}_0 = \mathsf{tr}_0'$ the corresponding UKE keys k_0 and k_0' are chosen at random such that $k_0 = k_0'$ holds. Otherwise, k_0 and k_0' are computed as in G_0.

Claim. $\Delta_1(\kappa) \leq \mathsf{Adv}_{\mathsf{KE}}^{\mathsf{UKE},\mathcal{B}}(\kappa)$. *Proof.* For client and server instances that do not share matching UKE transcripts both games are identical. Any $\mathcal{A}$ that can distinguish between G_1 and G_0 with non-negligible probability can be used to break the KE security from Definition 4. The description of the UKE adversary is exactly the same as in G_1 from the proof of Theorem 1. Hence, $\Delta_1(\kappa) \leq \mathsf{Adv}_{\mathsf{KE}}^{\mathsf{UKE},\mathcal{B}}(\kappa)$, as claimed.

G_2 This game proceeds as G_1, except that the simulator aborts if in the server-side tPkA session a client instance $[C, s]$ with tag t_C and (partial) transcript $\mathsf{tr}_{\alpha+\beta+\gamma+1}$ accepts server S but there exists no instance of S with matching (partial) transcript $\mathsf{tr}_{\alpha+\beta+\gamma+1}'$ and tag $t_S = t_C$, and sk_S is not corrupted.

Claim. $\Delta_2(\kappa) \leq \mathsf{Succ}_{\mathsf{CAuth}}^{\mathsf{tPkA},\mathcal{B}}(\kappa)$. *Proof.* As already described in games G_2 through G_4 in proof of Theorem 1 if $\mathcal{A}$ can distinguish between the two games, it can be immediately used to break CAuth-security of tPkA. (In this game CAuth-security is understood as a security property of PkA in case where the authenticating party is the server S. Recall that PkA offers single-side authentication and was defined from the perspective of an authenticating client. In this game the authenticating party is S but the notion of CAuth-security remains as defined.) Hence, $\Delta_2(\kappa) \leq \mathsf{Succ}_{\mathsf{CAuth}}^{\mathsf{tPkA},\mathcal{B}}(\kappa)$, as claimed.

Remark 7. Note that if the simulation does not abort in G_2 then it is guaranteed that for each client instance $[C, s]$ that is entering the confirmation round with partial transcript $\mathsf{tr}_{\alpha+\beta+\gamma+1}$ and tag t_C, there exists a server instance $[S, s']$ with partial transcript $\mathsf{tr}'_{\alpha+\beta+\gamma+1} = \mathsf{tr}_{\alpha+\beta+\gamma+1}$ and $t_S = t_C$ if neither C nor $S = \mathsf{pid}([C, s])$ has been fully corrupted.

G_3 This game proceeds as G_2, except that simulation aborts if an instance $[C, s]$ enters the confirmation round with partial transcripts tr_0 and $\mathsf{tr}_{\alpha+\beta+\gamma+1}$ and there exists $[S, s']$ with partial transcripts tr'_0 and $\mathsf{tr}'_{\alpha+\beta+\gamma+1}$ such that $\mathsf{tr}_{\alpha+\beta+\gamma+1} = \mathsf{tr}'_{\alpha+\beta+\gamma+1}$ but $\mathsf{tr}_0 \neq \mathsf{tr}'_0$.

Claim. $\Delta_3(\kappa) \leq q_{\mathcal{H}_T}^2 2^{-\kappa}$. *Proof.* G_2 already ensures that if $[C, s']$ accepts in the server side tPkA sub-protocol then there exists a server instance with $t_S = t_C$. The only difference between the two games is that G_3 may still abort even if G_2 does not. If $\mathcal{A}$ can distinguish between the games then $\mathcal{A}$ must have successfully caused the simulator to abort in G_3, in which case $[C, s]$ and $[S, s']$ hold tags $t_C = t_S$ but $\mathsf{tr}_0 \neq \mathsf{tr}'_0$. We can thus output a collision for $\mathcal{H}_T$. Since $\mathcal{H}_T$ is a random oracle we get $\Delta_3(\kappa) \leq q_{\mathcal{H}_T}^2 2^{-\kappa}$, as claimed.

G_4 This game proceeds as G_3, except that on behalf of an instance $[C, s]$ for which neither C nor $S = \mathsf{pid}([C, s])$ is fully corrupted the simulator computes $\nu_C \leftarrow \mathcal{H}'_S(s')$ using a private random oracle $\mathcal{H}'_S$, and sets $\nu_S = \nu_C$ for the corresponding $[S, s']$ that has matching UKE transcript and matching server-side tPkA sub-protocol transcript.

Claim. $\Delta_4(\kappa) \leq q \cdot q_{\mathcal{H}_S} \cdot 2^{-\kappa}$. *Proof.* Considering that in the previous game, confirmation values of $[C, s]$ were derived through the random oracle $\mathcal{H}_S$ on input k_0 (which is random as ensured by G_1) and the transcript s, any $\mathcal{A}$ that can distinguish between the games must ask at some point a query for $\mathcal{H}_S$ containing k_0 and s for any of the q invoked sessions as input. Therefore, $\Delta_4(\kappa) \leq q \cdot q_{\mathcal{H}_S} \cdot 2^{-\kappa}$, as claimed.

Assume that $\mathcal{A}$ wins in G_4. Then, $[C, s]$ must have received $\nu_S = \nu_C$ without having a partnered server instance. That is, $\mathcal{A}$ must have asked a Send query containing a value that matches a uniformly distributed ν_C. This happens with probability at most $q \cdot 2^{-\kappa}$ for up to q invoked client instances. We thus get

$$\mathsf{Succ}_{\mathrm{SAuth}}^{(\alpha,\beta,\gamma)\text{-MFAKE},\mathcal{A}}(\kappa) = \sum_{i=1}^{4} \Delta_i(\kappa) + q \cdot 2^{-\kappa}$$
$$\leq \mathsf{Adv}_{\mathrm{KE}}^{\mathsf{UKE},\mathcal{B}}(\kappa) + \mathsf{Succ}_{\mathrm{CAuth}}^{\mathsf{tPkA},\mathcal{B}}(\kappa) + (q_{\mathcal{H}_T}^2 + q(q_{\mathcal{H}_S} + 1)) \cdot 2^{-\kappa},$$

which is negligible by assumptions on UKE and tPkA.

$\square$

4   Conclusion

The proposed framework for multi-factor authentication and key exchange protocols enables black-box constructions from existing, better-understood single-factor authentication-only schemes. Our generic construction of the (α, β, γ)-MFAKE protocol avoids undesirable interactions amongst the different factors and bears optimization potential since messages of tag-based authentication sub-protocols can be interleaved or communicated over different channels. Thanks to its modularity the framework can easily be extended in the future to accommodate other authentication factors, e.g. based on friend-of-friend or social authentication [11,15].

Acknowledgments. Nils Fleischhacker was supported by the German Federal Ministry of Education and Research (BMBF) through funding for the Center for IT-Security, Privacy, and Accountability (CISPA, www.cispa-security.org). Mark Manulis was supported by the German Research Foundation (DFG), project PRIMAKE (MA 4957).

References

1. PCI Data Security Standard, Ver.2 (2010),
   http://www.pcisecuritystandards.org/
2. NIST Special Publication 800-63, Rev.1 (2011),
   http://csrc.nist.gov/publications/
3. Abdalla, M., Chevassut, O., Pointcheval, D.: One-Time Verifier-Based Encrypted Key Exchange. In: Vaudenay, S. (ed.) PKC 2005. LNCS, vol. 3386, pp. 47–64. Springer, Heidelberg (2005)
4. Abdalla, M., Fouque, P.-A., Pointcheval, D.: Password-Based Authenticated Key Exchange in the Three-Party Setting. In: Vaudenay, S. (ed.) PKC 2005. LNCS, vol. 3386, pp. 65–84. Springer, Heidelberg (2005)
5. Abdalla, M., Pointcheval, D.: Simple Password-Based Encrypted Key Exchange Protocols. In: Menezes, A. (ed.) CT-RSA 2005. LNCS, vol. 3376, pp. 191–208. Springer, Heidelberg (2005)
6. Bellare, M., Pointcheval, D., Rogaway, P.: Authenticated Key Exchange Secure against Dictionary Attacks. In: Preneel, B. (ed.) EUROCRYPT 2000. LNCS, vol. 1807, pp. 139–155. Springer, Heidelberg (2000)
7. Bellare, M., Rogaway, P.: Entity Authentication and Key Distribution. In: Stinson, D.R. (ed.) CRYPTO 1993. LNCS, vol. 773, pp. 232–249. Springer, Heidelberg (1994)

8. Benhamouda, F., Pointcheval, D.: Verifier-based password-authenticated key exchange: New models and constructions. Cryptology ePrint Archive, Report 2013/833 (2013), http://eprint.iacr.org/2013/833

9. Blake-Wilson, S., Johnson, D., Menezes, A.: Key Agreement Protocols and Their Security Analysis. In: Darnell, M. (ed.) Cryptography and Coding 1997. LNCS, vol. 1355, pp. 30–45. Springer, Heidelberg (1997)

10. Boyen, X., Dodis, Y., Katz, J., Ostrovsky, R., Smith, A.: Secure Remote Authentication Using Biometric Data. In: Cramer, R. (ed.) EUROCRYPT 2005. LNCS, vol. 3494, pp. 147–163. Springer, Heidelberg (2005)

11. Brainard, J.G., Juels, A., Rivest, R.L., Szydlo, M., Yung, M.: Fourth-Factor Authentication: Somebody You Know. In: ACM CCS 2006, pp. 168–178. ACM (2006)

12. Bringer, J., Chabanne, H.: An Authentication Protocol with Encrypted Biometric Data. In: Vaudenay, S. (ed.) AFRICACRYPT 2008. LNCS, vol. 5023, pp. 109–124. Springer, Heidelberg (2008)

13. Bringer, J., Chabanne, H., Izabachène, M., Pointcheval, D., Tang, Q., Zimmer, S.: An Application of the Goldwasser-Micali Cryptosystem to Biometric Authentication. In: Pieprzyk, J., Ghodosi, H., Dawson, E. (eds.) ACISP 2007. LNCS, vol. 4586, pp. 96–106. Springer, Heidelberg (2007)

14. Canetti, R., Krawczyk, H.: Analysis of Key-Exchange Protocols and Their Use for Building Secure Channels. In: Pfitzmann, B. (ed.) EUROCRYPT 2001. LNCS, vol. 2045, pp. 453–474. Springer, Heidelberg (2001)

15. De Cristofaro, E., Manulis, M., Poettering, B.: Private Discovery of Common Social Contacts. International Journal of Information Security 12(1), 49–65 (2013)

16. Dodis, Y., Reyzin, L., Smith, A.: Fuzzy Extractors: How to Generate Strong Keys from Biometrics and Other Noisy Data. In: Cachin, C., Camenisch, J.L. (eds.) EUROCRYPT 2004. LNCS, vol. 3027, pp. 523–540. Springer, Heidelberg (2004)

17. Eisenbarth, T., Kasper, T., Moradi, A., Paar, C., Salmasizadeh, M., Shalmani, M.T.M.: On the Power of Power Analysis in the Real World: A Complete Break of the KEELOQ Code Hopping Scheme. In: Wagner, D. (ed.) CRYPTO 2008. LNCS, vol. 5157, pp. 203–220. Springer, Heidelberg (2008)

18. Federal Financial Institutions Examination Council. Authentication in an Internet Banking Environment (2005),
http://www.ffiec.gov/pdf/authentication_guidance.pdf

19. Fleischhacker, N., Manulis, M., Azodi, A.: A Modular Framework for Multi-Factor Authentication and Key Exchange. Cryptology ePrint Archive, Report 2012/181 (2012), http://eprint.iacr.org/2012/181.pdf (last updated in 2014)

20. Garcia, F.D., de Koning Gans, G., Muijrers, R., van Rossum, P., Verdult, R., Schreur, R.W., Jacobs, B.: Dismantling MIFARE classic. In: Jajodia, S., Lopez, J. (eds.) ESORICS 2008. LNCS, vol. 5283, pp. 97–114. Springer, Heidelberg (2008)

21. Garcia, F.D., van Rossum, P., Verdult, R., Schreur, R.W.: Dismantling SecureMemory, CryptoMemory and CryptoRF. In: ACM CCS 2010, pp. 250–259. ACM (2010)

22. Gentry, C., MacKenzie, P.D., Ramzan, Z.: Password authenticated key exchange using hidden smooth subgroups. In: ACM CCS 2005, pp. 299–309. ACM (2005)

23. Gentry, C., MacKenzie, P.D., Ramzan, Z.: A Method for Making Password-Based Key Exchange Resilient to Server Compromise. In: Dwork, C. (ed.) CRYPTO 2006. LNCS, vol. 4117, pp. 142–159. Springer, Heidelberg (2006)

24. Hao, F., Clarke, D.: Security Analysis of a Multi-factor Authenticated Key Exchange Protocol. In: Bao, F., Samarati, P., Zhou, J. (eds.) ACNS 2012. LNCS, vol. 7341, pp. 1–11. Springer, Heidelberg (2012)

25. Jager, T., Kohlar, F., Schäge, S., Schwenk, J.: Generic Compilers for Authenticated Key Exchange. In: Abe, M. (ed.) ASIACRYPT 2010. LNCS, vol. 6477, pp. 232–249. Springer, Heidelberg (2010)
26. Kiefer, F., Manulis, M.: Zero-Knowledge Password Policy Checks and Verifier-Based PAKE. In: Kutyłowski, M., Vaidya, J. (eds.) ESORICS 2014, Part II. LNCS, vol. 8713, pp. 295–312. Springer, Heidelberg (2014)
27. LaMacchia, B.A., Lauter, K., Mityagin, A.: Stronger Security of Authenticated Key Exchange. In: Susilo, W., Liu, J.K., Mu, Y. (eds.) ProvSec 2007. LNCS, vol. 4784, pp. 1–16. Springer, Heidelberg (2007)
28. Lee, Y., Kim, S., Won, D.: Enhancement of Two-Factor Authenticated Key Exchange Protocols in Public Wireless LANs. Computers & Electrical Engineering 36(1), 213–223 (2010)
29. Li, C.-T., Hwang, M.-S.: An Efficient Biometrics-Based Remote User Authentication Scheme Using Smart Cards. Journal of Network and Computer Applications 33(1), 1–5 (2010)
30. Li, X., Niu, J.-W., Ma, J., Wang, W.-D., Liu, C.-L.: Cryptanalysis and Improvement of a Biometrics-Based Remote User Authentication Scheme Using Smart Cards. Journal of Network and Computer Applications 34(1), 73–79 (2011)
31. Park, Y.M., Park, S.K.: Two Factor Authenticated Key Exchange (TAKE) Protocol in Public Wireless LANs. IEICE Transactions on Communications E87-B(5), 1382–1385 (2004)
32. Paterson, K.G., Stebila, D.: One-Time-Password-Authenticated Key Exchange. In: Steinfeld, R., Hawkes, P. (eds.) ACISP 2010. LNCS, vol. 6168, pp. 264–281. Springer, Heidelberg (2010)
33. Pointcheval, D., Zimmer, S.: Multi-factor Authenticated Key Exchange. In: Bellovin, S.M., Gennaro, R., Keromytis, A.D., Yung, M. (eds.) ACNS 2008. LNCS, vol. 5037, pp. 277–295. Springer, Heidelberg (2008)
34. Song, R.: Advanced Smart Card Based Password Authentication Protocol. Computer Standards & Interfaces 32(5-6), 321–325 (2010)
35. Stebila, D., Udupi, P., Chang, S.: Multi-Factor Password-Authenticated Key Exchange. In: Eighth Australasian Information Security Conference (AISC 2010), vol. 105, pp. 56–66 (2010)
36. Tapiador, J.E., Hernandez-Castro, J.C., Peris-Lopez, P., Clark, J.A.: Cryptanalysis of Song's Advanced Smart Card Based Password Authentication Protocol. arXiv.org, Cryptography and Security (2011), http://arxiv.org/abs/1111.2744v1
37. Wang, X., Zhang, W.: An Efficient and Secure Biometric Remote User Authentication Scheme Using Smart Cards. In: Pacific-Asia Workshop on Computational Intelligence and Industrial Application (PACIIA 2008), vol. 2, pp. 913–917. IEEE (2008)

Improving the ISO/IEC 11770 Standard for Key Management Techniques

Cas Cremers and Marko Horvat

University of Oxford, Oxford, UK

Abstract. We provide the first systematic analysis of the ISO/IEC 11770 standard for key management techniques [18,19], which describes a set of key exchange, key authentication, and key transport protocols. We analyse the claimed security properties, as well as additional modern requirements on key management protocols, for 30 protocols and their variants. Our formal, tool-supported analysis of the protocols uncovers several incorrect claims in the standard. We provide concrete suggestions for improving the standard.

1 Introduction

The International Organisation for Standardisation (ISO) develops and promotes international standards, which include a wide variety of security mechanisms. Many large vendors aim to support ISO standards, for example because they are mandated by oversight bodies [15] or to prevent trade barriers. Hence, it is critical that the ISO standards for security mechanisms are thoroughly scrutinised. However, most previous analyses of the ISO security standards have been very limited in scope, e.g., [10,16,23,24]. One exception is the analysis of Basin et al. of the ISO/IEC 9798 standard for entity authentication [4] in 2012. Their analysis uncovered a series of issues that led to an updated version of the 9798 standard.

In this paper we focus on the ISO/IEC 11770 standard for key management protocols, in particular on parts 2 and 3 of this standard. In the most recent version as of June 2014, these two parts together describe 30 base protocols for key exchange, key agreement, and key transport. Many of the standard's protocols are based on protocols such as Diffie-Hellman, variants of MQV, and the TLS handshake. For many of the protocols, at least two variants are described. Thus, analysing these two parts is a significant undertaking.

In positive contrast to other security protocol standards [5], the ISO/IEC 11770 standard explicitly specifies security properties for each of its protocols. Two of these properties are structural properties, i.e., key control and replay detection. Additionally, there are four security properties that relate to active adversaries, namely key authentication, key confirmation, entity authentication, and forward secrecy.

In this work, we use tool-supported formal methods to determine if the protocols indeed satisfy their claimed properties. Additionally, we analyse the

L. Chen and C. Mitchell (Eds.): SSR 2014, LNCS 8893, pp. 215–235, 2014.

protocols for modern key exchange security properties, such as resilience against Key Compromise Impersonation (KCI) and Unknown Key Share (UKS) attacks.

Contributions. We perform the first comprehensive analysis of parts 2 and 3 of the ISO/IEC 11770 standard. Our analysis uncovers multiple previously unreported errors and weaknesses. For each of the discovered issues, we provide concrete recommendations for improving the standard.

Our protocol models and tools used are available for download from http://www.cs.ox.ac.uk/people/cas.cremers/scyther/iso11770/.

Overview In Section 2 we give some background on ISO/IEC 11770 and illustrate some of its protocols. We describe our analysis approach in Section 3 and present the results in Section 4. We provide concrete recommendations for improving the standard in Section 5, discuss related work in Section 6, and conclude in Section 7.

2   Background on ISO/IEC 11770

The ISO/IEC 11770 standard describes key management techniques. According to the standard, the purpose of key management is to provide procedures for handling cryptographic keying material to be used in symmetric or asymmetric mechanisms. Effectively, the standard describes a large number of key agreement, key transport, and key exchange protocols. We will therefore use the terms *mechanism* and *protocol* interchangeably.

The standard is currently divided into five parts. Part 1 was originally released in 1996 and has been updated over the years. It describes the context and framework. Parts 2 and 3 describe mechanisms based on symmetric and asymmetric techniques. Part 4 describes mechanisms based on weak secrets, such as password-based key exchange protocols. Part 5 describes group key management mechanisms. A part 6 on key derivation functions is currently under development.

2.1   Protocols

In this work we focus on part 2 [18] and part 3 [19] of the ISO/IEC 11770 standard. Part 2 describes 13 key establishment mechanisms. Part 3 describes 11 key agreement mechanisms and 6 key transport mechanisms. Many of these 30 mechanisms in parts 2 and 3 have optional message parts and message flows, giving rise to a large number of variants.

Additionally, the mechanisms produce keying material that must be used with a key derivation function to form a key for use in further messages. The standard does not specify a single key derivation function; instead it gives examples of various possible key derivation functions. Thus, using a single mechanism with different key derivation functions can be regarded as multiple variants of the same base mechanism. As we will see in Section 4.3, the choice of a key derivation function can influence the security of a mechanism.

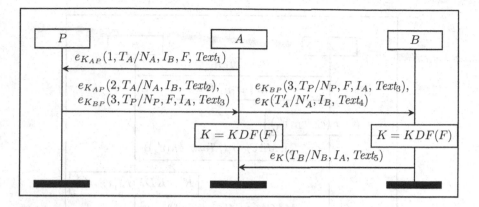

Fig. 1. Protocol 2-12 with optional fields

Naming conventions. We provide a unique name for each base mechanism in the considered parts of the standard. We refer to the thirteen key establishment mechanisms from part 2 as protocol 2-1, 2-2, ..., 2-13. We refer to the key agreement mechanisms as 3-KA-1, ..., 3-KA-11 and to the key transport mechanisms as 3-KT-1, ..., 3-KT-6.

We next describe two protocols from the standard. This enables us to introduce notation and provide an indication of the protocols contained in the standard.

Key Exchange Mechanism 12 (2-12).   We give an example of a protocol described in part 2 [18], referenced in the standard in Section 7.2 as Key Establishment Mechanism 12. The protocol is stated to be derived from, but not fully compatible with, the four-pass mutual authentication mechanism specified in ISO/IEC 9798-2 [17]. The protocol has several variants. For this example, we consider the variant with all optional parts included, depicted using a Message Sequence Chart (MSC) in Figure 1.

In the figure, T_A/N_A is either a time stamp T_A or sequence number N_A of A. I_A and I_B respectively identify entities A and B. $e_K(m)$ denotes the encryption of the message m with the key K. The protocol assumes that entities A and B respectively share long-term symmetric keys K_{AP} and K_{BP} with a trusted third party P. $Text_1$ through $Text_5$ are text fields whose contents are not specified by the standard. F denotes keying material.

The protocol proceeds as follows. When a party A wants to communicate with another party B, it contacts trusted third party P. A generates fresh keying material F and includes it in the message encrypted for P, who responds with two encrypted messages. They are respectively encrypted with K_{AP} and K_{BP}. Both encrypted messages are sent to A, who forwards the second encryption to B. B decrypts the message and obtains the keying material F. A and B now both use a key derivation function to compute the session key K from F. We are only considering the protocol variant with optional fields, so the protocol

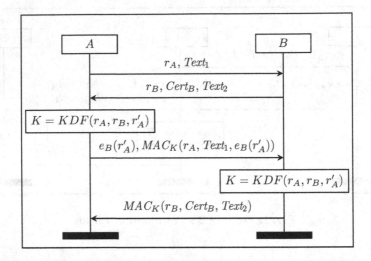

Fig. 2. Protocol 3-KA-11

proceeds with two messages that allow both entities to confirm to the other entity that they have successfully computed the key.

For the key derivation function, we consider two extremes from the KDFs described in the standard: at the one end, some KDFs take as input only F, whereas others include additional parameters, such as the identities I_A and I_B.

Key Agreement Mechanism 11 (3-KA-11). Key Agreement Mechanism 11 from part 3, shown in Figure 2, establishes a key shared by entities A and B. First, A generates a random value r_A and sends it to B. B responds with his own random value r_B and his certificate. Upon receiving this message, A generates a new random value r'_A. r'_A is used with the other two random values to derive a session key K. Then r'_A is encrypted using B's public key, and sent to B along with a message authentication code (MAC) keyed with K that includes the earlier randomness r_A. B decrypts the message, computes K, and checks the MAC. B then responds with his own MAC of r_B and his certificate.

According to the standard, this protocol is derived from the TLS handshake protocol [14]. In particular, since only B uses his private key (to decrypt the message) and the random values are directly input to the key derivation function, the protocol resembles TLS's unilaterally authenticated RSA mode, where A corresponds to the client and B to the server. The random value r'_A in 3-KA-11 plays the same role as TLS's *pre-master secret* and the two text fields are used in TLS for the cipher suite negotiation.

2.2 Security Properties and Threat Model of the Standard

Most standards for security protocols do not specify threat models or intended security properties [5]. In this respect, ISO/IEC 11770 is an exception since it explicitly specifies a set of security properties, and states for each protocol which of these properties it satisfies. ISO/IEC 11770 defines the following properties [18, 19]:

Implicit key authentication from entity A to entity B. Assurance   for entity B that A is the only other entity that can possibly be in possession of the correct key.

Explicit key authentication from entity A to entity B. Assurance   for entity B that A is the only other entity that is in possession of the correct key.

Key confirmation from entity A to entity B. Assurance for entity B that entity A is in possession of the correct key.

Entity authentication of entity A to entity B. Assurance of the identity of entity A to entity B.

Forward secrecy with respect to entity A. Property that knowledge of entity A's long-term private key subsequent to a key agreement operation does not enable an opponent to recompute previously derived keys.

Forward secrecy with respect to both entity A and entity B. Property that knowledge of entity A's long-term private key or knowledge of entity B's long-term private key subsequent to a key agreement operation does not enable an opponent to recompute previously derived keys.

Mutual forward secrecy. Property that knowledge of both entity A's and entity B's long-term private keys subsequent to a key agreement operation does not enable an opponent to recompute previously derived keys.

For example, regarding the protocols described in the previous section, the standard claims the following: protocol 2-12 with optional parts satisfies mutual explicit key authentication, mutual key confirmation and mutual entity authentication, and protocol 3-KA-11 provides mutual explicit key authentication, mutual key confirmation, entity authentication to B and mutual forward secrecy.

The standard does not specify an explicit threat model. However, the security properties stated above are not claimed for all protocols. Because some protocols apparently do not meet the above properties, we can conclude that the adversary is considered to have at least the following capabilities:

Injecting network messages. Entity authentication is claimed for some, but not all mechanisms. Entity authentication can only be effectively violated if the adversary is able to inject or tamper with network messages.

Eavesdropping on network messages. If the adversary cannot eavesdrop on messages, we would need no complex key management mechanism, and could exploit simple authentication mechanisms.

Compromising long-term private keys. For some protocols, perfect forward secrecy is claimed. The adversary can only violate perfect forward secrecy by compromising the long-term private keys of some entities.

3 Formally Modelling the Protocols and Their Properties

We analyse all 30 protocols specified in the standard, along with their described variants, by using formal methods. In particular, we use the Scyther framework [13] for the automatic symbolic analysis of security protocols. The Scyther tool [11] has built-in support for compromising adversaries [2], and is therefore especially suitable for analysing security notions that are common in the domain of protocols for key agreement, exchange, and transport.

3.1 Protocol Specification

Within the Scyther framework, protocols are specified using so-called role scripts. A protocol can have any finite number of roles, and is run by entities who execute those roles. Entities may execute each role multiple times, and every role can be executed by any entity. We call each such role instance a session. We assume that, prior to protocol execution, every entity has generated or securely received a long-term asymmetric key pair consisting of a public and a private key, it has authentic and secret copies of all its long-term symmetric keys shared with other entities, and authentic copies of the public keys of all other entities.

Roles are specified as sequences of send, receive and claim events. Events have term parameters, where terms are constructed from role names, function names, variables, and constants. Receive events correspond to pattern matching on incoming messages, and may therefore contain nonces generated in previous send steps and variables to store incoming payloads. Send events can contain freshly generated nonces and variables that have been previously initialised in receive steps. We specify intended security properties using claim events.

For example, we give in Figure 3 the input for the Scyther tool to describe protocol 2-12 from Section 2.1. Send, receive and claim events are respectively specified with send, recv, and claim. Freshly generated nonces are declared with fresh, variables with var, user-defined types with usertype, hash functions with hashfunction. Every function, constant and variable can have a different type, such as Nonce or a user-defined type such as Integer, KeyingMaterial, or String—the types are used to restrict the pattern matching in the execution of a receive event. The keyword macro can be used to define shorthands.

3.2 Specifying Security Properties

We model the following properties from the standard: *key authentication, key confirmation, entity authentication,* and *forward secrecy.* Additionally, we model *key compromise impersonation (KCI)* and *unknown key share (UKS)* attacks.

Key Authentication.   According to the standard, both implicit and explicit key authentication require that if an entity A uses a protocol to establish a key K with entity B, then only A and B will learn the key. We model this by analysing the secrecy of K whilst allowing the adversary to impersonate any entity except for A and B. The possibility of impersonation is modelled by allowing the adversary to learn the long-term private key of any entity except for A and B. Additionally, explicit key authentication requires that entities in fact compute the key. We cover this requirement in our modelling of key confirmation.

Key Confirmation.   This and the following property correspond to authentication properties in Lowe's hierarchy [22]. Key confirmation from A to B corresponds to non-injective data agreement on the key, which we model with two

```
 1  option "--partner-definition=2";          35  role B
 2                                             36  {
 3  usertype KeyingMaterial;                   37    var TNP,TNA2: Nonce;
 4  usertype String;                           38    var F: KeyingMaterial;
 5  usertype Integer;                          39    var Text3,Text4: String;
 6                                             40    fresh TNB: Nonce;
 7  hashfunction KDF;                          41    fresh Text5: String;
 8  const N1,N2,N3: Integer;                   42
 9                                             43    recv_3(A,B,{N3,TNP,F,A,Text3}k(B,P),
10  macro key = KDF(F);                        44             {TNA2,B,Text4}key);
11  macro sid = (A,B,key);                     45    claim(B,SID,sid);
12                                             46    claim(B,Running,A,key);
13  protocol 2-12-withOptional(A,B,P)          47    send_4(B,A,{TNB,A,Text5}key);
14  {                                          48
15    role A                                   49    claim(B,SKR,key);
16    {                                        50    claim(B,Commit,A,key);
17      fresh TNA,TNA2: Nonce;                 51    claim(B,Alive,A);
18      fresh F: KeyingMaterial;               52  }
19      fresh Text1,Text4: String;             53  role P
20      var Text2,Text5: String;               54  {
21      var T: Ticket;                         55    var TNA: Nonce;
22      var TNB: Nonce;                        56    var F: KeyingMaterial;
23                                             57    var Text1: String;
24      claim(A,SID,sid);                      58    fresh Text2: String;
25      claim(A,Running,B,key);                59    fresh TNP: Nonce;
26      send_1(A,P,{N1,TNA,B,F,Text1}k(A,P));  60    fresh Text3: String;
27      recv_2(P,A,{N2,TNA,B,Text2}k(A,P),T);  61
28      send_3(A,B,T,{TNA2,B,Text4}key);       62    claim(P,SID,P);
29      recv_4(B,A,{TNB,A,Text5}key);          63    recv_1(A,P,{N1,TNA,B,F,Text1}k(A,P));
30                                             64    send_2(P,A,{N2,TNA,B,Text2}k(A,P),
31      claim(A,SKR,key);                      65             {N3,TNP,F,A,Text3}k(B,P));
32      claim(A,Commit,B,key);                 66  }
33      claim(A,Alive,B);                      67  }
34    }
```

Fig. 3. Scyther input file for 2-12 with confirmation messages and claimed properties

claims: a Running claim in the A role and a Commit claim in the B role. If the Commit claim is executed, we require that the corresponding Running claim is executed as well: it must have the entities in reverse order, and the same contents (the entities are said to *agree* on the contents). It is called *non-injective* data agreement because replays are not considered.

Entity Authentication. Entity authentication from A to B corresponds to aliveness [22]: an Alive claim of A is placed in the specification of role B. Whenever the claim is executed, the entity assumed to be performing the A role is required to have executed some event.

Forward Secrecy. There are several definitions of forward secrecy in the literature, and it is not clear from the standard which property is intended. The *mutual forward secrecy (MFS)* notion from the standard seems to be closest to two common formal definitions. *Weak Perfect Forward Secrecy (wPFS)* [12, 20] requires that the adversary does not actively inject messages (and thus is passive) with respect to the session that he attacks. In contrast, *(strong) Perfect Forward Secrecy (PFS)* allows the adversary to actively interfere with the messages received by the session under attack. Scyther directly supports checking for both properties through its support of the LKRaftercorrect and LKRafter rules [2].

Our analysis revealed that the majority of protocols for which MFS is claimed in fact only achieve wPFS, and we therefore interpret MFS as wPFS.

Key Compromise Impersonation (KCI).   Resilience to KCI attacks is a desirable property of key exchange protocols [6]. KCI attacks are attacks in which the adversary exploits his knowledge of the long-term private key of Alice to impersonate any entity in subsequent communication with Alice.

This property is modelled in Scyther by a session key secrecy claim of an entity whose long-term private keys the adversary is allowed to reveal. These attacks can be seen as a broader class than unilateral forward secrecy attacks because KCI attacks allow for dynamic usage of the compromised keys: the adversary can use them during protocol execution to inject messages or otherwise tamper with the communication.

Unknown Key Share (UKS).   Unknown key share attacks are attacks in which only Alice and Bob know the session key K; however, Alice and Bob disagree on who they share K with [7]. For example, Alice correctly thinks K is shared with Bob, but Bob might think that K is shared with Charlie. Even though the adversary does not learn the key in such attacks, using the key is not sufficient to authenticate subsequent messages: if Alice sends a message encrypted with K or accompanied by a MAC keyed by K, Bob will assume that the message came from Charlie. Similarly, Bob will send messages intended for Charlie that will be received by Alice.

We model UKS attacks in the standard way, i.e., if the assumptions on the partner identities of the attacked session s do not match the assumptions of a session s', we allow the adversary to reveal the session key of s'. This causes UKS attacks to manifest as violations of secrecy of the session key computed by s. Note that false positives can also occur, where the revealed session key is used for more than computing the session key of s', e.g., for injecting messages.

We specify session identifiers (SIDs) manually in Scyther input files by including `option "--partner-definition=2"` and annotating each role with `SID` claims, in which the SID is specified for the role instance. For example, in Figure 3 we enable the manual specification of a partner session on line 1, define the session identifier on line 11 and insert it into role specifications on lines 24 and 45. When session key secrecy is analysed for a session s, and Scyther's SKR adversary rule (Session-Key Reveal) is enabled in the GUI or `--SKR=1` is provided as a command-line option, the adversary is able to obtain the session keys computed by any session whose identifier differs from that of s.

4   Results of the Formal Analysis

We analyse the protocol models described in the previous section with respect to their claimed properties, and afterwards consider KCI and UKS attacks.

4.1   Claimed Properties

We give an overview of our results when analysing the protocols with respect to their claimed properties in Table 1. The contents of this table are directly taken from the tables in [18,19], with the difference that we added notes and used red and bold to mark incorrect statements. We classify the incorrect claims in the standard into five categories AT1...AT5, which we describe below.

Note that the table in [18] only has an (explicit) key authentication column with "yes" or "no" in the cells, but this information has to be combined with NOTE 2 [18], which states that all protocols in part 2 achieve implicit key authentication, and that "yes" should be interpreted as explicit key confirmation.

AT1: Entity Authentication Failures for 2-8, 2-9, 2-12, and 2-13. We find several possible entity authentication failures for protocols in part 2 that are derived from protocols in an earlier version of the ISO/IEC 9798-2 standard for entity authentication [17].

These attacks are closely related to the attacks on the corresponding protocols from the 9798 standard as presented in [4]. The attacks work in all implementations where a single entity can perform not only the role of the trusted third party but also another role. In the attacks, the adversary can cause A to complete the protocol, apparently with B, even though B is not present. Thus, the attacks violate even the weakest form of entity authentication. We show an example of such an attack on protocol 2-12 in Appendix A.

Fixes for these protocols have been proposed in [4], which have been integrated into the ISO/IEC 9798 standard. As a result, these attacks no longer work on ISO/IEC 9798, but since no changes have been made to the derived protocols in ISO/IEC 11770, similar attacks are still possible on this standard.

AT2: 3-KA-11 Key Authentication/Confirmation Failure for B. According to the standard this mechanism (depicted in Figure 2) offers mutual explicit key authentication and mutual key confirmation. However, there is an attack on entity authentication on the B role that violates both of these claimed properties. In the attack, the adversary performs the A role, pretending to be Alice, and sends messages to Bob in the B role. Because executing the A role does not require the use of any long-term secrets, the adversary can simply claim to be anybody. The entity performing the B role therefore cannot obtain any authentication guarantees about its communication partner or about the secrecy of the key.

As said before, 3-KA-11 is derived from the unilaterally authenticated RSA mode of the TLS handshake [14]. In this mode of TLS, the server obtains no guarantee about whether the client is who he claims to be or not. The same issue occurs here for the B role of 3-KA-11.

AT3: Failure of MFS for 3-KA-11. Because protocol 3-KA-11 is derived from the RSA mode of TLS, it provides no forward secrecy. The adversary only needs to observe a regular session. If he afterwards obtains the long-term private key of B, he can decrypt $e_B(r'_A)$ and learn r'_A. Since r_A and r_B have been sent in plaintext, the adversary now has all the ingredients he needs to recompute the key K.

AT4: Failure of key authentication for 2-11. Depending on the implementation, it may be possible for an agent to misinterpret an agent identity

Table 1. Claimed properties Security properties claimed for the protocols in parts 2 and 3 of the standard. Our analysis revealed that some claims are incorrect, and we mark them using bold and red.

Mechanism in part 2	Key Authentication	Key Confirmation	Entity Authentication
2-1	implicit	no	no
2-2	implicit	no	no
2-3	explicit	no	A
2-4	explicit	no	A
2-5	explicit	no	A & B
2-6	explicit	no	A & B
2-7	implicit	no	no
2-8	explicit (AT1)	opt. (AT1)	opt. (AT1)
2-9	explicit (AT1)	opt. (AT1)	opt. (AT1)
2-10	explicit	no	no
2-11	explicit (AT4)	no	no
2-12	explicit (AT1)	opt. (AT1)	opt. (AT1)
2-13	explicit (AT1)	opt. (AT1)	opt. (AT1)

Mechanism in part 3	Implicit Key Authentication	Key Confirmation	Entity Authentication	Forward Secrecy
3-KA-1	A,B	no	no	no
3-KA-2	B	no	no	A
3-KA-3	A,B	B	A	A
3-KA-4	no	no	no	MFS
3-KA-5	A,B	opt	no	A,B
3-KA-6	A,B	opt	B	B
3-KA-7	A,B	A,B	A,B	MFS
3-KA-8	A,B	no	no	A
3-KA-9	A,B	no	no	MFS
3-KA-10	A,B	A,B	A,B	MFS
3-KA-11	**A,B (AT2)**	**A,B (AT2)**	B	**MFS (AT3)**
3-KT-1	B	no	no	A
3-KT-2	B	B	A	A
3-KT-3	B	B	A	A
3-KT-4	A	A	B	B
3-KT-5	A,B	(A),B	A,B	no
3-KT-6	A,B	**A,B (AT5)**	A,B	no

as (random) keying material, for example if both are the same bit length. If an implementation of 2-11 cannot tell the difference between these, it can be vulnerable to a type-flaw attack on key authentication.

The 2-11 protocol assumes pre-shared symmetric keys and a trusted third party P. In a regular execution of the protocol, A sends a request to P for a ticket to forward to B. The request is a triplet $(I_B, F, Text_1)$ encrypted with the key shared between A and P. P then returns a triplet $(F, I_A, Text_2)$ encrypted with the key shared between B and P, which A forwards to B.

The adversary Charlie can attack a session of Bob which assumes to be talking to Alice, even though Alice and Bob are not compromised. Charlie encrypts a message for the trusted third party Pete, requesting a key for Alice. However, instead of generating new keying material F, Charlie instead includes Bob's identity in the keying material field. Pete's response therefore is the triplet $(I_{Alice}, I_{Charlie}, Text_2)$ encrypted with $K_{Pete,Bob}$. Charlie re-sends this message to Pete. There is nothing in the standard that prevents Pete from accepting this message as a valid request. Now, Pete responds with the triplet $(I_{Charlie}, I_{Alice}, Text_2)$ encrypted with $K_{Bob,Pete}$. Upon receiving this message, Bob will assume that it is a valid message and that $I_{Charlie}$ is secure keying material for communicating with Alice. The adversary can now compute the session key that Bob computes.

MSCs of the protocol and the attack are provided in the appendix.

AT5: Failure of Key Confirmation for 3-KT-6.   There is a complex attack possible on some implementations of 3-KT-6 that meet three conditions. The 3-KT-6 protocol is a three-pass protocol that transfers two secret keys. After the exchange, a session key can be computed from either or both of these keys. We give the full attack description and the preconditions in the appendix.

4.2   Key Compromise Impersonation (KCI) Results

All of the protocols in part 2 use symmetric cryptography and hashing only. Hence, they are necessarily vulnerable to KCI attacks, which is implied by the impossibility result from [3]. All the key transport and public key transport protocols from part 3 are KCI resilient.

The automatic analysis shows that four of the eleven key agreement protocols in part 3 are vulnerable to KCI attacks: 3-KA-1, 3-KA-3, 3-KA-6, and 3-KA-8. Mechanisms 3-KA-1 and 3-KA-3 are variants of the unsigned Diffie-Hellman protocol. Mechanism 3-KA-1 is the static Diffie-Hellman protocol, so as expected the session key is not secret when one of the static keys is known to the adversary. Similarly, 3-KA-3 is a one-pass Diffie-Hellman variant where A's ephemeral and B's static half keys are used: if the adversary gets B's static private key, he can use A's half key to infer the session key. In 3-KA-6, the fact that the input to the key derivation function is only protected by the private key of B allows an adversary who knows B's key to impersonate A in subsequent communications that are only protected with the established session key.

3-KA-8 is derived from one-pass MQV, and is described in Appendix C. The adversary can construct a message with his own injected randomness, send it to B, and use B's key to infer the session key.

Note that all (unilateral) forward secrecy attacks can be considered to be KCI attacks with late occurrences of long-term key compromise. The converse does not hold: forward secrecy does not imply KCI resilience in general, because KCI attacks may require knowledge of the long-term keys before the attacked session ends. In this specific set of protocols, no KCI attacks require early knowledge of the relevant long-term keys, because the long-term keys are only needed for session key computation. Hence, the protocols in the standard that satisfy forward secrecy are also KCI resilient.

Therefore, in order to obtain KCI resilience guarantees while preserving the required properties, we turn to the Forward Secrecy column in Table 1. We replace each protocol vulnerable to KCI attacks with one that achieves all the already satisfied security guarantees, plus forward secrecy with respect to both entities:

- 3-KA-1 can be replaced by 3-KA-5 (optionally, key confirmation can be enabled),
- 3-KA-3 and 3-KA-6 can be replaced by 3-KA-7 (if entity authentication is required) or 3-KA-5 (otherwise), and
- 3-KA-8 can be replaced by 3-KA-9.

4.3 Unknown Key Share (UKS) Results

We used Scyther to analyse all protocols for which key authentication was claimed for UKS attacks. We found that two protocols are vulnerable to UKS for any implementation, and that some implementations of several other protocols are vulnerable to UKS attacks.

We first explain the unknown key share attack on the 3-KA-11 protocol in detail. A graphical representation is given in Figure 4. In the attack, the adversary does not modify the content of any messages, but only changes the implicit sender/recipient fields. When Alice executes role A with intended partner Bob, she sends out her first message. The adversary modifies the sender field to "Charlie" and forwards the message to Bob. Bob assumes Charlie wants to communicate with him and starts to execute the B role, and sends the response message to Charlie. The adversary redirects this message to Alice. The protocol continues as usual except that the adversary continues to modify the sender fields and redirecting the responses. There is nothing in the messages that allows the entities to check each other's beliefs about the communication partner. In the end, Alice and Bob compute the same key K. Although the adversary does not know this key, Bob will believe that any subsequent messages he receives, which are encrypted or authenticated using K, are coming from Charlie, where in fact they come from Alice, leading to a serious authentication flaw [8, p. 139].

Although 3-KA-11 is derived from the TLS protocol, the TLS protocol is not vulnerable to unknown key share attacks. The reason for this is that the TLS

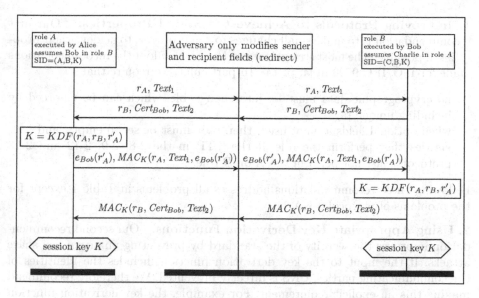

Fig. 4. Unknown key share attack on protocol 3-KA-11. Alice shares a key K with Bob as she expects, but Bob mistakenly assumes he is sharing K with Charlie.

protocol performs confirmation on all previously received messages, which in TLS contain the identities of the sender and recipient. This confirmation will fail if the parties have different views on their communication partners. In some sense, 3-KA-11 can be regarded as a stripped down version of the unilateral TLS-RSA handshake where security-relevant information (the identities of the participants) has been removed.

The second UKS attack is possible on the 2-10 protocol, which suffers from a role-mixup attack in which Alice and Bob both perform the A role and compute the same session key. This can lead to reflection attacks and misinterpretation attacks when the session key is later used to encrypt payloads. In implementations in which entities can perform multiple roles, protocols 2-2, 2-8, 2-9, 2-11, and 2-12 are also vulnerable to UKS attacks.

Fortunately, UKS attacks can be prevented by choosing a key derivation function that includes the identifiers (I_A and I_B) of the involved entities [7, 8]. For example, this is required by the NIST SP-800-56A key derivation [1], which is included in part 3 of the standard. We modelled the use of this KDF and used automated analysis to confirm that this prevents the UKS attacks. Intuitively, including the identities in the KDF ensures that entities that have different beliefs about their intended peers compute different keys, which prevents UKS attacks.

5 Recommendations

We provide three recommendations to improve the ISO/IEC 11770 standard.

1. Improving Protocols to Achieve the Stated Properties. Our first recommendation is to make small changes to the protocols to achieve their properties, if possible. The most straightforward way is to adopt the recommendations made for ISO/IEC 9798 in [4, p. 14]. In particular, we require that

- no cryptographic data must be interchangeable, which can be enforced by including unique tags,
- when optional fields are not used, then they must be set to empty, and
- entities that perform the role of the TTP in the 2-8, 2-9, 2-12 and 2-13 protocols must not perform the A or B role.

Following these recommendations addresses all problems in Table 1 except for the problems of protocol 3-KA-11.

2. Using Appropriate Key Derivation Functions. Our second recommendation improves the security of the standard by preventing unknown key share attacks. If the input to the key derivation function includes the identities of the communicating parties, UKS is directly prevented. We therefore recommend making this an explicit requirement. For example, the key derivation function from NIST SP-800-56A [1], which is described in the standard, meets this requirement.

3. Addressing Remaining Issues with 3-KA-11. 3-KA-11 inherently does not offer perfect forward secrecy or mutual authentication. Switching to a protocol that does, such as mutually authenticated TLS-DHE_RSA, substantially changes the environmental assumptions, including the pre-distribution of keys.

A simpler solution is to adapt the statements made about the protocol. In particular, it should *not* be claimed in the overview table [19, p. 42] that 3-KA-11 achieves implicit key authentication for both entities, that it achieves key confirmation for both entities, or that it achieves MFS. Similarly, the running text [19, p. 26] should not claim that 3-KA-11 achieves mutual explicit key authentication.

6   Related Work

In 1998, Horng and Hsu presented an attack on an early version of the 3-KT-6 protocol [16]. Their attack violated key confirmation and showed that the protocol did not offer any strong mutual authentication. In the same year, Mitchell and Yeun proposed a fix [24] that was later introduced in the standard. It essentially involves adding more identifiers to the messages, similar to Lowe's fix of the Needham-Schroeder protocol.

In 2004, Cheng and Comley presented two attacks on a previous version of the 2-12 protocol [10]. Their first attack is a replay attack. The second attack is a type flaw attack based on the possibility of interpreting an identity field as a fresh key. Cheng and Comley presented a fixed version of the protocol. Initially, protocol 2-12 was withdrawn from the standard, and it was later updated in 2008 with a new version that does not suffer from these attacks.

Mathuria and Sriram used Scyther to discover in 2008 [23] more complex type-flaw attacks on protocol 2-13 and on Cheng's and Comley's proposed fixed protocol. The attacks rely on the possibility that complex fields (concatenations, encryptions) can be interpreted as atomic fields (random values, keys, identities) in some implementations.

In 2010, Chen and Mitchell [9] generalised some of the concepts occurring in this class of type-flaw attacks and presented countermeasures, some of which found their way into later versions of ISO standards.

7   Conclusions

In retrospect, though we found all attacks through automatic analysis, it is clear that some attacks should have been found by manual inspection. This holds especially for 3-KA-11, which is based on TLS's unilaterally authenticated RSA handshake: it is clear that this protocol cannot offer key authentication or confirmation for both parties, since only one party is authenticated.

One way in which standardisation bodies could be more proactive is by being aware of analysis of standards on which they build. For example, many protocols in ISO/IEC 11770 are mentioned to be derived from authentication protocols in ISO/IEC 9798. In 2012, the ISO/IEC 9798 standard was analysed, several problems were identified [4], and it was subsequently updated to fix the identified problems. However, it seems that no attempt was made to determine if the derived protocols inherited these problems. Our analysis shows that this was in fact the case, implying that the attacks on protocols from part 2 could have been identified earlier. Our analysis shows that applying the recommendations for ISO/IEC 9798 as described in [4] to ISO/IEC 11770 would have prevented all issues in Table 1, except for those on 3-KA-11.

The standard currently does not claim resilience to UKS or KCI attacks. One could consider identifying those protocols that achieve these properties or improve the others. For example, all UKS attacks that we found can easily be prevented, at negligible cost, by using key derivation functions that include the identities of the participants. We therefore recommend including the identities in the input to the KDF.

Compared to other security protocol standards, ISO standards have been less analysed in the academic literature. A possible reason for this difference is that people who are not members of the working groups can only access the standards by purchasing the final versions. One possible way to promote the external analysis of ISO standards is to publish early drafts of proposed changes or new standards. Parties that are interested in applying the standards will still need to purchase the final versions to ensure they comply. However, interested parties can freely analyse the designs from the early drafts, which may help identify and prevent problems before the standards are deployed.

Acknowledgements. This work builds on, and extends abstract protocol models originally developed for an earlier analysis of ISO/IEC 11770 by Lara Schmid [25].

References

1. Barker, E., Johnson, D., Smid, M.: NIST SP 800-56: Recommendation for pair-wise key establishment schemes using discrete logarithm cryptography (revised) (2007)
2. Basin, D., Cremers, C.: Modeling and analyzing security in the presence of compromising adversaries. In: Gritzalis, D., Preneel, B., Theoharidou, M. (eds.) ESORICS 2010. LNCS, vol. 6345, pp. 340–356. Springer, Heidelberg (2010)
3. Basin, D., Cremers, C., Horvat, M.: Actor key compromise: Consequences and countermeasures. In: Proc. of the 27th IEEE Computer Security Foundations Symposium (CSF) (to appear, 2014)
4. Basin, D., Cremers, C., Meier, S.: Provably repairing the ISO/IEC 9798 standard for entity authentication. Journal of Computer Security 21(6), 817–846 (2013)
5. Basin, D., Cremers, C., Miyazaki, K., Radomirovic, S., Watanabe, D.: Improving the security of cryptographic protocol standards. IEEE Security & Privacy (2014)
6. Blake-Wilson, S., Johnson, D., Menezes, A.: Key agreement protocols and their security analysis. In: Darnell, M. (ed.) Cryptography and Coding 1997. LNCS, vol. 1355, pp. 30–45. Springer, Heidelberg (1997)
7. Blake-Wilson, S., Menezes, A.: Unknown Key-Share Attacks on the Station-to-Station (STS) Protocol (1999)
8. Boyd, C., Mathuria, A.: Protocols for Authentication and Key Establishment. Information Security and Cryptography. Springer (2003)
9. Chen, L., Mitchell, C.J.: Parsing ambiguities in authentication and key establishment protocols. Int. J. Electron. Secur. Digit. Forensics 3(1), 82–94 (2010)
10. Cheng, Z., Comley, R.: Attacks on an ISO/IEC 11770-2 key establishment protocol. I. J. Network Security 3(3), 290–295 (2006)
11. Cremers, C.J.F.: The Scyther Tool: Verification, falsification, and analysis of security protocols. In: Gupta, A., Malik, S. (eds.) CAV 2008. LNCS, vol. 5123, pp. 414–418. Springer, Heidelberg (2008), Available for download at http://www.cs.ox.ac.uk/people/cas.cremers/scyther/index.html
12. Cremers, C., Feltz, M.: Beyond eCK: Perfect forward secrecy under actor compromise and ephemeral-key reveal. Designs, Codes and Cryptography, 1–36 (2013)
13. Cremers, C., Mauw, S.: Operational Semantics and Verification of Security Protocols. Information Security and Cryptography. Springer (2012)
14. Dierks, T., Rescorla, E.: The Transport Layer Security (TLS) protocol version 1.2. IETF RFC 5246 (August 2008)
15. European Payments Council. Guidelines on algorithms usage and key management. Technical report, EPC342-08 Version 1.1 (2009)
16. Horng, G., Hsu, C.-K.: Weakness in the Helsinki protocol. Electronics Letters 34, 354–355(1) (1998)
17. International Organization for Standardization, Genève, Switzerland. ISO/IEC 9798-2:2008, Information technology – Security techniques – Entity Authentication – Part 2: Mechanisms using symmetric encipherment algorithms, 3rd edn. (2008)
18. International Organization for Standardization, Genève, Switzerland. ISO/IEC 11770-2:2008, Information technology – Security techniques – Key Management – Part 2: Mechanisms using Symmetric Techniques, 2009. Incorporating corrigendum (September 2009)
19. International Organization for Standardization, Genève, Switzerland. ISO/IEC 11770-3:2008, Information technology – Security techniques – Key Management – Part 3: Mechanisms using Asymmetric Techniques, Incorporating corrigendum (September 2009)

20. Krawczyk, H.: HMQV: A high-performance secure Diffie-Hellman protocol. Cryptology ePrint Archive, Report 2005/176 (2005), http://eprint.iacr.org/ (retrieved on June 1, 2014)
21. Law, L., Menezes, A., Qu, M., Solinas, J., Vanstone, S.: An efficient protocol for authenticated key agreement. Designs, Codes and Cryptography 28, 119–134 (2003)
22. Lowe, G.: A hierarchy of authentication specifications. In: Proc. 10th IEEE Computer Security Foundations Workshop (CSFW), pp. 31–44. IEEE (1997)
23. Mathuria, A., Sriram, G.: New attacks on ISO key establishment protocols. IACR Cryptology ePrint Archive, 2008:336 (2008)
24. Mitchell, C.J., Yeun, C.Y.: Fixing a problem in the Helsinki protocol. SIGOPS Oper. Syst. Rev. 32(4), 21–24 (1998)
25. Schmid, L.: Improving the ISO/IEC 11770 standard, Bachelor's thesis, ETH Zurich, Switzerland (2013)

A   Attack on Protocol 2-12

The attack on entity authentication claimed for protocol 2-12 is depicted in Figure 5. It depends on the fact that the entity running role A does not check the contents of the message encrypted for entities running roles B and P. In fact, normally such a check is impossible because all three roles are run by different entities. Seeing the payload of that particular message would be the only way for Pete to detect that something is wrong: he could see that the message contains I_{Alice} where I_{Pete} should be. Pete then gladly confirms the session key to Bob in role B, who falsely thinks that Alice just confirmed it.

B   AT5: 3-KT-6 Attack

There are three preconditions for the attack, which will not be met by most implementations. However, there is nothing in the standard that ensures that they are not met.

The first precondition is that the implementation must not implement any of the optional text fields except for $Text_1$. Second, nonces must be acceptable values for the $Text_1$ field. Third, entities must be able to perform both the A and the B role of the protocol, which occurs in many implementations.

If an implementation meets these conditions, the adversary can attack an instance of the A role by exploiting three instances of the B role. We give a graphical representation in Figure 7 in the appendix. The adversary redirects each sent message into the first receive of a new instance of the B role, and swapping the entity assumptions for the next B instance. This is possible since entities can perform multiple roles, and enabled by the fact that the nonces in messages sent by instances of the B role can be accepted into the $Text_1$ field. After three instances of the B role, the final message is then rerouted back to the final receive of the A role. Consequently, there is no instance of B that agrees with the A instance on *both* of the private keys. Thus, when the session key is computed from both these keys, key confirmation fails for the A instance.

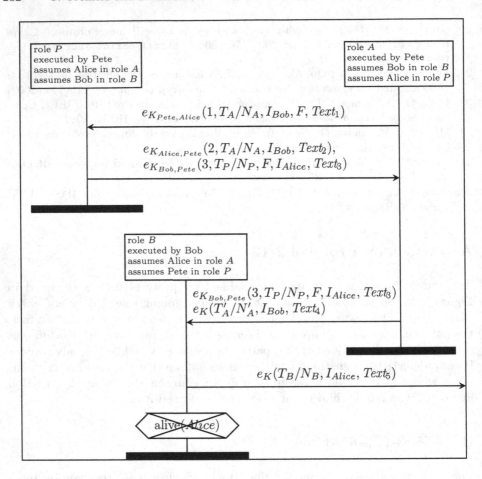

Fig. 5. Entity authentication attack on protocol 2-12 with optional fields

C Key Agreement Mechanism 8 (3-KA-8)

We next give an example from part 3 of the standard, referred to as Key Agreement Mechanism 8. This one-pass mechanism is derived from the one-pass variant of the MQV protocol [21]. It uses an elliptic curve agreed upon by entities A and B to establish a shared secret key. As shown in Figure 10, A generates an ephemeral secret and uses it to transmit to B a public value, modelled as a point on the elliptic curve. B then computes a fresh session key with his static private key and the public values of A (similarly, A uses her secrets and B's public values).

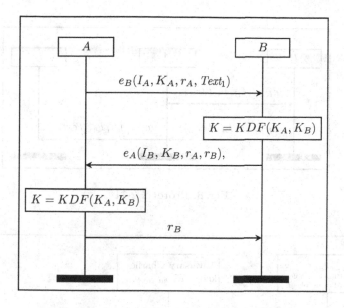

Fig. 6. Protocol 3-KT-6 combined key variant

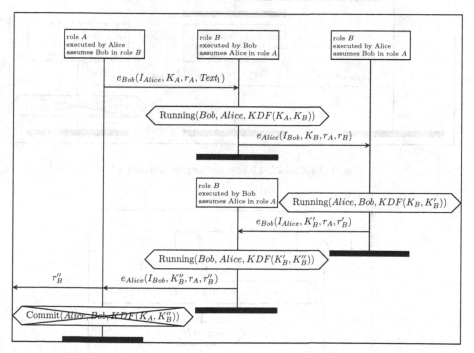

Fig. 7. 3-KT-6 combined key variant key confirmation attack

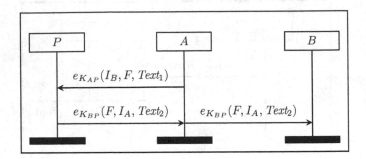

Fig. 8. Protocol 2-11

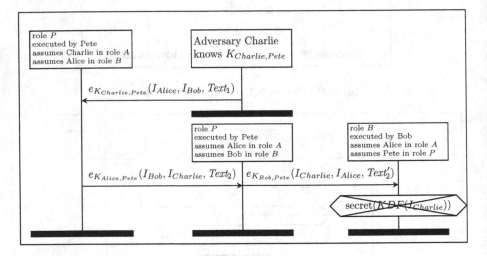

Fig. 9. Protocol 2-11 key authentication attack

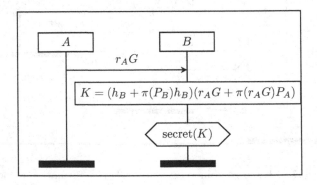

Fig. 10. Protocol 3-KA-8

Formally, let H be an elliptic curve over a finite field $\mathbb{F}_q$ and $G \in H$ of prime order n. The parameters H, q, G and n are known to both entities. It is assumed that each entity X has a private key $h_X \in \mathbb{Z}_n^*$, a public key $P_X = h_X G$, and an authenticated copy of the other entity's public key. For the function π defined for every point P on H by $\pi(P) = (P_x \bmod 2^{\lceil \frac{\rho}{2} \rceil}) + 2^{\lceil \frac{\rho}{2} \rceil}$, where P_x is the x-coordinate of P and $\rho = \lceil \log_2 n \rceil$, the protocol is executed as follows: A randomly generates r_A in $\mathbb{Z}_n^*$, computes $r_A G$ and sends it to B. B computes the shared key as $K = (h_B + \pi(P_B)h_B)(r_A G + \pi(r_A G)P_A)$.

Computationally Analyzing the ISO 9798–2.4 Authentication Protocol

Britta Hale and Colin Boyd

Norwegian University of Science and Technology (NTNU), Trondheim, Norway
{britta.hale,colin.boyd}@item.ntnu.no

Abstract. We provide a computational analysis of the ISO 9798–2.4 mutual authentication standard protocol in the model of Bellare and Rogaway. In contrast to typical analyses of standardized protocols, we include the optional data fields specified in the standard by applying the framework of Rogaway and Stegers. To our knowledge this is the first application of the Rogaway–Stegers technique in a standardized protocol. As well as a precise definition of the computational security properties achieved by the protocol, our analysis supplies concrete security requirements for the cryptographic primitive applied, which are absent from the protocol standard. We show that a message authentication code can be used to replace the encryption primitive if desired and that if authenticated encryption is applied it must be strongly unforgeable.

Keywords: ISO 9798, Bellare–Rogaway model, real-world protocol analysis.

1 Introduction

Because it is widely agreed that authentication protocols are difficult to design correctly, standardized authentication protocols are very useful for practitioners. Today, there are many such protocols available from a variety of different standards bodies; some of these, such as the well known TLS and SSH protocols, are widely deployed. Among its 9798 series of standards, the ISO have standardized a suite of authentication protocols. Like most standardized authentication protocols, the 9798 protocols are not defined in a fully formal way. Effectively, this can lead to a number of undesirable consequences, such as difficulty in establishing exactly what properties the protocols aim to achieve, doubts regarding whether the achieved aims are actually achieved, and uncertainty about how to correctly implement the protocols securely.

Recognizing the value of a formal analysis, Basin *et al.* [2] analyzed protocols in the ISO 9798–2 standard and found a number of potential weaknesses leading to a revision of the standard. However, such a symbolic analysis omits consideration of the specific cryptographic primitives used, instead assuming an idealized encryption function. Accordingly, implementors cannot be sure whether any chosen cryptographic primitive will satisfy the requirements for security. One of the

L. Chen and C. Mitchell (Eds.): SSR 2014, LNCS 8893, pp. 236–255, 2014.

motivations for our work is to provide computational proofs for one of the 9798–2 protocols which have so far been lacking.

In this paper we focus on the ISO 9798–2.4 protocol (9798–2, section 6.2.2 Mechanism 4 of the standard). This protocol is particularly interesting because it aims at an advanced level of authentication while at the same time has the potential to fit in the Bellare–Rogaway '93 model. First, it is a mutual authentication protocol, as opposed to a unilateral one. Second, unlike some other ISO 9798 protocols, it does not rely upon a time-stamps for freshness – instead using nonces. Finally, it does not require that confidence be placed in a third party (TTP), whereas several of the other ISO 9798 protocols do. Hence, the ISO 9798–2.4 standard is more suited than other 9798 protocols to modeling in this manner. ISO 9798–2.4 is presented in §2.

CHOICE OF CRYPTOGRAPHIC PRIMITIVES. ISO 9798–2.4 protocol makes use of an encipherment algorithm with a shared symmetric encipherment key. Per the standard, the encipherment algorithm used is required to be able to detect "forged or manipulated data" and authenticated encryption is recommended for its implementation. However, any formal definition or technical description of such properties is missing from the standard. We observe that in order to achieve entity authentication there is no requirement to use encryption at all. Therefore, so as to obtain security under maximal efficiency, we will show in our analysis that a message authentication code (MAC) algorithm can be safely implemented in place of the protocol's encipherment function. Using a MAC is arguably an improvement on the standardized protocol recommendation since it will generally result in a more efficient protocol than when applying authenticated encryption. Simultaneously, we recognize that, strictly speaking, such an improved protocol no longer conforms to the standard definition. Therefore we later show that authenticated encryption, in a formally defined sense, can also provide the required properties by a simple reduction in which we only use the authentication properties of the authenticated encryption algorithm.

OPTIONAL TEXT FIELDS. Like most of the protocols in the ISO 9798–2 standard, the ISO 9798–2.4 protocol includes optional text fields which can be chosen in any way desired by the protocol implementor. Potentially, this flexibility is a very useful feature since it allows users to include data which is authenticated by the protocol as an additional service to obtaining entity authentication. However, computational models for protocol analysis do not usually allow such flexibility in the protocols which they analyze. In fact, any change to the analyzed protocol can potentially introduce weaknesses and any security proof will become invalid. In 2009, Rogaway and Stegers [13] introduced the notion of a *partially specified protocol* in order to deal with exactly this problem. Concisely, their model allows the adversary to actively choose the extra data, but the adversary only wins if it changes the data while the parties still accept at the end of the protocol. We apply this technique to the ISO 9798–2.4 protocol to obtain a computational proof of security no matter how the free text is chosen. Rogaway and Stegers illustrated their technique with an academic protocol – as far as we are aware ours is the first example of application of the technique in a standardized protocol.

Contributions. We regard the following as the main contributions of this paper:

- a computational proof for the ISO 9798–2.4 protocol;
- the first application of the Rogaway–Stegers framework to a standardized protocol;
- concrete advice on appropriate primitives to ensure that the ISO 9798–2.4 protocol is provably secure.

Outline. The rest of this paper is structured as follows. In the next section we explain informally the ISO 9798–2.4 protocol, using the language employed in the standard. Section 3 describes the formal Bellare–Rogaway model used in this paper. Section 4 presents the analysis of ISO 9798–4.2 in the BR model on the assumption that the primitive used is a MAC rather than an authenticated encryption algorithm. Extending this analysis, Section 5 includes consideration of associated data into the security assessment by applying the framework of Rogaway and Stegers. In Section 6 we show that our security results will hold when using authenticated encryption, as informally stated in the standard, instead of a MAC.

2   ISO 9798–2.4

Notationally, let $Text_i$ be an optional text field, $\mathcal{E}_K$ an "encipherment function" between A and B [6, p. 4], d_K the corresponding decipherment function, I_B a unique identifier of the initiating party, and R_i a random nonce. In implementation situations where a reflection attack on the protocol is impossible, the distinguisher I_B is optional [6, p. 7]. Moreover, the symbol $||$ is employed to denote the concatenation of strings when order is specified (see [7] for further details on implementation). As presented in the standard, Figure 1 shows ISO 9798–2.4 with two-party three-pass authentication.

$$B \to A : R_B || Text_1$$

$$A \to B : Text_3 || \mathcal{E}_K(R_A || R_B || I_B || Text_2)$$

$$B \to A : Text_5 || \mathcal{E}_K(R_B || R_A || Text_4)$$

Fig. 1. ISO 9798–2 Protocol Mechanism 4 Three Pass Authentication

Per the ISO 9798 standard, $\mathcal{E}_K$ must have the property that "enables the recipient ... to detect forged or manipulated data" [6, p. 4]. Furthermore, it is recommended that authenticated encryption is used [6, p. 4].

Trade-offs between security and efficiency demand heavy consideration and it is desirable to find the least computationally costly implementation of $\mathcal{E}_K$

for which the protocol is secure. The chosen encipherment function will be a critical factor in the security proof presented in §4. While it is recommended that authenticated encryption (AE) be used for $\mathcal{E}_K$, this may in fact not provide optimal efficiency.

Predominantly, many popular implementations of authenticated encryption use a composition of a symmetric encryption scheme and a message authentication code (MAC) [3, p. 3]. Schemes which apply this composition method have, until recently, precluded the authentication of associated data, such as that which appears in the fields $Text_2$, $Text_4$ above, and even these non-composition AE schemes contain some MAC function in computation [11,8,12]. Accordingly, any AE scheme used on a message m will be no more efficient than a MAC on the same message. Consequently, we consider $\mathcal{E}_K$ concretized as a MAC function. Although this will not preserve confidentiality, just integrity, the scheme is designed for authentication only and not key exchange. Hence it turns out that a MAC is sufficient for security.

Of special note for consideration is the unique identifier I_B, which is addressed in the ISO standard by the following remark:

> When present, distinguishing identifier I_B is included...to prevent a so-called reflection attack. Such an attack is characterized by the fact that an intruder 'reflects' the challenge R_B to B pretending to be A. The inclusion of the distinguishing identifier I_B is made optional so that, in environments where such attacks cannot occur, it may be omitted. The distinguishing identifier I_B may also be omitted if unidirectional keys ...are used.

Analysis of the protocol in this paper will consider the protocol version which includes the unique identifier I_B. The alternative protocol with I_B omitted can still be proven secure in both the core and Rogaway–Stegers frameworks. Details of the required adjustments to the proofs can be found in Appendix B.

3   BR Model

In the seminal model introduced by Bellare and Rogaway [4] (henceforth referred to as the BR model), the security of a mutual authentication scheme is established on the session individuality of matching conversations. Basically, oracles for principals A and B should acquire matching conversations if and only if they both accept.

3.1   Adversary

In BR model, immense power is allowed to the adversary [4]. He is allowed to read, modify, replay, and delete messages – he is also allowed to provide his own messages to corresponding parties. Principals may engage in multiple sessions at once and the adversary may start up new sessions at his choosing.

Oracle calls allowed to the adversary are as follows:

- **Send.** Adversarial ability to request any instance $\Pi^s_{A,B}$ of a principal A to send a message to an instance of another principal B. In addition to learning the outgoing message, the adversary also learns whether or not it was accepted [4, p. 9].
- **Corrupt.** Adversarial ability to take over any principal A, obtain all of its private keys, and compute $\mathcal{E}_K$ under any symmetric key K that belongs to it.

While the **Send** query is used in the BR model, it should be noted that the query **Corrupt** is not. However, this research will employ a **Corrupt** query because it is reasonably within the realm of a real adversary's capabilities. Actually, it has become a generally accepted practice to allow this query since BR was published. Any instance will be considered *fresh* if neither its nor its partner's principal, if the partner exists, have been the subject of a **Corrupt** query by an adversary, and the instance has accepted.

Since ISO 9798–2.4 is designed for mutual authentication in a symmetric setting, there is no need for a (**Session**) **Key Reveal** query – the **Corrupt** query allows the adversary to access the symmetric key when such a query is desired.

3.2   Matching Conversations

Below is the definition of matching conversations, per the BR model. In short, matching conversations will be the requirement for the definition of a secure mutual authentication scheme in the model being considered for ISO 9798–2.4. Alternative definitions for determining uniqueness of a session have been applied in other research since the BR model was introduced, including using unspecified session identifiers [10]. Still later, more fully specified session identifiers that capture similar information to that of BR (e.g. [9]) have been utilized. Due to the simplicity of matching conversations and the natural session-identifier format that they epitomize, they will be employed in this work to capture partnering between sessions.

Definition 1. Matching Conversations [4]. *Fix a number of moves $R = 2\rho - 1$ and an R-move protocol Π. Run Π in the presence of an adversary E and consider two oracles, $\Pi^s_{i,j}$ and $\Pi^t_{j,i}$ that engage in conversations K and K', respectively. Let τ_l be time increments, α_l be messages sent by $\Pi^s_{i,j}$, and β_l be messages sent by $\Pi^t_{j,i}$.*

1. *Responder oracle has a conversation matching the conversation of an initiator oracle.*
   K' is in a matching conversation with K if there exists $\tau_0 < \tau_1 < \ldots < \tau_R$ and $\alpha_1, \beta_1, \ldots, \alpha_\rho, \beta_\rho$ such that K is prefixed by

$$< \tau_0, \lambda, \alpha_1 >, < \tau_2, \beta_1, \alpha_2 >, < \tau_4, \beta_2, \alpha_3 >, \ldots, < \tau_{2\rho-4}, \beta_{\rho-2}, \alpha_{\rho-1} >,$$
$$< \tau_{2\rho-2}, \beta_{\rho-1}, \alpha_\rho >$$

   and K' is prefixed by

$$< \tau_1, \alpha_1, \beta_1 >, < \tau_3, \alpha_2, \beta_2 >, < \tau_5, \alpha_3, \beta_3 >, \ldots, < \tau_{2\rho-3}, \alpha_{\rho-1}, \beta_{\rho-1} >.$$

2. *Initiator oracle has a conversation matching the conversation of a responder oracle.*

   K *is in a matching conversation with* K' *if there exists* $\tau_0 < \tau_1 < \ldots < \tau_R$ *and* $\alpha_1, \beta_1, \ldots, \alpha_\rho, \beta_\rho$ *such that* K' *is prefixed by*

$$< \tau_1, \alpha_1, \beta_1 >, < \tau_3, \alpha_2, \beta_2 >, < \tau_5, \alpha_3, \beta_3 >, \ldots, < \tau_{2\rho-3}, \alpha_{\rho-1}, \beta_{\rho-1} >,$$
$$< \tau_{2\rho-1}, \alpha_\rho, * >,$$

   and K *is prefixed by*

$$< \tau_0, \lambda, \alpha_1 >, < \tau_2, \beta_1, \alpha_2) >, < \tau_4, \beta_2, \alpha_3 >, \ldots, < \tau_{2\rho-4}, \beta_{\rho-2}, \alpha_{\rho-1} >,$$
$$< \tau_{2\rho-2}, \beta_{\rho-1}, \alpha_\rho >.$$

3.3   Secure Mutual Authentication

Concisely, the BR model prescribes that entities accept if and only if the session transcripts (conversations) match. This is presented below.

Definition 2. *Secure Mutual Authentication [4]. Let* $\mathrm{No - Matching}^E(k)$ *be the event that there exists an uncorrupted oracle* $\Pi_{i,j}^s$ *which accepted and there is no uncorrupted oracle* $\Pi_{j,i}^t$ *which engaged in matching conversation with* $\Pi_{i,j}^s$*. The protocol* Π *is a secure mutual authentication protocol if for any polynomial time adversary* E,

1. *Matching conversations* $\Rightarrow$ *acceptance.*
   If oracles $\Pi_{i,j}^s$ *and* $\Pi_{j,i}^t$ *have matching conversations, then both oracles accept.*
2. *Acceptance* $\Rightarrow$ *matching conversations.*
   The probability of $\mathrm{No - Matching}^E(k)$ *is negligible.*

In the BR model, the network is viewed as a 'benign adversary' whose actions are restricted to choosing an initiator oracle $\Pi_{i,j}^s$ and responder oracle $\Pi_{j,i}^t$, "faithfully conveying each flow from one oracle to the other" [4, p. 10], starting with the initiator. Such an adversary is deterministic. Thus the adversary has the power to determine i, j, s, and t, but must use these in any protocol execution with parameter k. While this is mostly of interest in a key-exchange setting, it is noted here to highlight strength of the model; i.e. if adversary behaves according to the protocol, with eavesdropping, it gains no additional advantage.

Pursuant to the definition of matching conversations is the uniqueness of matching partners. Bellare and Rogaway [4, p. 13] show that the probability of having multiple matching partners is negligible in this model .

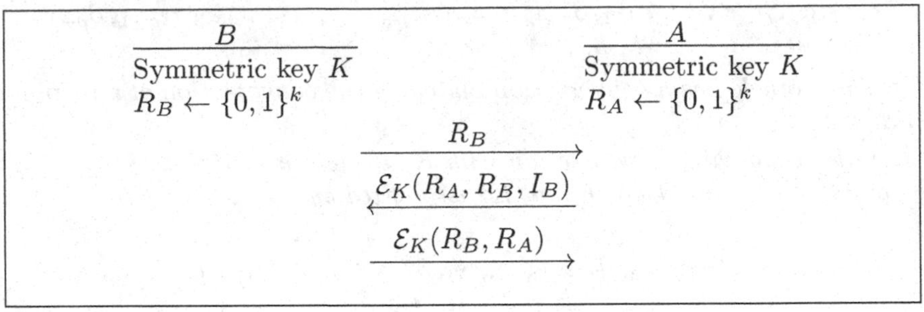

Fig. 2. ISO 9798–2.4 Protocol Core

4   Security of ISO 9798–2.4

In the following section, the security of Π will be considered in the case where $\mathcal{E}_K$ is implemented as a MAC function, with $\mathcal{E}_K(m) = (m, \text{MAC}(m))$, for a message m. Assessment will be performed in the BR model. Additionally, this proof focuses on the protocol core – the associated data in the ISO 9798–2.4 protocol, $Text_i$, will not be considered until §5. Figure 2 summarizes the core.

Theorem 1. *Let Π be the core of the ISO 9798–2.4 protocol implemented with a strongly unforgeable MAC algorithm[1] $\mathcal{E}_K(M) = \text{MAC}_K(M) = (M, T)$, as in Definition 4. Let E be a polynomial-time adversary against the mutual authentication scheme, running in time t and asking q queries. Then the advantage of E can be reduced to the advantage of an adversary against the MAC, running in time $t_F \approx t$ and asking $q_F = q$ queries:*

$$\mathbf{Adv}_{\Pi}^{MA}(E) \leq 2p^2 S \cdot \mathbf{Adv}_{\Pi}^{MAC}(F) + \frac{q^2}{2^{k+1}}.$$

where S is the number of sessions and p is the number of principals.

Proof (Proof with $\mathcal{E}_K$ implemented as a MAC).

Ideas from this proof follow from other proofs for entity authentication [4,5]. When Definition 2 is satisfied, the proof will be complete.

If the principals possess matching conversations, then they will both accept, by the protocol definition – hence satisfying the first condition of the definition of a secure mutual authentication protocol is trivial. Correspondingly, the remainder of this proof will target the second case; that acceptance implies matching conversations.

Adversarial advantage, $\mathbf{Adv}_{\Pi}^{MA}(E)$, will be defined as the probability that it can succeed in persuading an oracle to accept without a matching conversation.

---

[1] Strong-unforgeability is required since an adversary's ability to produce a different, yet valid, MAC on a protocol message flow would trivially result in principals accepting without matching conversations. Essentially, the tag would still verify correctly even though each instance held a different conversation transcript.

Let NC represent the event that two different instances accept with the same nonce pair. Then we can derive the following:

$$\mathbf{Adv}^{MA}(E) \leq \Pr[\neg \text{ Match.Conv }]$$
$$\leq \Pr[\text{NC}] + \Pr[\neg \text{ Match.Conv } | \neg \text{ NC}]$$

Let E be an adversary that attempts to inveigle acceptance from an oracle without that oracle being in matching conversation with a partner oracle. Let q be a polynomial bound on the number of oracle calls allowed to E.

Nonce Collision

As q calls are allowed to E and nonces are selected independently, the birthday bound yields

$$\Pr[\text{NC}] \leq q^2/2^{k+1}. \tag{1}$$

Matching Conversations without Nonce Collision

E will succeed with probability equal to the sum of the probability of success against at least one initiator oracle (i.e. gets an initiator oracle to accept without any other oracle in matching conversation with it) and the probability of success against at least one responder oracle (but no initiator oracle). Thus the proof follows two cases.

DESCRIPTION OF F: In the protocol game, E will get an oracle $\Pi_{i,j}^s$ for a principal i to accept with non-negligible probability, without the existence of an oracle $\Pi_{j,i}^t$ in matching conversation with $\Pi_{i,j}^s$. Using this fact, F's goal is to compute a valid MAC for a message m where m has not been queried from the MAC oracle.

F starts the game and initiates E on input 1^k. F selects a pair i, j at random from the set of all principals $\{1, \ldots, p\}$ as well as a session $s \in_R \{1, \ldots, S\}$ – thus F is selecting $\Pi_{i,j}^s$ as its guess for the initiator oracle against which E will succeed. F has a MAC oracle, per definition 4, that runs on a key K chosen randomly from $\{0,1\}^k$ which it will use to calculate the tag for messages between i and j. For all principals in the set $\{1, \ldots, p\} \setminus \{i, j\}$, F also selects keys k_l for each pair of principals; these keys will be used to calculate MAC_{k_l} and MAC.ver_{k_l} on message flows between all principals other than i and j.

F answers all of E's Send and Corrupt queries, according to the protocol. However, should E ask a Corrupt query on the principals corresponding to either of the instances i or j, F will give up. If E asks a Send query, for $\Pi_{i,j}^l$ or $\Pi_{j,i}^m$ for any l or m, F will compute the response with its MAC generation oracle and, if necessary, F checks incoming MACs using its MAC verification oracle

Against Initiator: Suppose that E succeeds against at least one initiator oracle with non-negligible probability of success. From E an adversary F will be constructed against the MAC.

Note: if E never calls on i to initiate a protocol run, F gives up.

Now suppose that E does call on i to initiate a protocol run. Then at some time τ_0, $\Pi_{i,j}^s$ will send out a flow R_i. For some time $\tau_2 > \tau_0$, $\Pi_{i,j}^s$ must receive a flow of the form $\mathcal{E}_K(R_j, R_i, I_i)$, else F gives up. If F has already received $\text{MAC}_K(R_j, R_i, I_i)$ from its oracle, then it gives up; else it returns $\text{MAC}_K(R_j, R_i, I_i) = ((R_j, R_i, I_i),\ Tag_1)$ as its guess for $\mathcal{E}_K(R_j, R_i, I_i)$.

GAME OF F: If we assume that E succeeds on the instance guessed by F, then the oracle $\Pi_{i,j}^s$ accepts. Given also that there are no collisions, F cannot have previously obtained the flow $\mathcal{E}_K(R_j, R_i, I_i)$. Therefore F outputs a valid forgery for $\mathcal{E}_K(R_j, R_i,\ I_i)$ (i.e. a valid forgery for the MAC per Definition 4).

Ergo (assuming that E always succeeds against at least one initiator oracle),

$$\mathbf{Adv}_\Pi^{\text{MAC}}(F) \geq \frac{1}{p^2 S} \cdot \Pr[\neg\ \text{Match.Conv} \mid \neg\ \text{NC}] \tag{2}$$

Against Responder: Suppose that E succeeds against at least one responder oracle but no initiator oracle with non-negligible probability of success. Similarly to the initiator case above, an adversary F of the MAC will be constructed.

Note: if E never calls on j as a responder to a protocol run, or if E succeeds against some initiator oracle, F gives up.

Now suppose that E does call on j as a responder oracle. Then at some time τ_1, $\Pi_{j,i}^t$ must receive a flow R_i and respond with a flow $\mathcal{E}_K(R_j, R_i, I_i)$. At time $\tau_3 > \tau_1$, $\Pi_{j,i}^t$ must receive a flow $\mathcal{E}_K(R_i, R_j)$, else F gives up. If F has already calculated $\text{MAC}_K(R_i, R_j)$, then it gives up; else it computes $\text{MAC}_K(R_i, R_j) = ((R_i, R_j),\ Tag_2)$ and returns this as its guess for $\mathcal{E}_K(R_i, R_j)$.

GAME OF F: As above, if we assume that the probability that E succeeds on the instance guessed by F then the oracle $\Pi_{j,i}^t$ accepts. Given also that there are no collisions, F cannot have previously obtained the flow $\mathcal{E}_K(R_i, R_j)$. Therefore F outputs a valid forgery for $\mathcal{E}_K(R_i, R_j)$ (i.e. a valid forgery for the MAC).

Therefore (under the assumption that E always succeeds against at least one responder oracle but no initiator oracle),

$$\mathbf{Adv}_\Pi^{\text{MAC}}(F) \geq \frac{1}{p^2 S} \cdot \Pr[\neg\ \text{Match.Conv} \mid \neg\ \text{NC}]. \tag{3}$$

Combining equations (2) and (3), and taking into account the two mutually exclusive cases, we have:

$$\mathbf{Adv}_\Pi^{\text{MAC}}(F) \geq \frac{1}{2}\left(\frac{1}{p^2 S} \cdot \Pr[\neg\ \text{Match.Conv} \mid \neg\ \text{NC}]\right)$$

or

$$2p^2 S \cdot \mathbf{Adv}_\Pi^{\text{MAC}}(F) \geq \Pr[\neg\ \text{Match.Conv} \mid \neg\ \text{NC}]. \tag{4}$$

Negligible Probability of Success

By equations (1) and (4), the probability that E secures its goal of oracle acceptance, while maintaining the absence of another oracle in matching conversation, is negligible. Particularly,

$$\mathbf{Adv}^{MA}(E) \leq \Pr[\neg \text{ Match.Conv }]$$
$$\leq \Pr[\text{NC}] + \Pr[\neg \text{ Match.Conv } |\neg \text{ NC}]$$
$$\leq q^2/2^{k+1} + 2p^2 S \cdot \mathbf{Adv}_{\Pi}^{MAC}(F).$$

Moreover, if E runs in time t and asks q queries, then F runs in time $t_F \approx t$ and asks $q_F = q$ queries. Thus, the protocol is secure with $\mathcal{E}_K$ implemented as a MAC function, where $\mathcal{E}_K(M) = \text{MAC}(M) = (M, T)$, for a message M.

5   Analysis with Associated Data – RS Model

While the analysis above demonstrates security of the ISO 9798–2.4 protocol core, it omits an important aspect of the original protocol: optional text fields. As with most protocols, these fields allow additional data to be sent – sometimes authenticated – during the mutual authentication process. However, the addition of this data would nullify the security statement in §4 since the inclusion of additional fields was not considered.

Rogaway and Stegers [13] introduced a model that addresses this issue by splitting the protocol into two parts: the partially specified protocol core (PSP) and the protocol details (PD). In essence, the protocol details selects the optional text fields to be added. Yet, since there is no restriction on the data that is sent in these fields, it is necessary to maintain the perspective that data choice could weaken the protocol. Fundamentally, this weakness is modeled by allowing the adversary itself to choose the optional text fields; thus, not only does the adversary call the security game but the game also calls the adversary.

Data fields in the model fall into two categories: associated data (AD) which are authenticated by the protocol and, by protocol security, should be guaranteed to be mutually held by all parties, and ancillary but unauthenticated data. While the RS model addresses both categories of data, the former is of salient concern. Even though the unauthenticated data fields are relevant to the security of the protocol, as they may influence the selection of the AD fields, they are also subject to being changed by an adversary en route. Consequently, no authenticity claims can be made on the non-authenticated fields.

Succinctly, the ISO 9798–2.4 protocol has text fields $Text_l$ for $l \in \{1, \ldots, 5\}$. Data fields $Text_1$, $Text_3$, and $Text_5$ are not authenticated and can therefore be modified en route later in the protocol. Hence they cannot be classified as AD. Likewise, since $Text_4$ is sent in the last message by the initiator, there is no guarantee that it will be received by the responder and is consequently also not AD, although it is authenticated. Ultimately, this leaves field $Text_2$ as the only AD. Applying the Rogaway–Stegers (RS) framework, it is our goal to demonstrate that ISO 9798–2.4 is still secure even when the AD selection is under adversarial control.

Rogaway and Stegers combine their AD framework with a particular mutual authentication model, using session IDs, and apply it to a variant of the

Needham-Schroeder-Lowe protocol [13, p. 7]. For application of the Rogaway–Stegers AD framework, we use the BR mutual authentication model with matching conversations, as in §4. To avoid trivial breaks of the matching conversations by an adversary, conversation transcripts will not include unauthenticated text fields – i.e. $Text_1$, $Text_3$, and $Text_5$

Notably, the RS model is a public-key mutual authentication model, whereas ISO 9798–2.4 is a symmetric-key protocol. As a result of these details, slight model adaptations are required. Nonetheless, security in RS framework for associated data can be summarized as shown below.

Definition 3. RS Framework for Associated Data with BR Mutual Authentication Model [4,13]. *Let* $\mathrm{No-Matching}^E(k)$ *be the event that there exists an uncorrupted oracle* $\Pi_{i,j}^s$ *which accepted and there is no uncorrupted oracle* $\Pi_{j,i}^t$ *which engaged in matching conversation with* $\Pi_{i,j}^s$. *The protocol* Π *is a secure mutual authentication protocol if for any polynomial time adversary* $\mathcal{A}$,

1. *Matching conversations* $\Rightarrow$ *acceptance.*
   If oracles $\Pi_{i,j}^s$ *and* $\Pi_{j,i}^t$ *have matching conversations, then both oracles accept.*
2. *Acceptance* $\Rightarrow$ *matching conversations.*
   The probability of $\mathrm{No-Matching}^E(k)$ *is negligible.*
3. *Matching Conversations* $\Rightarrow$ *Matching AD.*
   If oracles $\Pi_{i,j}^s$ *and* $\Pi_{j,i}^t$ *have matching conversations, then the associated data in the protocol is guaranteed to be mutually held.*

Theorem 2. *Let* Π *be the ISO 9798–2.4 protocol implemented with a strongly unforgeable MAC algorithm* $\mathcal{E}_K(M) = (m, \mathrm{MAC}_K(M))$, *including the optional text fields* $Text_l$ *and the associated data* $Text_2$. *Then advantage of an polynomial-time adversary against the mutual authentication scheme can be reduced to the adversarial advantage against the MAC:*

$$\mathbf{Adv}^{MA}(\mathcal{A}) \le 2p^2 S \cdot \mathbf{Adv}_\Pi^{\mathrm{MAC}}(F) + q^2/2^{k+1}.$$

Proof. Succinctly, this proof will build on that of §4 and the following previously used notation will continue, with the addendum of matching AD. For conciseness, text fields $Text_l$ will be denoted T_l.

– NC: Two different instances accept with the same nonce pair.
– Match.Conv: Two different instances are in matching conversation.
– Match.AD: The AD held by both parties at the end of the protocol matches.
– q: number of calls allowed to the adversary.
– S: number of sessions.
– p: number of principals.
– 1^k: security parameter.

Simply, the first requirement for mutual authentication in definition 3 follows from the protocol description. It remains to be shown that acceptance still implies matching conversations even with the optional text fields included and that this in turn guarantees the associated data is mutually held by both parties at termination. Adversarial advantage against the mutual authentication scheme thus complies with the following inequalities. The final reduction will serve as a triad proof infrastructure.

$$
\begin{aligned}
\mathbf{Adv}^{MA}(\mathcal{A}) &\leq \Pr[(\neg \text{ Match.Conv}) \vee (\neg \text{ Match.AD} \mid \text{Match.Conv})] \quad (5)\\
&\leq \Pr[\neg \text{ Match.Conv}] + \Pr[\neg \text{ Match.AD} \mid \text{Match.Conv}]\\
&\leq \Pr[\text{NC}] + \Pr[\neg \text{ Match.Conv} \mid \neg \text{ NC}]\\
&\quad + \Pr[\neg \text{ Match.AD} \mid \text{Match.Conv} \wedge \neg \text{ NC}]
\end{aligned}
$$

Nonce Collision

As q calls are allowed to $\mathcal{A}$ and nonces are selected independently, the birthday bound yields

$$
\Pr[\text{NC}] \leq q^2/2^{k+1}. \quad (6)
$$

Acceptance Implies Matching Conversations

Immediately, this proof is in parallel to that in §4. Furthermore, nonce collision has already been accounted for. Correlatively, the following are addenda to the proof and reduction statement presented in §4:

Case 1: (continued from §4.) Let F be an adversary against the MAC, having a MAC oracle that runs on a key K chosen randomly from $\{0,1\}^k$. Suppose that the probability that $\mathcal{A}$ succeeds in having an initiator oracle accept without being in matching-conversation is non-negligible.

When the PSP requires a choice of text fields, it calls on the PD, answered by $\mathcal{A}$, to select T_l. If the responder exists, at time τ_1, the PSP calls the PD which responds with its selection for all text fields in the second message flow while also setting $AD = T_2$ for the responder. Regardless of the responder's view, when the initiator receives the flow $T_3 \| \mathcal{E}_K(R_j, R_i, I_i, T_2)$ at time τ_2, the PD sets AD $= T_2$ for the initiator.

Even though $\mathcal{A}$ has power over the PD and is therefore able to choose the AD, it is deterministic in its selection. Essentially, $\mathcal{A}$ is not allowed to simply change T_2 at a later date; once chosen, $\mathcal{A}$ may not attempt to insure that principals i and j hold different AD values by simply reselecting T_2 via the PD when setting the AD from the initiator's view. Thus, any attempt by $\mathcal{A}$ to ensure that conversations do not match by changing T_2, and therefore the AD, must be made by exchanging the flow with a previous one or by a valid forgery.

Consequently, as in §4, F has previously calculated the flow $\mathcal{E}_K(R_j, R_i, I_i, T_2)$ or F outputs a valid forgery for it (i.e. a valid forgery for the MAC). Since there are no nonce collisions, F has not previously calculated the flow. Therefore it must output a valid forgery for the MAC.

Case 2: (continued from §4.) Let F be an adversary against the MAC, having a MAC oracle that runs on a key K chosen randomly from $\{0,1\}^k$. Suppose that the probability that $\mathcal{A}$ succeeds against a responder oracle but no initiator oracles is non-negligible.

When the PSP requires a choice of text fields, it calls on the PD, answered by $\mathcal{A}$, to select T_l. At time τ_1, the PD sets $AD = T_2$ for the responder. When it receives the flow $T_5 \| \mathcal{E}_K(R_j, R_i, I_i, T_2)$, the PD again sets $AD = T_2$ from the initiator's view at time τ_2.

As in Case 1, the AD is chosen deterministically in the PD call and may not simply be changed later by the PD to ensure non-matching conversations. Consequently, in order to get the responder to accept at time τ_3, F has either previously calculated the flow $\mathcal{E}_K(R_i, R_j, T_4)$ or F outputs a valid forgery for it (i.e. a valid forgery for the MAC). Since there are no nonce collisions, F has not previously calculated the flow. Therefore it must output a valid forgery for the MAC.

Combining Case 1 and 2, the reduction is summarized as

$$\mathbf{Adv}_{\Pi}^{\text{MAC}}(F) \geq \tfrac{1}{2} \left(\tfrac{1}{p^2 S} \cdot \Pr[\neg \text{ Match.Conv} \mid \neg \text{ NC}] \right).$$

Hence,

$$\Pr[\neg \text{ Match.Conv} \mid \neg \text{ NC}] \leq 2p^2 S \cdot \mathbf{Adv}_{\Pi}^{\text{MAC}}(F). \tag{7}$$

Associated Data Agreement

Compactly, it can be assumed that an instance and its partner are in matching conversations and that there are no nonce collisions. It remains to show that the same AD, T_2, is equally held by both sides.

Since i and j are in matching conversation, at some time τ_1 the responder sent the message $T_{3_j} \| \mathcal{E}_K(R_j, R_i, I_i, T_{2_j})$ and at some time τ_2 the initiator received a message

$$T_{3_i} \| \mathcal{E}_K(R_j, R_i, I_i, T_{2_i}),$$

where T_{3_i} and T_{3_j} may or may not be equal. Moreover, this was authenticated under the symmetric key K, $T_{2_i} = T_{2_j}$. Therefore $AD_i = AD_j$.

Ergo,

$$\Pr[\neg \text{ Match.AD} \mid \neg NC \wedge \text{ Match.Conv}] = 0. \tag{8}$$

Combining the reductions from equations 6–8 with equation 5 yields the full reduction of security for ISO 9798–2.4 with inclusion of associated data.

Remark: As previously observed, there is no guarantee that the final message flow is received, which limits T_4 from inclusion in the AD. However, assuming matching conversations, a responder that receives a flow $T_5 \| \mathcal{E}_K(R_i, R_j, T_4)$ can be assured that T_4 has been authenticated by the sender. Namely, this follows from the assumptions above – if $T_{5_j} \| \mathcal{E}_K(R_i, R_j, T_4)$ is received by an instance

$\Pi_{j,i}^t$ under a symmetric key, then the flow $T_{5_i}||\mathcal{E}_K(R_i, R_j, T_4)$ was sent by a partner instance $\Pi_{i,j}^s$ of principal i and $\mathcal{E}_K(R_i, R_j, T_4)$ must also form part of the conversation transcript of $\Pi_{i,j}^s$.

6   Using Authenticated Encryption

As previously stated, the ISO 9798–2 standard currently does not specify the primitive to be used as the encipherment function $\mathcal{E}_K$. Likewise, while the standard concurs that its integrity requirements "can be achieved in many ways" [6, p. 4], authenticated encryption per the ISO/IEC 19772 standard is recommended. Consequently, it is desirable to check that a protocol implemented with an AE primitive will have security traceable to that of a protocol under a MAC primitive, proven in §4 and §5.

For the lemmas and theorem below, *strongly-unforgeable authenticated encryption* (SUF-AE) is defined as in Definition 5. The reader should note that this definition may be more commonly referred to as INT-CTXT$_{\mathcal{AE}}$ [3] – however, the term SUF-AE is used here for the sake of clarity with its relationship to SUF-CMA.

Lemma 1. *Suppose that $(\mathcal{K}, \mathcal{E}, \mathcal{D})$ is a strongly-unforgeable authenticated encryption algorithm according to Definition 5. Then the MAC algorithm with $\text{MAC}_K(M) = (M, \mathcal{E}(K, M))$ is SUF-CMA secure, according to Definition 4.*

Proof. Suppose that E is an adversary that succeeds against the MAC with advantage $\mathbf{Adv}_{\text{MAC}}^{\text{SUF-CMA}}(E)$, which is non-negligible; from E an adversary F will be constructed against the authenticated encryption algorithm.

Let MAC be a strongly-unforgeable message authentication code. F starts the game, chooses $K \xleftarrow{\$} \mathcal{K}$ and initiates adversary E on 1^n.

Provide F with an oracle for $\text{MAC}_K(\cdot) = (\cdot, \mathcal{E}_K(\cdot))$ which it will use to answer E's MAC queries. Since E forges the MAC under key K, it can output a valid pair (M, C) such that $\mathcal{D}_K(C) = M$ and E did not ask a query $\text{MAC}_K(M)$ such that $C = \mathcal{E}_K(M)$.

Now, if F wishes to forge an authenticated encryption on message $M_l \in$ Message, $l \in \{0, \dots, w\}$, it will call E on M_l. Respectively, E will output the message-tag pair (M_l, C_l) where $\mathcal{D}_K(C_l) = M_l$ and C_l was not previously the answer to an oracle call $\text{MAC}_K(M_l)$. Since E succeeds with non-negligible probability $\mathbf{Adv}_{\text{MAC}}^{\text{SUF-CMA}}(E)$, F will also succeed in forging the authenticated encryption with $\mathcal{E}_K(M_l) = C_l$, where C_l has not previously been produced as a ciphertext on M_l, with non-negligible probability. Thus,

$$\mathbf{Adv}_{\text{MAC}}^{\text{SUF-CMA}}(E) \leq \mathbf{Adv}_{(\mathcal{K}, \mathcal{E}, \mathcal{D})}^{\text{SUF-AE}}(F).$$

Lemma 2. *Let Π be the 9798–2.4 protocol implemented with a strongly unforgeable authenticated encryption algorithm $(\mathcal{K}, \mathcal{E}, \mathcal{D})$. Let Π' be the 9798–2.4 protocol implemented with the MAC as in Lemma 1. An efficient adversary against Π can be efficiently converted into an adversary against Π'.*

Proof. Let $\mathcal{A}$ be an efficient adversary against the mutual authentication protocol Π operating with advantage $\mathbf{Adv}_{\Pi}^{\text{MA-AE}}(\mathcal{A})$; from $\mathcal{A}$, an adversary $\mathcal{B}$ will be constructed against Π'. Let the advantage of $\mathcal{B}$ be denoted $\mathbf{Adv}_{\Pi'}^{\text{MA-MAC}}(\mathcal{B})$.

Starting the protocol game, $\mathcal{B}$ chooses $K \xleftarrow{\$} \mathcal{K}$ and initiates adversary $\mathcal{A}$ on 1^k. Let n be a polynomial bound on the number of queries allowed to $\mathcal{A}$.

$\mathcal{B}$ will answer $\mathcal{A}$'s first Send query in the open. Thereafter, all Send queries will be answered by submitting them to $\text{MAC}_K(M)$, $\mathcal{B}$'s oracle for the MAC, and only passing on the tag part $\mathcal{E}(K, M)$ of the answer pair $(M, \mathcal{E}(K, M)) \leftarrow \text{MAC}_K(M)$ to $\mathcal{A}$.

When $\mathcal{B}$ wishes to succeed against Π' it must convince an instance to accept without matching conversations. To do this, $\mathcal{B}$ will pass all messages to $\mathcal{A}$, which will answer each Send query, outputting encryptions C_i on message queries $i = 2, 3$ (second and third protocol flows). $\mathcal{B}$ will then relay the pair (M_i, C_i) to its respective instances in the second and third protocol flows as required.

Since $\mathcal{A}$ can succeed against Π and therefore get one of its instances to accept without matching conversations with non-negligible probability, relaying the message-tag pair will also ensure that one of the instances in Π' will accept without matching conversations, so long as $\mathcal{A}$ has not made a forgery on the AE. Should $\mathcal{A}$ forge the AE, then the message added by $\mathcal{B}$ to create the message-tag pair will not match the decryption of the ciphertext output by $\mathcal{A}$. However, with n queries allowed to $\mathcal{A}$ and a probability of forgery $\mathbf{Adv}_{(\mathcal{K},\mathcal{E},\mathcal{D})}^{\text{SUF-AE}}(F)$, this only occurs with negligible probability. Thus,

$$\mathbf{Adv}_{\Pi}^{\text{MA-AE}}(\mathcal{A}) \leq \mathbf{Adv}_{\Pi}^{\text{MA-AE}}(\mathcal{A} \text{ succeeds} \mid \mathcal{A} \text{ does not forge}) + \Pr(\mathcal{A} \text{ forges})$$
$$\leq \mathbf{Adv}_{\Pi'}^{\text{MA-MAC}}(\mathcal{B}) + n \cdot \mathbf{Adv}_{(\mathcal{K},\mathcal{E},\mathcal{D})}^{\text{SUF-AE}}(F)$$

Theorem 3. *Let Π be the 9798-2.4 protocol implemented with a strongly unforgeable authenticated encryption algorithm $(\mathcal{K}, \mathcal{E}, \mathcal{D})$. Let Π' be the 9798-2.4 protocol implemented with the MAC as in Lemma 1. In the Rogaway–Stegers framework with associated data considered, an efficient adversary against Π can be efficiently converted into an adversary against Π'.*

Proof. Applying the reductions from Lemma 6, Lemma 2, and §5 Game 5, it follows that

$$\mathbf{Adv}_{\Pi}^{\text{MA-AE}}(\mathcal{A}) \leq \mathbf{Adv}_{\Pi'}^{\text{MA-MAC}}(\mathcal{B}) + n \cdot \mathbf{Adv}_{(\mathcal{K},\mathcal{E},\mathcal{D})}^{\text{SUF-AE}}(F)$$
$$\leq 2p^2 S \cdot \mathbf{Adv}_{\text{MAC}}^{\text{SUF-CMA}}(E) + q^2/2^{k+1} + n \cdot \mathbf{Adv}_{(\mathcal{K},\mathcal{E},\mathcal{D})}^{\text{SUF-AE}}(F)$$
$$\leq 2p^2 S \cdot \mathbf{Adv}_{(\mathcal{K},\mathcal{E},\mathcal{D})}^{\text{SUF-AE}}(F) + q^2/2^{k+1} + n \cdot \mathbf{Adv}_{(\mathcal{K},\mathcal{E},\mathcal{D})}^{\text{SUF-AE}}(F)$$
$$= (2p^2 S + n) \cdot \mathbf{Adv}_{(\mathcal{K},\mathcal{E},\mathcal{D})}^{\text{SUF-AE}}(\mathcal{A}) + q^2/2^{k+1}$$

7   Conclusion

Ultimately, these results underscore the security of the ISO 9798-2.4, a real-world mutual authentication standard. Basing security on matching conversations, the protocol core was first analyzed in the Bellare–Rogaway model. Being more efficient than an authenticated encryption scheme, a MAC function was used in the security assessment of the protocol and shown to be sufficient. While this no longer yields privacy, it attests to the security of the mutual authentication scheme in the most fundamental cases – when merely integrity and authenticity are required.

Integrated into the proof of security for the protocol core is a polynomial-time reduction to the security of the MAC. Furthermore, we have shown that while a strongly-unforgeable MAC is sufficient for security, the current recommendation of ISO 9798-2 of authenticated encryption will also result in a secure protocol, albeit a less efficient one. Strong unforgeability is required for both the MAC and the AE since an adversary's ability to produce a different, yet valid, encipherment for a message flow would trivially result in principals accepting without matching conversations.

Subsequently, the full protocol, inclusive of associated data, was analyzed in the RS model. With additional data fields included and adversarial selection of the data for those fields allowed, the protocol was again demonstrated to be secure under the MAC implementation.

Ad interim, a parallel symbolic analysis of the protocol for juxtaposition was also performed using Scyther, albeit omitted from this paper. In comparison to the symbolic analysis by Basin *et al.* [2] which applied symmetric encryption and checked for aliveness and weak-agreement, we reconnoitered security while implementing a MAC for the encipherment function as well as demanding non-injective agreement and non-injective synchronization; thereby twinning restrictions with those used in the computational analysis. Results from the separate symbolic analysis affirmed those in this research.

Collectively, these results demonstrate a clarification and efficiency improvement to the ISO standard's current requirements, while also validating the protocol's security in the computational model. Even though authenticated encryption is recommended for implementation of the standard, it is confirmed in this research that security is achievable for the mutual authentication scheme using only a MAC. If authenticated encryption is applied, our security analysis demands that it be strong unforgeable.

While this research addresses a specific ISO 9798 protocol, the standard covers several variants for use in authentication. Of these, some highlight particular aspects that would need to be taken under consideration, should similar analyses be performed. Notably, the BR model would require adjustment in the case of one-pass authentication, as this is not addressed in the traditional model. Likewise, some ISO 9798 protocols utilize timestamps instead of nonces for freshness and care would be required in modeling these if an analysis is to be performed in the manner of our work.

References

1. Abdalla, M., Bellare, M., Rogaway, P.: The Oracle Diffie-Hellman Assumptions and an Analysis of DHIES. In: Naccache, D. (ed.) CT-RSA 2001. LNCS, vol. 2020, pp. 143–158. Springer, Heidelberg (2001)

2. Basin, D., Cremers, C.J.F., Meier, S.: Provably repairing the ISO/IEC 9798 standard for entity authentication. Journal of Computer Security 21(6), 817–846 (2013)
3. Bellare, M., Namprempre, C.: Authenticated encryption: Relations among notions and analysis of the generic composition paradigm. In: Okamoto, T. (ed.) ASIACRYPT 2000. LNCS, vol. 1976, pp. 531–545. Springer, Heidelberg (2000)
4. Bellare, M., Rogaway, P.: Entity Authentication and Key Distribution. In: Stinson, D.R. (ed.) CRYPTO 1993. LNCS, vol. 773, pp. 232–249. Springer, Heidelberg (1994)
5. Blake-Wilson, S., Johnson, D., Menezes, A.: Key agreement protocols and their security analysis. In: Darnell, M.J. (ed.) Cryptography and Coding 1997. LNCS, vol. 1355, pp. 30–45. Springer, Heidelberg (1997)
6. ISO. Information technology – security techniques – entity authentication – part 2: Mechanisms using symmetric encipherment algorithms. ISO ISO/IEC 9798-2:2008, International Organization for Standardization, Geneva, Switzerland (2008)
7. ISO. Information technology – security techniques – entity authentication – part 2: Mechanisms using symmetric encipherment algorithms. ISO ISO/IEC 9798-2:2008/Cor 1:2010, International Organization for Standardization, Geneva, Switzerland, Technical Corrigendum 1 (2010)
8. Jutla, C.S.: Encryption Modes with Almost Free Message Integrity. In: Pfitzmann, B. (ed.) EUROCRYPT 2001. LNCS, vol. 2045, pp. 529–544. Springer, Heidelberg (2001)
9. LaMacchia, B.A., Lauter, K., Mityagin, A.: Stronger Security of Authenticated Key Exchange. In: Susilo, W., Liu, J.K., Mu, Y. (eds.) ProvSec 2007. LNCS, vol. 4784, pp. 1–16. Springer, Heidelberg (2007)
10. Canetti, R., Krawczyk, H.: Analysis of key-Exchange protocols and Their Use for Building Secure Channels. In: Pfitzmann, B. (ed.) EUROCRYPT 2001. LNCS, vol. 2045, pp. 453–474. Springer, Heidelberg (2001)
11. Rogaway, P.: Authenticated-Encryption with Associated-Data. In: Ninth ACM Conference on Computer and Communications Security (CCS-9). ACM Press (2002)
12. Rogaway, P., Bellare, M., Black, J.: OCB: A Block-Cipher Mode of Operation for Efficient Authenticated Encryption. In: Eighth ACM Conference on Computer and Communications Security (CCS-8), pp. 365–403. ACM Press (2003)
13. Rogaway, P., Stegers, T.: Authentication without Elision: Partially Specified Protocols, Associated Data, and Cryptographic Models Described by Code. In: Proceedings of the 2009 22nd IEEE Computer Security Foundations Symposium, pp. 26–39. IEEE Computer Society (2009)

A   Definitions for Security of MAC and Authenticated Encryption

Definition 4. *([5,1], modified.) A strongly-unforgeable message authentication code is a probabilistic polynomial-time algorithm* $\mathrm{MAC}_{(.)}(\cdot)$. *Let* $\mathtt{Message} = \{0,1\}^*$, $\mathtt{mKey} = \{0,1\}^k$ *for some number* k, *and* $\mathtt{Tag} = \{0,1\}^{tLen}$ *for some number* $tLen$.

A message authentication code is defined by a pair of algorithms $(\mathrm{MAC}_{(.)}(\cdot), \mathrm{MAC}.ver_{(.)}(\cdot, \cdot))$. To compute the MAC, $\mathrm{MAC}_{(.)}(\cdot)$ takes a key $K \in mKey$ and a message $M \in \mathtt{Message}$ and computes:

$$(M, T) = \text{MAC}_K(M).$$

The authenticated message is the pair (M, T); T is called the tag on M.
  To verify a purported message-tag pair (M, τ), any entity with key K computes

$$\text{MAC}.ver_K(M, \tau),$$

which returns either 0 (message unauthentic) or 1 (message authentic). It is required for all $K \in mKey$ and $M \in$ Message, $\text{MAC}.ver_K(MAC_K(M)) = 1$.

  An adversary F (of the MAC) is a probabilistic polynomial-time algorithm which has access to an oracle that computes MACs under a randomly chosen key K'. The output of F is a pair (M, T) such that (M, T) was not previously output by the MACing oracle.

  A MAC is secure if, for every adversary F of the MAC, the function $\epsilon(k)$ defined by

$$\epsilon(k) = P[K' \leftarrow \{0, 1\}^k; (M, T) \leftarrow F : (M, T) = MAC_{K'}(M)]$$

is negligible. Note that F wins even if it outputs a different tag on a previously queried message.

Definition 5. *([11], modified.) A strongly-unforgeable authenticated encryption scheme (SUF-AE scheme) is a three-tuple $(\mathcal{K}, \mathcal{E}, \mathcal{D})$, with associated set Message $\subseteq \{0, 1\}^*$, satisfying $M \in$ Message $\Rightarrow M' \in$ Message for any M' of equal length to M.*

  Algorithm $\mathcal{E}$ is probabilistic, returning a string $C \overset{\$}{=} \mathcal{E}_K(M)$ on input of a string $K \in \mathcal{K}$ and $M \in$ Message.

  Algorithm $\mathcal{D}$ is deterministic, taking in a string $K \in \mathcal{K}$ and $C \in \{0, 1\}^*$, and returning $\mathcal{D}_K(C)$ which is either a string in Message or a symbol $\perp$. Moreover, it is required that $\mathcal{D}_K(\mathcal{E}_K(M)) = M$ for all $K \in \mathcal{K}$ and $M \in$ Message.

  Let $\$(\cdot)$ be an oracle that, on input of M, returns a random string of length $l(|M|)$ where l is the length function of $(\mathcal{K}, \mathcal{E}, \mathcal{D})$. Let $\mathcal{A}$ be an adversary. Define $\mathbf{Adv}_{(\mathcal{K}, \mathcal{E}, \mathcal{D})}^{IND\$-CPA}(\mathcal{A}) = P[K \overset{\$}{\to} \mathcal{K} : \mathcal{A}^{\mathcal{E}_K(\cdot)} = 1] - P[\mathcal{A}^{\$(\cdot)} = 1]$. In this instance, IND\$-CPA is the indistinguishability from random bits under a chosen-plaintext attack.

  Let $(\mathcal{K}, \mathcal{E}, \mathcal{D})$ be a SUF-AE scheme. Choose $K \overset{\$}{\leftarrow} \mathcal{K}$ and run the adversary $\mathcal{A}$, providing it with an oracle for $\mathcal{E}_K(\cdot)$. We say that adversary $\mathcal{A}$ forges, under key K for the particular run, if $\mathcal{A}$ outputs a ciphertext C where $\mathcal{D}_K(C) \neq \perp$ and $\mathcal{A}$ did not ask a query $\mathcal{E}_K(M)$ such that $C = \mathcal{E}_K(M)$. However, $\mathcal{A}$ is allowed to have previously queried $\mathcal{E}_K(M)$, such that $C_1 = \mathcal{E}_K(M)$, as long as $C \neq C_1$. Let $\mathbf{Adv}_{(\mathcal{K}, \mathcal{E}, \mathcal{D})}^{SUF-AE}(\mathcal{A})$ be the probability that $\mathcal{A}$ forges against the authentication. The probability is over the random choice of K.

B   Security Revisited without Unique Identifier

Per the ISO 9798–2.4 protocol, the unique identifier of the initiator I_B may be excluded if either uni-directional keys are used or the protocol environment precludes reflection attacks. These two cases will be discussed below for the security proofs of §2 and §5.

B.1   Analysis of Core Security Proof Revisited

The security argument for the core proof of §4 remains largely unchanged for an environment where the unique identifier I_B is excluded. In particular, in the case of uni-directional keys, F has two MAC oracles, per definition 4, that run on keys K_i and K_j chosen randomly from $\{0, 1\}^k$ which it will use to calculate the tags for messages sent to instances of i and j, respectively. Thus the tag in the case of $\mathcal{A}$'s success against an initiator is calculated as $\text{MAC}_{K_i}(R_j, R_i)$ and the tag in the case against a responder is likewise changed to $\text{MAC}_{K_j}(R_i, R_j)$. Thus, as in the original proof, F must win by producing a forgery

Furthermore, in an environment without reflection attacks, the case of F against an initiator remains unchanged. Against a responder oracle, $\mathcal{A}$ cannot win by reflecting back to $\Pi_{j,i}^t$ the flow $\mathcal{E}_K(R_j, R_i)$. Consequently, F must again win as in §4 by producing a forgery.

B.2   Analysis with Associated Data Revisited

Theorem B1. *Let Π be the ISO 9798–2.4 protocol implemented with a strongly unforgeable MAC algorithm $\mathcal{E}_K(M) = (m, \text{MAC}_K(M))$, including the optional text fields $Text_1$ and the associated data $Text_2$, uni-directional keys in an environment that precludes reflection attacks, and no unique identifier I_B. Then advantage of an polynomial-time adversary against the mutual authentication scheme can be reduced to the adversarial advantage against the MAC:*

$$\mathbf{Adv}^{MA}(\mathcal{A}) \leq 2p^2 S \cdot \mathbf{Adv}_\Pi^{\text{MAC}}(F) + q^2/2^{k+1}.$$

Proof. Essentially, the theorem follows from the proof in §5 with some alterations as noted below.

Nonce Collision

Proof as shown in §5.

$$\Pr[\text{NC}] \leq q^2/2^{k+1}. \tag{9}$$

Acceptance Implies Matching Conversations

Proof as in §5 with the added notes of §B.1.

$$\Pr[\neg \text{ Match.Conv} \mid \neg \text{ NC}] \leq 2p^2 S \cdot \mathbf{Adv}_\Pi^{\text{MAC}}(F). \tag{10}$$

Associated Data Agreement

Compactly, it can be assumed that an instance and its partner are in matching conversations and that there are no nonce collisions. It remains to show that the same AD, T_2, is equally held by both sides.

Uni-Directional Keys Since i and j are in matching conversation, at some time τ_1 a responder sent the message $T_{3_j} || \mathcal{E}_{K_i}(R_j, R_i, T_{2_j})$ and at some time τ_2 an initiator received a message $T_{3_i} || \mathcal{E}_{K_i}(R_j, R_i, T_{2_i})$, where T_{3_i} and T_{3_j} may or may not be equal. Then the responder must be an instance $\Pi_{j,i}^t$ of j, and the initiator must be an instance $\Pi_{i,j}^s$ of i. Moreover, the data T_2 has been authenticated under the key K_i, so $\Pi_{i,j}^s$ can be assured of the integrity of T_2. Therefore $AD_i = AD_j$.

Ergo,

$$\Pr[\neg\ \text{Match.AD} \mid \neg NC \wedge\ \text{Match.Conv}] = 0. \tag{11}$$

Reflection Attacks Disallowed Proof as in §3, with unique identifier I_B removed.

Combining the reductions from equations 9–11 with equation 5 yields the full reduction of security for ISO 9798–2.4 with inclusion of associated data.

Author Index